THE JOBLESS FUTURE

The Jobless Future

Sci-Tech and the Dogma of Work

Stanley Aronowitz
and
William DiFazio

University of Minnesota Press
Minneapolis
London

Published by the University of Minnesota Press
111 Third Avenue South, Suite 290, Minneapolis, MN 55401-2520
Printed in the United States of America on acid-free paper
Second printing, 1995

Library of Congress Cataloging-in-Publication Data

Aronowitz, Stanley.
    The jobless future / Stanley Aronowitz and William DiFazio.
        p.   cm.
    Includes bibliographical references and index.
    ISBN 0-8166-2193-4 (hc : acid-free paper)
    ISBN 0-8166-2194-2 (pb : acid-free paper)
        1. Unemployment, Technological—United States.   2. Labor supply—
    Effect of technological innovations on—United States.   I. DiFazio,
    William.   II. Title.
    HD6331.2.U5A76   1994
    331.13'7042'0973—dc20   94-14102

The University of Minnesota is an
equal-opportunity educator and employer.

For Stanley's son Michael
For Bill's daughter Liegia Stella

Look at the noble savage whom the missionaries of trade and the traders of religion have not yet corrupted with Christianity, syphilis and the dogma of work, and then look at our miserable slaves of machines.

Paul Lafargue, *The Right to Be Lazy*

Contents

Preface

We live in hard times. The economic stagnation and decline that changed many lives after the stock market crash of 1987 and blossomed into a full-scale recession in 1990 lingers despite frequent self-satisfied statements by politicians and economists that the recovery has finally arrived (1992, 1993, 1994). Nevertheless, there are frequent puzzled statements by the same savants to the effect that although we are once again on the road to economic growth, it is a jobless recovery. Then, in the months in which the Department of Labor records job growth, we are dismayed to discover that most of the jobs are part-time, near the minimum wage.

Bemoaning the high costs of doing business in the 1990s, *Business Week* attributes some of the problem to increases in benefits such as health care. It is particularly concerned that "small business, the engine of job creation in the 1980s," is unable to overcome the steep job cuts that big business has imposed on hundreds of thousands of blue-collar, pink-collar, and white-collar employees. For there is no doubt that we have yet to feel the long-term effects on American living standards that will result from the elimination of well-paid professional, technical, and production jobs. At the same time, nearly everyone admits that many of these jobs are gone forever. Still, we are amazed that as experts, politicians, and the public become acutely aware of new problems associated with the critical changes in the economy—crime, poverty, homelessness, hunger, education downsizing, loss of tax revenues to pay for public services, and many other social issues—the solution is always the same: jobs, jobs, jobs.

The central contention of this book is that if jobs are the solution, we are in big trouble. We argue that the tendency of contemporary global economic life is toward the underpaid and unpaid worker. The characteristics of the industrializing era of world capitalist development—a highly differentiated workforce, strong unions, powerful states that guaranteed economic security and promoted economic growth—no longer describe our new era. Scientifically based technological change in the midst of sharpened internationalization of production means that there are too many workers for too few jobs, and even fewer of them are well paid. As we completed the final words of this book, there was considerable evidence that this point has reached deeply into the public imagination, but not into the political or cultural discourse. The aim of this work is to suggest political and social solutions that take us in a direction in which it is clear that jobs are no longer the solution, that we must find another way to ensure a just standard of living for all.

This book is the result of a collaboration that began in 1983 when the authors were asked by Lou Albano and Brad Smith of the New York Technical Guild (District Council 37, AFSCME) to perform a study of the possible effects of the introduction of computer-aided design and drafting on sections of the union's membership. As engineers and drafters, they were naturally concerned about the impact of the new technology on their job security and working conditions. The first results of our interviews and ethnography were published as a report to the union, *Time for Decision,* that appeared in August 1984. We were fortunate to have the enthusiastic and able assistance of several colleagues without whose participation this study and, hence, this book might never have been published. We want to thank Patricia D'Andrade, Dana Fenton, Eric Lichten, Bill Watson, Martha Ecker, and Ted Yanow. A second study, for the Communications Workers of America Local 1082, employees of the New Jersey Department of Transportation, became an indispensable verification of our intial conclusion that computer-aided design could be an extraordinary boon to engineers and technicians *only* if it were used in a context of equality and opportunity. Used as a management tool, it could become a powerful weapon against professional and technical knowledge. Eric Lichten and David Wagner were particularly important in bringing this study to fruition.

The next step in the process was the study group on science and technology that the authors participated in, together with David Chorlian, Margaret Yard, Louis Amdur, Patricia D'Andrade, Janet Biehl, and Jeff Schmidt. The result of this group study became the basis for Stanley Aronowitz's *Science as Power,* published by the University of Minnesota Press in 1988. *The Jobless Future* is, in many respects, a second volume of this earlier collective effort. Of course, it may be read as a self-contained work.

In the development of our general perspective on science, technology, and work, we have been informed and supported by a number of friends and colleagues, including Kimberly Flynn, Lynn Chancer, Paolo Carpignano, Carl Johnson, Maria Milagros Lopez, Bruce Vanacour, and Ellen Willis.

Judy Gregory, then of the Professional Department of the AFL-CIO, gave us an early opportunity to share our ideas and findings, published as "Unions, Technology and Computer-Aided Design," with a group of engineers. Professor Sam Bloom enthusiastically helped us in our work on biomedical scientists. William Kornblum, one of America's leading ethnographers, has been a source of unfailing support.

Sol Yurick's editing and intellectual toughness have been indispensable. He helped shape an often amorphous manuscript and put our noses to the grindstone, forcing us to make nascent ideas intelligible. Biodun Iginla and the copyeditors at University of Minnesota Press have provided unfailing support.

For DiFazio, Dan Barnestein was an inspiration. I would like to thank supportive colleagues at St. John's University: Richard Harris, Father Brian O'Connell, Rosalyn Bologh, Henry Lesieur, Susan Meswick, Dawn Esposito, Joseph Trumino, Judy Desena, and Michael Welch. Conversations with Susanne MacGregor, Danny Graham, Arthur Lipow, Frances Mannino, Slawomir Magala, and Arthur B. Shostak were very helpful. When I was overwhelmed, long telephone conversations with Donna Gaines always helped get me back on track. And, especially, thanks to Susanna Heller, who taught me to see in new ways. I would also like to thank my friends at St. John's Bread and Life Soup Kitchen.

We also want to thank all of the architects, biomedical scientists, drafters, engineers, and university professors who allowed us into their worlds.

Introduction

The nation's economy is staggering out of the recession, say most of the gauges that measure it, but people who are getting jobs tell a sobering story:

> Many good factory jobs and white-collar office jobs with good wages and benefits are giving way to unstable and mediocre jobs. That makes the recovery different from any other.
>
> Trends that started in the 1980s have produced a new look to working America. Part-time jobs, temporary jobs, jobs paying no more than the Federal minimum wage, jobs with no more benefits than a few vacation days are displacing permanent regular jobs that people would lose in past recessions and reclaim when business picked up.[1]

Less than three months after this report appeared in the *New York Times* (in December 1992), there was a variation on this theme: "Employment agencies call them contingent workers, flexible workers or assignment workers. Some labor economists, by contrast, call them disposable and throwaway workers." According to Audrey Freedman, an economist for Manpower, the big temporary-help company, "The labor market today, if you look at it closely, provides almost no long-term secure jobs."[2]

After nearly three years of recession, the painfully slow "recovery" that economists said occurred in late 1992 failed to reduce official joblessness below 7 percent, or 9 million people, throughout 1992 and early 1993. By March 1993 the January figures on industrial production, consumer buying, and other indicators showed a slight dip, leaving in their wake both consternation and doubt that the U.S. economy was finally breaking out of the doldrums. When February payrolls grew by 365,000, Secretary of Labor Robert Reich noted that unemployment was still two percentage points higher than before the recession.[3] Moreover, most employment gains had been part-time jobs, he said. In fact, if part-time employment was calculated as partial unemployment, if the military was excluded from the employed (as it had been until the Reagan Bureau of Labor Statistics revised the basis for computing the number of jobholders), and if discouraged workers—those who had stopped looking for work—were factored into the jobless figure, the numbers would be much higher, closer to 12 percent, which corresponds to the rate in most of Western Europe. Since the late 1980s unemployment rates around the globe seem intractable, and not directly influenced by the overall performance of a national economy.

The traditional "smokestack industries" have been in decline. The general decline has not been only in the manufacturing sectors but in the nonmanufacturing sectors as well. This process has been going on for twenty years as U.S. economic dominance has been transformed into the United States as one major player among others in the global economy. Barry Bluestone and Bennett Harrison have traced the transformation:

> During the decade of the 1970's we estimate that between 450,000 and 650,000 jobs in the private sector in both manufacturing and non-manufacturing were wiped out somewhere in the United States of both small and large runaway shops. But it turns out that such physical relocations are a tip of a huge iceberg. When the employment lost as a direct result of plant, store and office shutdowns during the 1970's is added to the job loss associated with runaway shops, it appears that more than 32 million jobs were destroyed. Together, runaways, shutdowns and permanent physical cutbacks short of complete closure may have cost the country as many as 38 million jobs.[4]

In 1982, 1,287,000 jobs were lost as the result of plant closings and layoffs. Between 1977 and 1982, 150,000 jobs were lost in the steel industry. There are millions of unemployed, underemployed, and discouraged workers in the United States. Unemployment in the major industrial sectors seems to be out of control, and employed workers wonder if they are next. Union membership declines, and workers who strike legally are threatened with "replacement workers," a sanitized term for scabs in an antiunion decade. Workers are replaced, and no one knows who is next. Will they go the way of longshoremen replaced by containerization, or auto workers replaced by robots, or steelworkers replaced by disinvestment and a restructured global steel industry? No one knows, but the predictions are ominous. As Leontief and Duchin report, a 1982 study from Japan "suggests that among the most

2

advanced robots currently in use, displacement rates of 2–4 workers per shift are possible."[5]

All of the contradictory tendencies involved in the restructuring of global capital and computer-mediated work seem to lead to the same conclusion for workers of all collars—that is, unemployment, underemployment, decreasing skilled work, and relatively lower wages. These sci-tech transformations of the labor process have disrupted the workplace and workers' community and culture. High technology will destroy more jobs than it creates. The new technology has fewer parts and fewer workers and produces more product. This is true not only in traditional production industries but for all workers, including managers and technical workers:

> Forecasts made by the BLS [Bureau of Labor Statistics] and for *Business Week* by Data Resources Incorporated (DRI) in fact show that the number of high-tech jobs created will be less than half the two million jobs lost in manufacturing in the past three years. While high-tech industries as defined by the BLS will generate 10 times the number of jobs expected from the rest of industry, it will still amount to only 730,000 to one million jobs. And most of those will be traditional occupations, not technical ones. Fewer than one-third will be for engineers and technicians, according to DRI, and the remainder will be managers, clerical workers, operators and other factory workers.[6]

Technological progress and capital accumulation seem to disrupt the social fabric in the United States. A weakened position in the international economy demands that American industry increase its productivity and cut its unit labor cost. As Carl G. Thor, president of the American Productivity Center in Houston, says, "The trick is to get more output without a surge in employment."[7] Technological change and competition in the world market guarantee that increasing numbers of workers will be displaced and that these workers will tend to be rehired in jobs that do not pay comparable wages and salaries.[8] Women and minorities will suffer the most as the result of these changes; the increased participation in an occupational sector by women and minorities is often an indicator of falling wages in that occupation.

The problems of plant shutdowns and technological change in production industries is fairly well known, but we lack reliable information concerning the fate of displaced workers, and our social knowledge of the effects of technological change and corporate restructuring in the private service sectors is virtually nonexistent. Until recently, most economists, business analysts, and sociologists assumed that the long-term decline of manufacturing employment was not a serious social issue because of the concomitant expansion of such industries as financial services, insurance, and retail and wholesale trades. The guiding assumption here was that the expansion of jobs in these industries would absorb those who lost their production jobs. For forty years after World War II, this assumption proved more or less valid for large numbers of younger workers. Now, however, contemporary trends in major financial corporations point to mass layoffs among middle managers as well as clerical employees: consolidation of computer services may cause skilled operators and programmers to seek jobs in other sectors; a glut of college graduates has saturated the job market, changing the criteria for employment in sectors that traditionally did not hire the educated; recent M.B.A.'s are being hired to replace senior managers and reduce payroll costs; and there has been a dramatic increase in part-time and contract employees replacing permanent staff. These changes have become apparent since the "crash of '87." Wall Street layoffs have become a daily occurrence, spreading to banks and other financial institutions. These layoffs are a result of three factors: corporate mergers that result in reorganization and allow for downsizing, that is, the elimination of workers who duplicate services; the rapid spread of technological innovations such as telecommunication systems; and more advanced computer networks that eliminate workers even as productivity rises. In banking, the combination of these changes has already led to layoffs among some of the country's leading banks, including Chase Manhattan, Bank of America, and Chemical Bank-Manufacturers Hanover Trust.

Prevailing economic wisdom explained the puzzlement of what became known as the "jobless recovery" in two ways: by ascribing the job cuts to the industrial restructuring necessary to make the economy more "efficient" in a highly competitive global market; and the more pessimistic view that real growth, including jobs, could not occur in a debt-ridden economy.

The optimists argued that global restructuring has forced U.S. industries to become meaner and leaner. America's weakened competitive position had to be improved through efficiencies such as technological change that reduced the size of the factory labor force; mergers and acquisitions that eliminated redundancy, which resulted in plant closings and consequent elimination of "excess" workforces; and applying a hatchet to "overhead" costs such as clerical workers and middle management. The large tent under which these changes were made is the term *productivity*. Everything, including jobs, had to be sacrificed to make the worker more productive. Included in these measures were company demands for wage concessions, substantial cuts in health and pension benefits, and conversion of many permanent jobs to temporary and part-time work. Many experts argue that "leaner and meaner" production is the right medicine for the U.S. economy. For example, many of the largest industrial corporations were converting to "flexible specialization," better known as "just-in-time" production methods. Instead of building huge inventories, a growing number of industrial companies were producing only enough to meet orders. This method yields savings in warehousing, truck fleets, and the labor needed to operate them. While advocates acknowledged that this innovation might result in smaller workforces, they contend that joblessness today might be the necessary prelude to full employment tomorrow.[9]

The second explanation, more sobering, emphasized the role of huge federal and consumer debt accumulated beween the late 1970s and the early 1990s that has drained public and private investment, inhibiting recovery and growth. According to this view, economic growth, including more jobs, may not be possible unless investors are encouraged to pour money into productive activity instead of into high-interest-bearing paper such as Treasury bills, municipal bonds, and instruments used by government to pay the huge debt service to banks and other large institutional investors. Following this analysis, President Bill Clinton proposed an economic plan to reduce government spending and raise taxes rather than create new jobs, except for some work repairing bridges and roads. According to many fiscal and economic analysts, any sustained economic growth requires reinstituting austerity, initiated first by the Carter administration in the late 1970s but disrupted by Ronald Reagan's subsequent massive military buildup, which expanded the credit system to the breaking point even as his administration tried, in the name of austerity, to cut entitlements to the bone. In this view, the party is over and Americans will have to learn to tighten their belts further. They can expect little from the federal government and may be required to give back some of the already enfeebled welfare-state benefits. Where Reagan and George Bush failed, Clinton can succeed because only a Democrat can persuade the working class to sacrifice its historical memory.

In the broad picture there is no shortage of capital, only shortages of capital investment in economic sectors associated with production, which, compared to other options, cannot deliver the maximum possible return. When, by electronic means, an investor can purchase, on margin, a large quantity of U.S. Treasury bills or bank money market funds in the morning and sell them at a substantial profit in the afternoon, choosing to sink money in a steel, chemical, or auto stock may not be a rational choice. The paper market promises to deliver rapid investment turnover, while industrial capital turnover rarely takes fewer than five years. Given the prevailing rationality according to which investment efficiency is measured in terms of the time it takes to yield maximum profits, industrial corporations have had an uphill battle attracting investors away from the bustling traffic in what might be called "spurious" capital formation. In the 1980s, real estate—both commercial development such as shopping malls and, especially, co-op and condo building and conversions—was a preferred investment. As the economy slowed down in the late 1980s, leaving many of the malls empty (especially in the overexpanded South and Southwest) and condos repossessed by banks because of default, the transfer of paper—which produces few, if any, jobs, and no white elephants—was the investment of choice for many institutional as well as private investors.

The paradox is that even when business investors pour substantial portions of their capital into machinery and buildings, these investments do not significantly increase the number of new permanent, full-time jobs. Computerized machines employ very little direct labor; most of it is devoted to setup, repair, and monitoring. The actual production process is almost laborless because control over it is now built into the machines by computer processes such as numerical controls, lasers, and robotics. In short, computer-based technology inherently eliminates labor. The more investment in contemporary technologies, the more labor is destroyed.

Computer-mediated work processes are now worldwide. In the early 1990s, Japan and Germany, whose national economies were second only to that of the United States, began to experience substantial reductions in production and workforces. In stark contrast to the 1980s hype of the invincibility of the Japanese, especially its effective corporatism, the early 1990s witnessed severe changes in the Japanese economic outlook. For the first time since the early postwar years, many of the largest corporations such as Mitsubishi laid off thousands of workers, cut production, and began to transfer work to less-developed countries. Suddenly, Japanese industrial and labor relations policies resembled those of Western Europe and the United States. Germany, whose economic "miracle" cluttered press reports for two decades, found itself in the doldrums; plant closings and layoffs dotted the industrial landscape and, not unexpectedly, the government faced deep budget deficits and debt.

Some of the decline can be attributed to the steep descent of consumption of basic industrial and consumer goods such as steel and autos in the United States, which, despite the relative deterioration of its world position, remains the largest market for both industrial and consumer goods. But we must consider the possibility that capital-intensive production processes that aggravated unemployment, the extensive use of overtime work as an alternative to new hiring, and relative wage and benefit reductions have deepened the recession and made it much harder to produce a vigorous recovery. Since the end of World War II, working people have been encouraged to mortgage everything, including their souls, on the assumption that, economic ups and downs notwithstanding, there was no real barrier to ever higher living standards. The historic demand for shorter hours that accompanied the introduction of labor-replacing technology has been ditched in favor of work without end. But, as Bill Clinton has reiterated as both candidate and president, there is a shortage of "quality" jobs in the restructured economy; we wish to add that there is shortage of quality pay, as well.

Of course, not all investment has been in manufacturing or money instruments. Some of it has fueled the enormous growth of informatics: the production of computer hardware and software; electronic communication equipment and processes; and the vast array of services attached to information, much of it channeled to finance. Informatics facilitates even more rapid money and other market exchanges. The various forms of labor-reducing informatics span a wide range of industries from goods production to retail and wholesale trades and nearly all services, large and small. Supermarkets use electronic devices for inventory control and to register prices on nearly all items. Small businesses use informatics as well. A liquor store, for example, is likely to use a computer to track inventory, eliminating the need to hire a full-time stock clerk. Computers are now part of the pharmacist's tools, and even some independent grocers use them. Although the information and computer "revolution" at first created millions of new jobs, it destroyed many others. Tens of thousands of workers once employed as retail clerks in "mom and pop" stores, albeit at low wages, are no longer needed. In the wake of the growth of the giant supermarket chains and multilocation department store chains that dot the retail landscape, one of the major weapons of survival for owners of small clothing, hardware, and grocery stores is cutting labor costs by working longer hours themselves and using machinery.

Even before the worldwide recession began with the stock market crash of November 1987, hundreds of companies entered into transnational mergers and acquisitions. In some cases, such as the automobile industry, these relationships predated the 1980s, but that decade was marked by several joint ventures between U.S. and Japanese corporations that reduced further the need for many of their respective parts plants. The practice of "outsourcing" meant that many corporations closed their parts plants and used the same contractors to produce everything from windshield wipers to engine parts to sheet metal components. It was the decade not only of the global car, but also of the global sweatshop. Americans were regularly plied with reports of products they used in everyday life being produced by workers in Mexico, Malaysia, and China who earned a small fraction of a U.S. or Western European wage. In many other industries such as machine tools, computer chips, and other hardware components, the idea of a national identity to a commercial product became an advertising fiction used for the purposes of neutralizing protectionism. In the communications industry itself, several major U.S. newspapers were purchased by Australian communications mogul Rupert Murdoch, who also acquired the *Times* of London.

Sony, Japan's leading electronics corporation, bought substantial portions of the U.S. record industry. Japanese and European capital formed conglomerates through takeovers of U.S. corporations; in turn, U.S. investment grew in countries like China, Mexico, and Taiwan, leaving substantial sectors of the U.S. economy in the hands of foreign companies and, equally important, laying waste to vast stretches of the industrial heartland. Pittsburgh and the Monongahela valley, the Youngstown area, and Buffalo, once booming steel towns, are industrial wastelands; only Pittsburgh managed to remain economically viable by becoming a regional financial center.

The effects of global restructuring are fairly well known by now. Awakened by the near demise of the Chrysler Corporation in 1979, Americans were sternly warned that unless they learned to tighten their belts, the consequences for the American dream would be dire. Indeed, by 1990 the once-proud U.S. lead in wage levels had been wiped out. The Bureau of Labor Statistics reported that U.S. wages eroded from first in 1970 to sixth, behind Germany, Sweden, Austria, Italy, and France. For example, the wages of German workers were nearly 50 percent higher and those of Swedish workers more than 40 percent higher than those of U.S. workers. In that twenty-year period, real wages actually deteriorated, while the wages for all other major industrial countries except the Soviet Union and Great Britain grew in real terms. The concerted assault by government and major corporations on workers' income, begun in 1981 when, in a dramatic gesture, President Reagan fired 11,000 striking air traffic controllers, managed to impose a virtual wage freeze on the overwhelming majority of manufacturing workers. Plant closings, tepid trade union response bordering on the supine, and an intense antiunion ideological environment conspired to weaken workers' resistance so that, by the mid-1980s, the nonunion sector drove labor relations.

This, in brief, is the context within which a severely reduced job "market" began to take shape, not only for U.S. workers, but potentially for all workers. In this book we argue that the progressive destruction of high-quality, well-paid, permanent jobs is produced by three closely related developments.

First, in response to pervasive, long-term economic stagnation and to new scientifically based technologies, we are experiencing massive restructuring of patterns of ownership and investment in the global economy. Fewer companies dominate larger portions of the world market in many sectors, and national boundaries are becoming progressively less relevant to how business is done, investment deployed, and labor employed. As the North American Free Trade Agreement illustrates, there is no reason other than political considerations to value the concept of the nation with respect to the production and distribution of goods and services. As much as machinery, organization at the level of the corporate boardrooms and the workplace is a crucial technology of labor destruction.

Second, the relentless application of technology has destroyed jobs and, at the same time, reduced workers' living standards by enabling transnational corporations to deterritorialize production. Today, plant and office locations are less dependent on geographic proximity to markets, except in the case of some services. Informatics, of which computer-mediated processes are the most common in financial, retail, and wholesale services, permit production and services to be dispersed throughout the globe with impunity. Increasingly, electronically transmitted information is the medium of business, and for the most part it does not depend on place.

Third, National Public Radio recently reported that a number of U.S. corporations were locating their design and development activities in India, where, they claimed, systems analysts, programmers, and engineers were highly competent and much lower paid than their Western (American) counterparts. Similarly, Du Pont is preparing to build a major synthetics fiber and petrochemical facility in Shanghai. In the past, U.S. scientific and technical experts were largely responsible for designing and supervising foreign plant construction, but Du Pont is employing Chinese engineers, chemists, and technicians, many of whom were trained in the United States and other advanced industrial countries and others who were not. These cases illustrate a second theme of this book: informatics not only displaces and recomposes manual labor but also displaces technical and scientific labor—a new and expanding frontier of global restructuring. As we shall see, because these are most of the new "quality" jobs about which economists and political leaders speak, we argue that there is absolutely no prospect, except for a fairly small minority of professional and technical people, to obtain good jobs in the future.

Consequently, whatever validity it had in the past, the neoclassical economic philosophy and the policies it engenders, according to which economic growth leads to relatively full employment and higher living standards, are rendered obsolete by recent developments. If the economy respects no national boundaries, the impact of nationally based fiscal, monetary, and industrial policies are severely limited, just as labor and welfare policies that are not truly international in scope and application are fated to be eroded. Forget the old social-democratic slogan of full employment in a humane welfare state. Abolish the welfare bureaucracy and decommodify work by separating work from income. Even with considerable political will, national governments by themselves can do next to nothing to overcome the new conditions of life and labor. The old slogan "international labor solidarity" is now a practical political proposal to achieve redistributive justice.

The era of "jobs, jobs, jobs" and all that this slogan implies is over. We suggest that if justice depends on employment and the good life depends on the rewards of hard work, there can be no justice, and the good life may be relegated to a dim memory. However, we renounce neither justice nor pleasure. In the final section of this book we propose alternatives to the long wave of the job culture as the substitution for the good life.

Part I

Technoscience and Joblessness

Chapter 1

The New Knowledge Work

Overview

In 1992 the long-term shifts in the nature of paid work became painfully visible not only to industrial workers and those with technical, professional, and managerial credentials and job experience but also to the public. During the year, "corporate giants like General Motors and IBM announced plans to shed tens of thousands of workers."[1] General Motors, which at first said it would close twenty-one U.S. plants by 1995, soon disclaimed any definite limit to the number of either plant closings or firings and admitted the numbers of jobs lost might climb above the predicted 70,000, even if the recession led to increased car sales. IBM, which initially shaved about 25,000 blue- and white-collar employees, soon increased its estimates to possibly 60,000, in effect reversing the company's historic policy of no layoffs. Citing economic conditions, Boeing, the world's largest airplane producer, and Hughes Aircraft, a major parts manufacturer, were poised for substantial cuts in their well-paid workforces. In 1991 and 1992 major retailers, including Sears, either shut down stores or drastically cut the number of employees; in late January 1993, Sears announced it was letting about 50,000 employees go. The examples could be multiplied. Millions, worldwide, were losing their jobs in the industrialized West and Asia. Homelessness was and is growing.

Also in 1992, twelve years of a Republican national administration came to an end. Presidential candidate Bill Clinton successfully made the economy the central issue, eradicating the seemingly unassailable popularity President George Bush won during the brief Gulf War of the previous year. Among the keystones of his campaign, Clinton promised federal action to create new jobs through both direct investment in roads and mass transit and a tax credit to encourage business to invest in machinery and plants. Did this signal a return to presumably antiquated Keynesian policies and fiscal policies to encourage private investment? An important part of Clinton's approach to recovery was more money for education and stepped-up training and retraining programs, the assumption of which is that development of "human capital" was a long ignored but important component of the growth of a technologically advanced economy.[2] According to this argument, a poor educational system and inadequate apprenticeship and retraining programs would inevitably result in a competitive disadvantage for the United States in an increasingly competitive global economy.

The peculiar feature of the latest economic recovery is that while the economic indicators were turning up, prospects for good jobs were turning down. Even before Clinton took office, the trend toward more low-paid, temporary, benefit-free blue- and white-collar jobs and fewer decent *permanent* factory and office jobs called into question many of the underlying assumptions of his campaign. For if good jobs were disappearing as fast as "unstable and mediocre"[3] jobs were being created, more education and training geared to a shrinking market for professional and technical labor might lead nowhere for many who bought the promise. As in many manufacturing sectors, labor-displacing technological change has reached the construction industry. Unless public and private investment is specifically geared to hiring labor rather than purchasing giant earth-moving machines, for example—which would imply changing a tax structure that permits write-offs for technology investments—it seems dubious that Clinton's plan to plow $20 billion a year into infrastructure could result in a significant net employment gain even as it generated orders for more labor-saving equipment. Machine tool and electrical equipment industries are leaders in the use of computer-mediated labor processes. Labor-saving technologies combined with organizational changes (such as mergers, acquisitions, divestitures, and consolidation of production in fewer plants, a cost saving made possible by the technology) yield few new jobs.

The central contention of this book lies somewhere between common sense and new knowledge. We discuss what should already be evident to all but those either suffering the political constraints of policy or still in the thrall of the American ideology and blind belief in the ineluctability of social mobility and prosperity in the American system. The two are linked: no politician who aspires to power may violate the unwritten rule that the United States is the Great Exception to the general law that class and other forms of social mobility are restricted for the overwhelming majority of the population. Our first argument—that the Western dream of upward mobility has died and it is time to give it a respectful funeral—may have at long last seeped into the bones of most Americans, even the most optimistic economist.

The dream has died because the scientific-technological revolution of our time, which is not confined to new electronic processes but also affects organizational changes in the structure of corporations, has fundamentally altered the forms of work, skill, and occupation. The whole notion of tradition and identity of persons with their work has been radically changed.

Scientific and technological innovation is, for the most part, no longer episodic. Technological change has been routinized. Not only has abstract knowledge come to the center of the world's political economy, but there is also a tendency to produce and trade in symbolic significations rather than concrete products. Today, knowledge rather than traditional skill is the main productive force. The revolution has widened the gap between intellectual, technical, and manual labor, between a relatively small number of jobs that, owing to technological complexity, require more knowledge and a much larger number that require less; as the mass of jobs are "deskilled," there is a resultant redefinition of occupational categories that reflects the changes in the nature of jobs. As these transformations sweep the world, older conceptions of class, gender, and ethnicity are called into question. For example, on the New York waterfront, until 1970 the nation's largest, Italians and blacks dominated the Brooklyn docks and the Irish and Eastern Europeans worked the Manhattan piers. Today, not only are the docks as sites of shipping vanishing, the workers are gone as well. For those who remain, the traditional occupation of longshoreman—dangerous, but highly skilled—has given way, as a result of containerization of the entire process, to a shrunken workforce that possesses knowledge but not the old skills.[4] As we show in chapter 3 and in the concluding section of this book, this is just an example of a generalized shift in the nature and significance of work.

As jobs have changed, so have the significance and duration of joblessness. Partial and permanent unemployment, except during the two great world depressions (1893–1898 and 1929–1939) largely episodic and subject to short-term economic contingencies, has increasingly became a mode of life for larger segments of the populations not only of less industrially developed countries, but for those in "advanced" industrial societies as well. Many who are classified in official statistics as "employed" actually work at casual and part-time jobs, the number of which has grown dramatically over the past fifteen years. This phenomenon, once confined to freelance writers and artists, laborers and clerical workers, today cuts across all occupations, including the professions. Even the once buoyant "new" profession of computer programmer is already showing signs of age after barely a quarter of a century. We argue that the shape of things to come as well as those already in existence signals the emerging proletarianization of work at every level below top management and a relatively few scientific and technical occupations.

At the same time, because of the permanent character of job cuts starting in the 1970s and glaringly visible after 1989, the latest recession has finally and irrevocably vitiated the traditional idea that the unemployed are an "industrial reserve army" awaiting the next phase of economic expansion. Of course, some laid-off workers, especially in union workplaces, will be recalled when the expansion, however sluggish, resumes. Even if one stubbornly clings to the notion of a reserve army, one cannot help but note that its soldiers in the main now occupy the part-time and temporary positions that appear to have replaced the well-paid full-time jobs.

Because of these changes, the "meaning" (in the survival, psychological, and cultural senses) of work—occupations and professions—as forms of life is in crisis.[5] If the tendencies of the economy and the culture point to the conclusion that work is no longer significant in the formation of the self, one of the crucial questions of our time is what, if anything, can replace it. When layers of qualified—to say nothing of mass—labor are made redundant, obsolete, *irrelevant*, what, after five centuries during which work remained a, perhaps *the*, Western cultural ideal, can we mean by the "self"? Have we reached a large historical watershed, a climacteric that will be as devastating as natural climacterics of the past that destroyed whole species?

Some of these new epochal issues have been spurred by a massive shift in the character of corporate organization. Beginning in the 1970s much of the vertical structure of the largest corporations began to be dismantled. New kinds of robber barons appeared; among them, Bernard Cornfeld and James Ling were perhaps the most prominent pioneers in the creation of the new horizontal corporate organizational forms we now call conglomerates. Ling's empire, for example, spanned aircraft, steel, and banking. Textron, once a prominent textile producer, completely divested itself of this product as its business expanded to many different sectors. U.S. Steel became so diversified that it replaced *steel* with X in its name shortly after it bought Marathon Oil in 1980. And Jones and Laughlin, a venerable steel firm, became only one entry in Ling's once vast portfolio of unconnected businesses. Every television watcher knows that Pepsi-Cola has gone far beyond producing soft drinks to operating a wide array of retail food services; by the 1980s it owned Pizza Hut and several other major fast-food chains.

New forms of organization such as mergers and acquisitions, which have intensified centralized ownership but also decentralized production and brought on the shedding of whole sections of the largest corporations, have spelled the end of the paternalistic bureaucracy that emerged in many corporations in the wake of industrialization and, especially, the rise of the labor movement in the twentieth century.

Large corporations such as IBM and Kodak, for example, have reversed historic no-layoffs policies that were forged during the wave of 1930s labor organization as a means of keeping the unions out. Equally important, layoffs as a mode of cost cutting have been expanded to include clerical, technical, and managerial employees, categories traditionally considered part of the cadre of the corporate bureaucracies and therefore exempt from employment-threatening market fluctuations. The turnover of ownership and control of even the largest corporations combined with technological changes undermines the very concept of job security. The idea of a lifetime job is in question, even in the once secure bastions of universities and government bureaucracies. Thus, the historic bargain between service workers and their employers, in which employees accepted relatively low wages and salaries in return for security, has, under pressure of dropping profit margins and a new ideology of corporate "downsizing," been abrogated.

For the corporate conglomerate, the particular nature and quality of the product no longer matter since the ultimate commodity, the one that subsumes and levels all others, is the designating language, the representing language in the terms of forms of credit. That is to say, along with the technical changes, which are knowledge changes, the changes in representation are not only parallel, but in conflict. Knowledge itself, once firmly tied to specific labor processes such as steelmaking, now becomes a relatively free-floating commodity to the extent that it is transformed into information that requires no productive object. This is the real significance of the passage from industry-specific labor processes to computer-mediated work as a new universal technology.

Science and technology (of which organization is an instance) alter the nature of the labor process, not only the rationalized manual labor but also intellectual labor, especially the professions. Knowledge becomes ineluctably intertwined with, even dependent on, technology, and even so-called labor-intensive work becomes increasingly mechanized and begins to be replaced by capital- and technology-intensive—*capitech-intensive*—work. Today, the regime of world economic life consists in scratching every itch of everyday life with sci-tech: eye glasses, underarm deodorant, preservatives in food, braces on pets. Technology has become the universal problem solver, the postmodern equivalent of deus ex machina, the ineluctable component of education and play as much as of work. No level of schooling is spared: students interact with computers to learn reading, writing, social studies, math, and science in elementary school through graduate school. Play, once and still the corner of the social world least subject to regimentation, is increasingly incorporated into computer software, especially the products of the Apple corporation. More and more, we, the service and professional classes, are chained to our personal computers; with the help of the modem and the fax we can communicate, in seconds, to the farthest reaches of the globe. We no longer need to press the flesh: by E-mail, we can attend conferences, gain access to library collections, and write electronic letters to perfect strangers. And, of course, with the assistance of virtual reality, we can engage in electronic sex. The only thing the computer cannot deliver is touch, but who needs it, anyway?[6]

Each intrusion of capitech-intensiveness increases the price of the product, not makes it cheaper, because the investment in machines has to be paid off. For example, getting on E-mail is not free; in 1993 dollars it costs about ten dollars a month for the basic service. Joining a conference or forum network might cost an additional ten or fifteen dollars, and each minute on the E-mail line carries an additional charge, although it is not as high as the charge for traditional voice-based telecommunications. Like the 900 number used to get vicarious sexual experience or esoteric information, computer-driven information is not free. Of course, as scientific, technical, academic, and other professionals feel that they "need" access to information that, increasingly, is available only through electronic venues, the cost of being a professional rises, but the privileges are also more pronounced.

The new electronic communication technologies have become the stock-in-trade of a relatively few people because newspapers, magazines, and television have simply refused to acknowledge that we live in a complex world. Instead, they have tended to *simplify* news, even for the middle class. Thus, an "unintended" consequence of the dissemination of informatics to personal use is a growing information gap already implied by the personal computer. A relatively small number of people—no more than ten million in the United States—will, before the turn of the century, be fully wired to world sources of information and new knowledge: libraries, electronic newspapers and journals, conferences and forums on specialized topics, and colleagues, irrespective of country or region around the globe. Despite the much-heralded electronic highway, which will be largely devoted to entertainment products, the great mass of the world's population, already restricted in its knowledge and power by the hierarchical division of the print media into tabloids and newspapers of record, will henceforth be doubly disadvantaged.

Of course, the information gap makes a difference only if one considers the conditions for a democratic, that is, a participatory, society. If popular governance even in the most liberal-democratic societies has been reduced in the last several decades to *plebiscitary* participation, the potential effect of computer-mediated knowledge is to exacerbate exclusion of vast portions of the underlying populations of all countries.

The problem is not, as many have claimed, that the U.S. economy has gone global. Not only has *every* economy become global, but in fact economies have been "global" for centuries.[7] International trade and investment were part of the impetus that brought Columbus to these shores more than five hundred years ago. The movement of both capital and labor across geographic expanses has been, since 1492, a hallmark of U.S., Caribbean, and Canadian economic development. From its days as a series of English, French, and Spanish colonies, the geographic expanse that, eventually, became the United States and later Canada was, because of fertile agricultural and horticultural production, a valuable source of food, raw materials, and tobacco for Europe. Only after 1850, more than three centuries after Columbus, did the U.S. economy enter the industrial era, the major products of which—textiles, iron, steel, and later machine tools, for example— were closely intertwined with the European economy.

In fact, the era of U.S. international economic dominance began, at the beginning of the twentieth century, with its leadership in the development of industrial uses for electricity, oil, and chemicals. The application of electricity to machinery and chemical manufacturing eliminated vast quantities of industrial labor and accelerated the process whereby science, rather than craft, drove the production process. The primacy of knowledge is reflected as well in the rationalization, by assembly-line methods, of automobile production even though, at least until the late 1970s, this type of mass production required vast quantities of semiskilled labor. Needless to say, under pressure from growing international competition, the introduction of computer-mediated robotics and numerical controls has, since the late 1970s, enabled auto corporations to substantially reduce their labor forces and impose other efficiencies in production.

In retrospect we can see how temporary U.S. domination of world industrial markets really was. Despite truly remarkable industrial development, agricultural products are still the most important U.S. export commodities today. In the wake of the rise of Middle Eastern, Soviet, and Latin American oil production in the interwar period, the rapid recovery of European industrial capacity by the 1960s, especially in conventional mechanized industries such as autos and steel, and the truly dramatic development of Japanese export industries in consumer durable goods such as cars, electronics, and computer hardware as well as basic commodities, the U.S. economy had little else besides agriculture and its substantial lead in the development of informatics to commend it to foreign markets. Only France and Canada were able to offer significant farm competition, and Japan was the only significant competitor in computer hardware, though the explosion of the personal-computer market was initiated by IBM. The major industrial commodities in which the United States still enjoys an uncomfortable lead are computer chips and software. Japanese-made IBM clones have outdistanced the parent in the production and sales of personal computers. And it has taken the alliance of Intel, the world's premier chip producer, and Microsoft, whose lead in software is quite wide, to fend off Japanese and European competition.

What is new is that after a century of expansion and then world economic, political, and military dominance after World War II, the U.S. lead remains only in these sectors—and in military production. The United States is now the only military superpower. Still a formidable economic and political force, the United States nevertheless no longer commands, in terms of either production or, increasingly, consumption, anything like its former authority. Moreover, the United States has lost its great historical leadership and ability to mobilize science in the service of technological change. Today, not only Japan, Korea, and Germany but also China and India are producing engineers and scientists who compare favorably with those employed in the West.[8] Consequently, U.S. companies engaged in the production of high technologies such as robotics, lasers, and other computer-mediated industries are either relocating in these regions—not for their cheap manual labor, but for their cheap technically and scientifically trained labor—or are in fact communicating with this labor.

New uses of knowledge widen the gap between the present and the future; new knowledges challenge not only our collectively held beliefs but also the common ethical ground of our "civilization." The tendency of science to dominate the labor process, which emerged in the last half of the nineteenth century but attained full flower only in the last two decades, now heralds an entirely new regime of work in which almost no production *skills* are required. Older forms of technical or professional knowledge are transformed, incorporated, superseded, or otherwise eliminated by computer-mediated technologies—by applications of physical sciences intertwined with the production of knowledge: expert systems—leaving new forms of knowledge that are *inherently* labor-saving. But, unlike the mechanizing era of pulleys and electrically powered machinery, which retained the "hands-on" character of labor, computers have transferred most knowledge associated with the crafts and manual labor and, increasingly, intellectual knowledge, to the machine. As a result, while each generation of technological change makes some work more complex and interesting and raises the level of training or qualification required by a (diminishing) fraction of intellectual and manual labor, for the overwhelming majority of workers, this process simplifies tasks or eliminates them, and thus eliminates the worker.

The specific character of computer-aided technologies is that they no longer discriminate between most categories of intellectual and manual labor. With the introduction of computer-aided software programming (CASP), the work of perhaps the most glamorous of the technical professions associated with computer technology—programming—is irreversibly threatened. Although the "real" job of creating new and basic approaches will go on, the ordinary occupation of computer programmer may disappear just like that of the drafter, whose tasks were incorporated by computer-aided design and drafting by the late 1980s. CASP is an example of a highly complex program whose development requires considerable knowledge, but when development costs have been paid and the price substantially reduced, much low-level, routine programming will be relegated to historical memory.

The universal use of computers has increased exponentially the "multiplied productive powers" of labor.[9] In this regime of production, the principal effect of technological change—labor displacement—is largely unmitigated by economic growth. That is, it is possible for key economic indicators to show, but only for a short time, a net increase in domestic product without significant growth of full-time employment. On the other hand, growth itself is blocked by two effects of the new look to working in America. Labor redundancy, which is the main object of technological change, is, indirectly, an obstacle to growth. In the wake of the shrinking social wage, joblessness, the growth of part-time employment, and the displacement of good full-time jobs by mediocre badly paid part-time jobs tend to thwart the ability of the economic system to avoid chronic overproduction and underconsumption.

The Global Metastate

Thus, for many employers the precondition of weathering the new international economic environment of sharpened competition is to ruthlessly cut labor costs in order to reverse the free fall of profits. The drop of profits over the past five years may be ascribed to a number of factors including declining sales; increased costs of nearly all sorts, but especially of borrowing; and the high price of expensive technologies used to displace even more expensive labor. But many corporations experience profit loss in terms of falling prices, a telltale sign of *overproduction* in relation to consumption. Along with labor-displacing technological change aimed at reducing the size of the labor force, wages must be reduced and benefits cut or eliminated, especially those that accrue on the basis of length of employment. And, wherever possible, employers are impelled to export production to areas that offer cheap labor and, like Mexico, free plants, water, and virtually no taxes.

These measures produce chronic overproduction of many commodities that formed the foundation of postwar domestic growth: cars, houses, and appliances. Consistent cost cutting leads to a domestic labor force that suffers short-term—that is, security-free—jobs. This situation is exacerbated by the accelerated globalization of production and the current international recession, so that raising the level of exports as a means to overcome the structural crisis within the national economy is much more difficult to achieve even as it plays a greater role in foreign policy. In fact, the very notion of "exports," just like the notion of a purely national working class in a global economy, is problematic if not already anomalous.

Here, from the economic perspective, we can observe the effective breakdown of the purely "national" state and the formation of what might be called the "metastate," in which the intersection of the largest transnational corporations and the international political directorates of many nations constitute a new governing class. Institutional forms of rule—multilateral trade organizations such as the General Agreement on Tariffs and Trade (GATT) and the North American Free Trade Agreement; proliferating international conferences on terrorism, technological change, and new forms of international economic arrangements in which business leaders, diplomats, academics, and other "experts" regularly consult; and increasingly frequent summits among government leaders of the key national states, usually flanked by trade representatives recruited from the international business establishment—are taking over.

Until recently, from the perspective of these metastates, to the extent that currency regulation remained a national affair, national states were important as the major means for valorization of capital. Labor was regulated within the framework of national law, and police forces and armies were raised in this way. Of course, the nation, with or without the state, remained the context within which culture and ideology are produced, itself an aspect of control, at least from the perspective of international business. Although these functions are still partially served by national states, we may discern, in the various forms of spurious capital formation made possible by informatics, a definite decline in the valorization functions of national treasuries; the emergence of a de facto international currency undermines the power of the dollar, the yen, and the mark as universal media of exchange. Further, international capital has forced many states to relax enforcement of protective labor codes if not the law itself, leaving employers freer to pay lower wages, export jobs, and import (undocumented) labor. The very idea of national "border" in all except its most blatant geographic connotation is becoming more dubious as labor flow becomes heavier between formally sovereign states. Finally, while elements of national culture remain, the past quarter century is definitively the era of media and cultural internationalization, precisely because of available technologies as well as the proliferation of transnational production and distribution companies for (primarily) U.S. cultural products. The international culture industry has destroyed all but a few national film industries (in Europe, in France and Germany but not in Britain and Italy; in Asia, in India; and in Latin America, Argentina's is dead, Mexico's weak, and Brazil's almost nonexistent). American television syndication has reached deeply into the world market, and only Great Britain and, to a lesser degree, France have achieved transnational dissemination in Western countries.

The pressure on profits and the imperative to subsume labor under the new global arrangements is the "rational" basis for the decimation of the industrial heartlands—of the United States as well as European countries such as France and Great Britain—manifested in plant closings, drastic workforce reductions, and the definitive end of the social compact that marked the relationship between a significant portion of industrial labor and corporations since the New Deal and the postwar European compromise between capital and labor. Ronald Reagan's dramatic and highly symbolic firing of 11,000 air traffic controllers in 1981 may be remembered as the definitive act that closed the book on the historic compromise between a relatively powerful, if conservative, labor movement and capital. As the American unions whimpered but offered little concrete resistance, employers' groups quickly perceived that it was possible to undertake a major frontal assault on labor's crucial practice, collective bargaining. The ensuing decade witnessed rapid deterioration in union power and *therefore* a decline in real wages (what income can actually buy) for a majority of workers. Millions of women entered the wage-labor force in part to mitigate the effects of a fairly concerted employer/conservative campaign to weaken unions and to reduce wages and salaries beginning in the 1970s.[10]

In the 1980s, the two-paycheck family became a commonplace. Of course, the entrance of large numbers of women into the wage-labor force was also a sign of their growing refusal to accept subordination within the male-dominated family. At the same time, as the computerization of the labor process accelerated, millions of well-paid industrial jobs were eliminated by technological change and others migrated to the global South, both within the United States and in other parts of the world.

Recall that one of the major terms of the compromise between labor and management forged in the 1930s and 1940s was the exchange of the job control inherent in traditional crafts for high wage levels, which, in the era of U.S. domination of world markets for autos and steel, for example, spread from the crafts to many categories of unqualified labor. When the U.S. labor force was about 60 million, most of the 20 million industrial workers belonged to unions that negotiated steadily increased wages and benefits. At the end of World War II, union strength had grown to nearly a third of the labor force. By 1990, however, when the nonfarm labor force had reached about 105 million and factory and transportation employment was about 22 million, unions represented only 16 percent of the labor force, 12 percent in the private sector. Although in key industries such as autos, communications, chemicals, oil, electrical, and steel unionization as a percentage of the nonsupervisory workforce had not significantly diminished, collective bargaining no longer determined general wage levels. In fact, union wages were being driven by the low-wage nonunion sector abroad as well as at home. Having surrendered job control, workers were unable, and often unwilling, to control the pace and effects of technological change lest the employer close up shop, and their diminished political power made unions virtually incapable of stemming capital flight. Needless to say, union concessions in the form of wage and benefits reductions in the 1980s failed to halt industrial migration. This resulted in lower living standards for many Americans.

As we have argued elsewhere, the surprisingly feeble union response to the concerted employer offensive on the social compact that drove labor relations for almost forty years may be attributed to one of the tacit provisions of that compact: that labor accept its role as merely another variable "factor" of the costs of doing business.[11] Fulfilling labor's historical social justice agenda became dependent on the health of American business and subordinate to the exigencies of U.S. foreign policy.

Organized labor's ideological subordination in the first decades after World War II seemed to serve American workers' interests well. Real wages rose nearly every year until 1967 and, with short disruptions, continued to improve until the late 1970s. During this period, the AFL-CIO was perhaps the most reliable and powerful nongovernmental organization that provided a social base for U.S. foreign policy, especially its periodic war aims, but also its program of intervention in Europe, Africa, and Latin America, where American unions supplied training and financial assistance to "free" (read anticommunist) trade unions.[12]

As important as this full-throated patriotic fervor was for disciplining American workers, at least politically, perhaps the most important result of labor regulation since the New Deal was the emergence of a highly autocratic—in some cases semifeudal—labor bureaucracy to administer the terms of regulation, especially the crucial task of keeping workers in line. The labor baronate is among the most stable of U.S. institutions. In fact, many large unions resemble, in their culture as much as their structure, the large corporations with which they bargain. Far from the social movement in which they had their roots, many unions became, in effect, the mirror image of the corporations with which they bargain collectively. Although,

24

since the 1930s, unions have modified the old doctrine of business unionism by adopting a distinct political agenda, their daily operation resembles that of an insurance company, and union leaders take on the aura of corporate executives.

Even before the debacle of the 1980s, unions in private production and service sectors had ceased to grow, even as they experienced enormous expansion at all levels of public employment after President Kennedy's 1961 executive order sanctioning union recognition and bargaining for federal employees. Yet in banking and financial services, in the crucial industrial South, and among the rapidly expanding scientific and technical categories, organized labor made almost no inroads, even during the booming 1960s when ostensibly prolabor Democratic administrations were in power. The 1947 Taft-Hartley amendments, especially the anticommunist provisions and restrictions on the use of the strike weapon, halted labor's forward march during the 1950s, and labor's vaunted legislative clout was unable to rescind this blight even under the most favorable political circumstances; the cold war continued to drive labor relations, but, perhaps significantly, except for teachers and other public employees, hospital workers, and agricultural workers, unions had ceased to be a dynamic force in U.S. society.

Computers and Economic Growth

In sectors run by computer-mediated work, productivity outstrips growth. Now, *productivity* is a hotly disputed term. The conventional economists' definition focuses on the formula of measuring the ratio of the price of goods and services, adjusted for inflation, to labor costs. In these terms labor productivity, which has stagnated since the recession, grew in 1992 by nearly 3.5 percent, while the gross domestic product (GDP) increased by less than half that percentage.[13] But in specific sectors such as industrial production, productivity is much higher, perhaps 5 percent, which, in view of the relative stagnation of manufacturing, accounts for the substantial reduction of the labor force in these sectors. Computer-mediated work has spread from offices and the industrial sectors where a high level of corporate concentration prevails (such as autos, machine tools, and steel) to light manufacturing and, more recently, to the professions. Economic growth, which has proceeded in the last decade at a very modest rate (1 to 3 percent a year), is structurally unable, in the long run, to overcome the job-reducing effects of technological and organizational changes. This relatively new development contrasts sharply with the two great periods of U.S. growth in this century: 1900 to 1925 and the years between 1938 and 1970. Although the average annual growth rate was about 3 percent during the period after 1938, the level of technological displacement was relatively low and offshore capital flight virtually absent, except in shoes and the needle trades. However, the internal capital flight began at the turn of the twentieth century, principally in textiles, garments, shoes, lumber, oil refining, and steel. While the distribution of industry changed within the United States from the Northeast to the Middle West and the Southeast, the national workforce grew substantially and so did income.

Of course, much of the prolonged era of general expansion of the economy and the proliferation of jobs corresponded to spurious growth associated with the permanent war economy. That is, much of U.S. economic activity in the last great expansion period is attributable to enormous state investments in military goods that do not circulate but remain within a self-contained defense sector. "Real" growth, that is, the accretions of capital stock for the civilian sectors of the economy, was severely limited by what might be termed "disaccumulation," defined here as investments in sectors such as public employment, military expenditures, and financial services that exist to get rid of surplus capital.[14] As we have noted, from the perspective of the individual or institutional investor, putting money in these activities may yield higher returns and are evaluated by neoclassical criteria as a "rational" choice. But spurious capital formation, favored since the 1950s over industrial production and transportation, conspired to accumulate debt rather than capital and contributed to the relatively slow growth rate.

From the Craft Era to Fordism

Once craftspersons controlled production. From the point of view of the labor process, the skilled worker of the early, manufacturing phase of capitalism was really no different from the artisan who owned as well as operated the shop. The artisan designed the product and made and chose the tools needed to shape the raw material and the tools needed to convert the raw materials into a finished product. In manufacturing, the main change was the shift in property relations. The craftsperson now worked for an employer who may or may not once have worked at the bench. More typically the employer was a merchant who supplied the building, some of the tools, the raw materials, and the means to bring the product to the market. The procedures of the traditional craft(s) still dominated the labor process, which, in that period, remained virtually unchanged.

The industrializing era of the late nineteenth century was marked by the growing rationalization of the labor process. The introduction of rationalized methods of production and scientifically designed machinery simultaneously reduced, but did not obliterate, the crafts. The machine was

26

designed to "transfer" craft knowledge to those who organized production, and to machines that were designed by engineers, who became technical adjuncts of management. Artisans began to lose their design function, except for the modifications required to adapt machines to specific production requirements. The crafts were progressively reduced to making tools and maintaining or fixing machinery designed by an engineer.[15]

Marxist theorists of the labor process have attempted to account for this rationalization process by citing the pressure of capitalist competition on profits and have tended to reduce explanation to its economic dimension. But concrete studies of the labor process reveal, time and again, the degree to which reducing the worker to an interchangeable part by means of "rational" organization and machine technology rationalization (which was, at that stage of development, a kind of analog to actual work and worker) is also paralleled and dominated by the struggle among the workers, management, and the owners for *power*. The chief weapons of craft resistance are the traditional skills shared by the community of craftspersons and their *culture*, which prescribes *how* things are to be done consistent with the moral economy implicit in their culture. As E. P. Thompson and Harry Braverman have suggested, the struggle of capital to dominate the labor process has for centuries been linked to breaking craft culture, not only by bitter tactics such as provoking strikes and lockouts to smash craft unions, but also by making mechanization and rationalization a "natural" process that is regarded as socially progressive. What could not be accomplished by Taylorism was finally achieved by computerization.[16]

A case in point is the transformation of the relatively "crude" division of labor in carriage production to the extreme rationalization of the Ford assembly line in the second decade of the twentieth century. Henry Ford was able to portray the introduction of the conveyor belt and the breakdown of assembly tasks to fine details as a function of "efficiency." By the 1920s, together with Frederick Winslow Taylor's reports of his own industrial engineering innovations at Midland Steel, Frank Gilbreth's work in "scientific rationalization," and Lillian Gilbreth's application of the Taylorist system to the household, the assembly line as an emblem of industrial efficiency became a new "common sense."[17] People debated whether goods made by such methods were equal or superior in quality to those made by the older craft methods, but there was virtually no debate about whether these commodities could be made at prices most working people could afford. Selling the idea of a car in every garage was one of the major rationales, and also the promise that, with the introduction of mass production, one could purchase almost any product, even a house, from a Sears catalog. Fordism——including mass production, relatively high wages, and mass consumption through a liberal credit system—quickly became the

American *cultural ideal*.[18] Thanks to mass production and easy credit, the prevailing ideology was that nearly every hard-working American could own a house, a car, and appliances without saving for a lifetime. Of course, this new common sense was frequently violated by the persistent evidence that many who had successfully internalized the common sense of hard, steady work as the path to success often had trouble paying the rent, let alone affording a late-model car.

Yet there is little doubt that the economic and social history of the past century may be identified, to a considerable degree, with the conjunction of a new regime of capital accumulation based on the expansion of consumer credit (debt) with the new cultural ideal in which the categories of want and need were melded so that absolute need became confused with want.[19] Or, to put it another way, the distance between want and need were so confused that large segments of the population were taught to experience a sense of deprivation as their inability to achieve what has been established as the distinctly "American" living standard grew. In this regime of regulated accumulation, there is little point to comparing U.S. living standards in terms of the conventional measures of the earlier industrializing era when the issue was whether or not workers had adequate food, clothing, and shelter (could meet their needs). Although there is growing evidence that for an important and growing minority of the populations of advanced industrial societies these "necessities" have been denied, the economic/cultural situation of perhaps of a majority of citizens of these countries is still framed by expectations created during the "Fordist" age. Despite determined efforts by ideologists as well as political leaders to complete a post-Fordist counterrevolution, selling, as it were, lowered expectations (new ideology, which began to be sold during the great oil "crisis" of the mid-1970s), large segments of the U.S. population have internalized a contrary, if materially antiquated, cultural ideal.

In the rise of the Taylorist-Fordist regime, the "mass," semiskilled worker or artificer is organized into a highly cooperative division of labor and assigned specific tasks that are closely integrated by managers and engineers who monopolize, and thus keep secret, as much knowledge as they can.[20] In the early mechanical era, the conflict between employers and their agents with the artisans, who still possessed considerable know-how, led to hard-fought strikes, machine breaking, and other forms of protest. The battle extended into the twentieth century, for even though artisans were being driven from their central place in the production process, they remained indispensable as long as tool and die making could not be performed by unaided machines, machines that could not yet fully build other machines and were unable to repair themselves. After the crafts were broken up into detailed and repetitive tasks, these detail workers (mass workers, in our usage) became interchangeable like standardized machine parts. This situation, felicitous for management, diminished but did not eliminate the capacity of the mass worker for self-organization. When the mass production workers organized after the turn of the twentieth century into strong industrial unions and were able to form alliances with the somewhat more privileged "skilled" workers (themselves performing only some of the segments that once constituted a traditional craft) and with workers in other plants within the same company and industry, they were able to exercise a substantial degree of power that mediated, if not reversed, management's monopoly over the labor process. Employers counterattacked, beginning in the 1920s, first by moving plants, especially in light industry, to nonunion regions of the country and then by globalizing industrial production so that the mass worker, who benefited from the concentration of production in a few large factories, was considerably weakened. And, lacking a truly international movement to coordinate labor's resistance, unions in industries such as textiles and shoes were decimated even as unions in other industries were still expanding.

The heart of management's power was its ability to control the labor process, especially to develop technologies of production and administration that progressively eliminated large quantities of labor and further transferred worker know-how to machines. By the middle of the twentieth century, most key industries, while still ensconced in the mechanical mode of production, were already highly rationalized. The conveyor-driven assembly line was only one of the methods of control. The fabrication of almost all major appliances, clothing, processed food, and machinery itself was organized according to the rules of detailed rationalization even when no conveyor belt was used. For example, a power press operator spent the entire shift stamping out a single sheet metal part of a television set, a section of an auto body or of a washing machine. Machine tool operators worked on lathes that had already been set up and whose cutting tools were presharpened, producing thousands of screws, nuts, and other small parts. In another part of the shop a drill press operator spent the entire day performing a single operation on a machine tool designed for the specialized function of cutting holes.

Specialization also arose within engineering: quality control, industrial, and design engineering became subspecialties, and managers and engineers at first recruited from the ranks of craftworkers were increasingly considered professionals, even when they did not have academic credentials. The professionalization of engineering signaled a regime in which large segments of industrial production became based on applications of natural science after about 1850; this process accelerated after the Civil War. By the beginning of the twentieth century, formal credentials increasingly became a condition for advancement beyond the level of foreman within large enterprises. Gradually, management was recruited from the growing ranks of those trained in business and professional schools, but the front lines still remained a shrinking site where a sliver of manual labor could enter the ranks of low-level management. Plant superintendents intimately familiar with most of the jobs because they had performed most of them during their years as manual workers began to be replaced by administrators with academic credentials who relied on a staff of production specialists. The plant manager became a kind of cost accountant and personnel manager and, above all, a corporate politician without shop floor background. As is evident from this description, the emergence of management and engineering is closely linked to the process of rationalization, that is, to the expanded requirements for control of labor.

This, of course, is a highly schematic account; the "pure" type never corresponded to the actual labor process in which the craftworker, the "mass worker," and the engineer modified each other's core functions through conflict as well as cooperation. This noncorrespondence between model and performance attests to the persistence of tacit as opposed to formal, abstract knowledge. To the extent that craft and other categories of manual labor had a primary relationship with "nature" throughout the industrializing era, they remained indispensable for transforming materials into products. The craftsperson's intimate knowledge of the "feel" of materials such as wood and metal is an integral feature of the labor process: the blacksmith "knows," tacitly and by "touch," how much heat to apply to metal in order to soften it for fabrication, just as the shoemaker can discern the texture of leather without referring to a manual. It may be argued that the power of scientific knowledge to transform the labor process in the image of capitalist rationality relies, historically, upon two developments: the invention of artificial materials based on the chemistries of hydrocarbons and petrochemicals, from which synthetic fibers and especially plastics may be produced; and the rationalization of the labor process via innovations such as those introduced by Fredrick Winslow Taylor and Henry Ford, the aim and result of which is the subordination of craft. Chemistry as the basis of synthetics technologies replaces natural products such as textiles, leather, and metals whose quality depends, crucially, on traditional craft-wrought knowledge. And crafts are broken up by techniques of organization.

These technological innovations occurred in the context of the dramatic improvement of communication and transportation technologies—the telegraph, the telephone, the airplane, the motor vehicle—that helped extend the market for goods and information beyond conventional borders. We do not claim a direct relationship between scientific knowledge, invention, and economic relations. Indeed, many scientifically based technologies were invented before the conditions for their wide dissemination were present. Yet we do not doubt that futuristic dreams, in both fiction and popular philosophy and science—the aspiration to fly, to reach the moon and the planets, to achieve the laborless factory—are conditioned by the *dialectic* between scientific discovery, economic development, and political struggle. Jules Verne's celebrated nineteenth-century journeys to the bottom of the sea, to the center of the earth, and to outer space presupposed the emergence of rudiments of scientific and technological possibility and the collective perception that the economic means were, at least hypothetically, available. Similarly, in 1914, H. G. Wells's novel *The World Set Free* foretold the discovery of nuclear chain reaction and the consequent development of an atomic bomb that changes the face of world politics and culture.

Of course, from Mary Shelley's *Frankenstein* to Fritz Lang's *Metropolis* to Phillip K. Dick's novels of simulacra *Do Androids Dream of Electric Sheep?* (made into the movie *Blade Runner*), *We Can Build You*, and *The Three Stigmata of Palmer Eldritch* (among many others), writers and filmmakers have envisioned the dystopian consequences of fulfilling what is, perhaps, the most vivid capitalist fantasy, the automatic factory and human-free work. Most recently, Alvin Toffler and Jacques Attali have offered futurologies that confirm prognostications of the workerless factory and office.[21] With computer applications in virtually every aspect of work and play, none of these once unimaginable dreams is beyond practical possibility.

The second consequence of the replacement of mechanical tools of both production and administration by computers is the transformation of the occupation structure of the old mechanizing era. In this transformation, every new technological generation transfers more skill to the computer, and the life of each generation of computer technology is, in terms of acquired knowledge, radically foreshortened. In fact, the point of the technological change is to elide a work regime fatally subject to human error in which tolerances are extremely narrow; in the mechanical era, hand tools such as micrometers and calipers were important verification devices.

Culture as habit formation includes language and discourse as well as algorithms and other practices. These practices often involve the habituation of the body as well as the "mind" in ways that render a practice largely second nature. For example, the skilled worker in a cold rolling mill adjusts the rollers by "feel" to attain the desired width of the sheet or wire. Similarly, "feel" is the major instrument of micromeasurement by the machinist on work in which tolerances are extremely narrow; micrometers and calipers are important verification devices. But as any experienced machinist or tool and die maker will attest, intuition plays the major role in meeting "close tolerances."

In the period of automatic production, the motions, practices, skills, and intuition of the mechanic are copied and programmatically simulated and mimicked by an instrument. As exemplified by computers, the digitalization of machinery signifies the final triumph of science and technology over craft, abstract over concrete labor, the "mind" over the "body," or, to be more exact, the ideology of the dominance of the intellect over the body-subject. Number is the language of science; the body is the language of craft. Or, put more precisely, although number is latent in older forms of analog machines, its full impact as a governor of labor is still mediated by traditional skill. In contrast, while it is true that the engineer, for example, working with a computer-aided drafting and design program still relies to some degree, on hunch and intuition and must understand mathematics in order to make choices among several options, the program allows only a finite menu of options. Although designing a switch or transformer, routine electrical engineering work, is never an entirely spontaneous activity, the three-dimensional graphics on the computer screen partially displace the imagination, at least for routine designs, while at the same time stimulating the imaginative dimension of more complex design work.

Writing programs and designing systems that displace or otherwise alter the character of intellectual labor is, at first, heady stuff for high-level computer engineers. One must reproduce, rationalize, and transform, in the logical language of computers, aspects of professional knowledge, rather than the classical manual task. But, after barely two decades of experiment with and perfection of these programs, impelled by the cost-cutting imperative, programmers found ways to displace their own work. Today computer-aided software programming eliminates much of the routine work of the programmer, just as, in the 1970s, numerical controls and robotics eliminated vast quantities of craft and unqualified labor. Knowledge, once the vehicle of the unprecedented expansion of our collective dominion over nature, including traditional skill, has turned on itself. Its *tendency*, if not a completed task, is toward the self-regulated workplace in which no category of labor is spared, even some categories of management.

Of course, not all paid work is obsolete, especially work that entails the perfection of control, but this is the tendency that results from the displacements currently in process. The new computer-mediated technologies bifurcate paid work in a new way. Much of this work, in both its "manual" and its "intellectual" incarnations, is now engaged in the labor of *control* of living labor or its machine form. In other words, where in the previous regime of mechanical production the skilled worker was partially disempowered by the rationalization of the traditional crafts and transformed into a mass worker, the technologies of many knowledge-based industries require a more qualified worker.

For example, production workers in oil refineries are required to understand and to monitor fairly complex computer-mediated production systems and be able to operate similarly intricate environmental and safety equipment. The environmental equipment conforms to state and federal antipollution requirements, but the safety equipment is installed to protect expensive machinery as much as the workers. And, to operate these systems properly requires the production worker to have some scientific and technical knowledge as distinct from mere familiarity with operational procedures and rules. Hence, production workers in this industry, perhaps the most sophisticated and costly of advanced technologically based industries, must undergo some degree of scientific education as opposed to manual or clerical training. But in 1991 there were fewer than 50,000 production and maintenance workers in U.S. oil refining as compared to nearly 200,000 in 1960. Technological change has produced similar reductions in steel, electrical equipment, auto, and even traditionally low-technology industries such as garments and textiles, where computers are now typically attached to sewing machines and to looms. This trend is, of course, much better known in offices, where computer-mediated electronic devices dominate the labor of both clerical workers and professionals.

Much of this book concerns the implications and consequences of the proliferation of scientifically based technology for the future of paid work. Our broad conclusion is that the highly publicized benefits of these technologies for work and its culture are vastly overblown. For an ever smaller number of people in virtually all occupations, the qualifications required by computers have created new opportunities for satisfying and economically remunerative work. But for the immense majority, computerization of the workplace has resulted in further subordination, displacement, or irrelevance. As we shall argue in the last section of this work, the transformation of production from mechanically based technologies to what some have termed the "mode of information" not only changes the notion of the "subject," as Mark Poster has observed, but also changes the prospects that paid work can remain the defining activity of human existence.[22] It is not a question of posing labor against "leisure," for in this dichotomy work remains dominant. Rather, we pose questions that, although they have been on the agenda of philosophers and cultural theorists for much of this century, have not (yet) intruded into the precincts of social policy—except welfare. To raise the question of the partial eclipse and decentering of paid work is to ask crucial questions concerning the purpose of education, the character of economic and social distribution, and, perhaps more profoundly, what it means to be human.

The Incorporation of Psychology

In chapter 2 we will discuss the by now widely accepted claim that the effect of computerization is to reintegrate work. Clearly, many students of computer-mediated work such as Shoshana Zuboff acknowledge that the long wave of industrializing capitalism was marked by the transformation of highly integrated, qualitatively oriented craft labor into a series. For much of the growth of the new technical and administrative middle class in this century is simply the other side of the growth of the mass worker. Designing machines and organizing production and distribution of goods were intimately connected to both the rationalization of labor and the struggle over knowledge. "Technology" was no longer an outgrowth of the older concept of "techne," which, in its Greek etymology, means the activity of "uncovering" or "disclosing" nature (including human nature).[23] Now, the process of "disclosing" was intertwined with the exploitation and commodification of nature. That is, the role of science and technology is to uncover and *select* for investigation those properties that are useful for specific ends, particularly production and profit.

The "human relations" (recently transmuted to "human resources") engineer became a corporate technician who sought to explore and exploit worker motivation. The primary work was to apply the insights of behavioral and Freudian-derived psychology to finding ways to improve the efficiency of the worker not only in the productive labor process, but also as an instrumentalized consumer. Methods to enhance worker "motivation" were by no means confined to wage incentives such as more humane piecework systems; in larger corporations, human relations experts, who are frequently professionally trained in psychology, may intervene in the workers' personal problems such as addiction to drugs and alcohol, family issues such as divorce, and children's school difficulties. Perhaps most important, they may assist the employer in motivating workers to identify with the company, its business prospects, its production problems, its personality.

The incorporation of psychology as an arm of industrial and labor relations exemplifies the long-range movement of science as a relatively free enterprise of curiosity ("techne") to science in the service of capital ("technician"); it is only a relatively late example of the general subsumption of science under capital. From an early emphasis on wage incentives to persuade workers to increase their productivity in quantitative terms, the new industrial psychology has been induced, by the increasingly competitive economic environment, to develop methods of improving the quality of products. Although initially developed in the 1920s as an employer-generated alternative to trade unionism, the human relations movement in industry has expanded to include such innovations as "quality circles" in which employees are encouraged to participate in finding ways to improve production methods, criticize arbitrary management practices, and raise the quality of the product through dedicated labor. Many unions were at first dubious about, even hostile to, management-generated worker participation schemes. In the current grim economic environment where every day brings further news of labor-force cuts and plant closings, however, many unions have revised their antagonism to quality circles and other productivity schemes. As a result, accident rates in industrial plants have jumped as traditional workplace safety and health protections and other restrictive work rules are modified or scrapped as too expensive to maintain the position of "their" employer in the world market.[24]

The Subordination of Scientific Knowledge

While modern science has always been indebted to some extent to capital, the beginning of the subordination of science to the requirements of capital can be traced to the development of electromagnetism, Babbage's calculating machine experiments, early genetic experiments, and Lyell's development of the science of geology, each of which exhibited characteristics of both the older science and the new technoscience. From Lyell's work in the 1820s and 1830s until the late nineteenth century, geology was constituted by its articulation with the development of the theory of evolution; it was the premier discipline of *natural* history, of which archaeology and evolutionary biology were, respectively, the study of the history of living things and, specifically, of human evolution. In 1858 Lyell himself presented the papers of Darwin and Wallace on natural selection to the Linneaus Society.[25] With the emergence of coal and oil as crucial fossil fuels in the iron and steel industries, for electricity, and for providing power to the internal combustion engine, however, geology became an industrially linked discipline. Like botany's dual role in evolution and commercial agriculture, Faraday's work on electromagnetism similarly led to new physical discoveries and to broad industrial uses in lighting, the invention of electrical motors, and production machinery. Faraday himself stood in both realms; he was the grand old man of British science by the age of forty but also patented his discoveries for commercial uses.[26]

There is no question of the direct subordination of science to industry until later in the nineteenth century when, as we have seen, chemistry was more directly integrated into large corporations. Rather, capital supported the allocation of substantial government funds for research, especially within the universities and independent research institutes. Faraday's employment by the British Royal Institution was an early indication of the end of the era of the so-called gentleman scientist of which Joseph Priestley, Robert Boyle, and Charles Darwin were exemplary figures. Increasingly, physical, chemical, and biological research required elaborate facilities that most larger corporations were unable (or unwilling) to provide because of the uncertain practical results of any basic research. It was not until some U.S. corporations, especially General Electric and American Telephone and Telegraph, became directly involved in transforming Faraday's discoveries into commercial uses that capital entered the research field. Yet despite the extraordinary sophistication of these examples of corporate-sponsored research, few of them were prepared to displace universities as the primary site of basic research.

Babbage's work on the difference engine and on the calculating machine reveals a similar ambiguity between the realms of pure science and practical science. Like his friend Faraday, Babbage was acutely aware of the relevance of his inventions to industry. Himself a prominent, even leading political economist, he made detailed studies of the labor process while working on his computer and saw its development as a potentially important addition to British industry. At the same time, he was interested in the progress of British science and participated actively in the intellectual debates within the Royal Society, the leading forum of the scientific community in Great Britain.

We may observe parallel developments in American science in the latter half of the nineteenth century. By this time, Faraday's discoveries had had an enormous impact on American physicists such as Willard Gibbs and inventors such as Thomas Edison and George Westinghouse. In the United States, the integration of science and technology first took hold in the development of agriculture, which, perhaps in advance of most of Europe, became scientifically based even before the end of the century. This rapid transformation was promoted by the federal government's grants of land to states that agreed to establish colleges devoted in large measure to research and education in agriculture and mechanical technology.

As David Noble has shown, the growth of American industry depended on the emergence of engineering and other technical education programs in these land-grant colleges and private universities.[27] Geological surveys conducted by engineers and scientists were crucially necessary for the discovery of coal and oil resources; experimental physics emerged in the wake of the widespread application of electricity in many sectors of industrial production; and civil engineering expanded with the growth of railroads and roads for wagon, truck, and automobile travel in a sprawling emerging national market. And in addition to these infrastructural aspects of industrial growth, as we have seen, "industrial" engineering became a recognized subdiscipline when, after the Great War, Taylorism and Fordism became crucial technological regimes of production. At the same time, many corporations, especially in the newer industries such as oil, chemicals, steel, communications, and electrical, established research laboratories. Three of the most famous—General Electric's Schenectady laboratory directed by Charles Steinmetz, the Edison labs at Menlo Park, and later the AT&T labs at Kearny and western New Jersey—were responsible for both basic physics research and applications to commercial and industrial uses. Increasingly, however, basic research was centered in universities while the corporations retained significant responsibility for applied science. By 1950 virtually every major corporation whose product was rooted in basic chemical and physical knowledge employed professional scientific and engineering personnel.

Despite much current discussion about the need to privatize basic science, the newest industrially oriented science, molecular biology and its applications to biomedical research, has emerged within universities. The federal government and corporations have funded some of the research on condition that they own the patents to life forms developed in university labs. Given the emerging austerity policies of most state legislatures and the high cost of scientific research, it is not surprising that many scientists have willingly accepted this bargain. The arrangement suggests not only that molecular biology is emerging as a leading technoscience but also that it will remain within universities as long as corporations can benefit directly from discovery and invention.[28]

As continuous-flow technologies came to dominate the workplace, some of the "older" industries such as oil and chemicals that for many years had been based on scientifically wrought knowledge reduced the mass worker and the skilled trades to a minority of the workforce by the 1960s. Similarly, in the telephone industry, computerized processes resulted in a rough parity between professionally and technically trained employees and the manual and clerical workforces. As a result, in both industries, supervisors and nonunion professionals and technicians can operate the facility for considerable periods of time during strikes. In an exemplary instance in the 1960s, during a strike by 3,600 refinery workers against Gulf Oil, six hundred salaried employees successfully operated the plant for a year. Similarly, the communications industry is virtually strikeproof (but not immune to internal industrial action).

These instances suggested, by the mid-1950s, a movement toward a permanent displacement of manual labor, at least within certain types of industries. More important than its relative decline in quantitative terms has been the shift of power over the production process from manual to intellectual labor. With the arrival of the cybernetic revolution (in technological terms) the tendencies already evident in continuous-flow and analog automation technologies are brought to fruition. However much the knowledge entailed in computer programming refers to its origins in crafts and older mechanical-era skills, cybernetics really signals a new era. These skills are now abstracted and converted into self-referential logical symbols. "Computer" languages translated and transmuted these skills so that they no longer appear to have an autonomous existence. Tacit knowledge has been transformed and incorporated into the mathematical system but remains invisible. More striking, intellect no longer appears as a derivation of concrete labor; the abstract symbolic system is now imposed on concrete labor as if from the outside.[29]

So it is not merely that scientifically based technical knowledge has become the dominant productive force; it appears to requalify, even as it continues to subordinate, manual labor. Consequently, for every unit of investment in technological innovations in the cybernetic era, the proportion of intellectual to manual labor increased in two ways: by the elimination of categories of manual labor and by increasing *in relative terms* the number of qualified intellectual workers while the number of manual workers, whose productive powers are magnified by cybernetic technologies, remains stationary or declines.

This historical reversal, marking the emergence of what various commentators have labeled a new middle class, a new working class, a professional-managerial class, or simply the expansion of a different kind of middle strata, has posed significant issues for the study of professions, for class and stratification analysis, and for the study of work. In 1857 Marx's notebooks for *Capital* published as the *Grundrisse* already asserted that scientific knowledge was rapidly becoming the main productive force: "The entire production process appears as not subsumed under the direct skilfulness of the worker, but rather as the technological application of science. It is hence a tendency of capital to give production a scientific character."[30] Veblen's fundamental insight—arrived at independently of Marx's statement since the *Grundrisse* was not published in German until more than a decade after Veblen's *Engineers and the Price System* (1921)—that technical knowledge had taken hold of the labor process led him to speculate that its bearers were, at least potentially, new historical agents. But Veblen warned that even though engineers were now at the center of the production system they might be subsumed, perhaps permanently, under corporate capital by inducements such as high salaries and easy routes to mobility. Of course, this is precisely what has happened throughout this century. If our thesis that computerization permits a hierarchical bifurcation of scientific and technical labor is right, however, low- and middle-range scientists and engineers may not look forward to occupying a privileged place in the corporate order.

Early Theories of the Knowledge Class

These days, the terms *crisis* and *catastrophe* are ordinary components of social communication. They are no longer consigned to revolutionary or dystopian futurity but are categories of the present and are employed with abandon by journalists and savants and on the street. Walter Benjamin's once bold remark that the catastrophe is not before us but has already occurred seems almost commonplace. If this is so, we live today amid the ruins of Western civilization, but the extent of the damage is hidden by the proliferation of material cosmetics, among them the miracle technologies that make possible the consumption of considerable quantities of cheap goods.

The ruins are not merely physical, although Mexico City and São Paolo, no less than New York, reveal the extent of the decay and destruction that have afflicted many of the world's largest cities. Perhaps more to the point is the palpable decline of the *political agents* associated with the industrializing era, even the old ruling groups. As the industrial working class is reduced and recomposed, the parties and movements linked to it suffer concomitant crisis: despite the unpopularity of Margaret Thatcher, the British Labour party was unable to assume governmental power; the French Socialists found that even structural reforms were difficult, if not impossible, to implement despite twelve years of discontinuous power that ended ignominiously in spring 1993; and the German Social Democrats, the government party of the 1970s, seem relegated to permanent opposition, suffering erosion from both the ecologists and the ultra-right-wing parties. In the United States, as we have seen, organized labor, safely subsumed under the Democratic party, plays an ever diminishing role in national politics, and even relatively modest demands such as a national health scheme and a federal jobs program to alleviate unemployment require herculean effort to enact.

The apparent eclipse of the radicalism that claimed the unfinished work of the French Revolution in the wake of World War I prompted those for whom technological development constituted a certain sign of "progress" to seek new change agents. As it turned out, the Russian Revolution was the last great event of the old radicalism. Some were provoked by the blatant evidence that the working class in the most advanced capitalist societies had cooperated with their own capitalists in pursuing *national* rather than class war aims, and began the search for new agents. For many theorists still tied to discovering agency in class relations that were ineluctably tied to production, Thorstein Veblen's argument that the "engineers" were the new subjects of advanced technology provided the basis for the judgment that technical intellectuals specifically tied to production rather than manual workers or the owners would hold in their hands the economic and political future of humankind.

Indeed, for many writers, the war was a watershed. The 1920s marked, in John Dewey's terms, the "eclipse of the public,"[31] Walter Lippmann wrote that the public had become a "phantom" in the conduct of day-to-day affairs of state, including the information media, which, more and more, were dominated by "experts."[32] The engineers of the polity, a position that corresponded, roughly, to Veblen's idea, had replaced the public as the real decision makers. According to Lippmann, the public was subject to mobilization in the service of keeping elites honest, but was no longer, if it ever was, a genuine political actor. Although Dewey disputed the ethical basis of Lippmann's judgment, he could scarcely disagree with its empirical veracity.

In the 1930s Veblen's ruminations had been followed, with some variation, by parallel if not identical theories put forward by writers such as Alfred Bingham, Emil Lederer, and Lewis Corey and, in a somewhat different vein, by James Burnham, who proclaimed at the dawn of the 1940s the "managerial" revolution.[33] Lederer, in concert with some strands of prewar German and Austrian Marxism, attempted, even before the war, to offer an account of the rise of the white-collar (administrative) worker, although he focused chiefly on the emergence of "organized" or "monopoly" capitalism by the turn of the century.

Those who retained hope for radical change even in the wake of the calumnies of the interwar decades tended to place an ambivalent bet that what Corey called the "crisis" of the middle class could provide for the working class the alliance without which any thought of profound social transformation was foreclosed. For by the 1930s it was evident to many that the emergence of what might be termed the technical—as opposed to traditional literary and philosophical—intellectuals associated with the expanded role of scientific and technical knowledge in the production process foreshadowed changes in the relations of economic and, eventually, political power in those countries where knowledge had become the crucial productive force. For Corey, the crisis consisted in the contradiction between their centrality to the new shape of economic life and the relentless proletarianization to which they were subjected. The new middle class of professional and technical employees was subject to the vicissitudes of the economic crisis: during the Depression of the 1930s, they suffered unemployment, wage cuts, and deteriorating working conditions even as they became increasingly necessary in the labor process and in administration.

These developments form the backdrop for the theme of the growing importance of the so-called new middle class, which has been repeated with considerable variation in subsequent years. The post-World War II economic expansion thrust this social category even more decisively to the middle of the new production and administrative processes, first with the development of the analog machine (the first phase of automation) and then with the introduction of cybernetic technologies in the administrative and goods-producing workplaces.

Fueled by the GI Bill of Rights, which guaranteed veterans income and tuition grants for secondary and higher education, the postwar era marks the decisive rise of a new stratum of technical intellectuals. Between 1945 and 1950 U.S. colleges and universities doubled their enrollments in the liberal arts, engineering, and the sciences. By the mid-1950s millions of degrees were awarded to men and women who would not otherwise have attended and graduated from college. These people became the cadre of the postwar knowledge class. They were sons and daughters of factory and service workers, truck drivers, and considerable sections of the old middle class of small shopkeepers and small farmers who perceived that their futures were, at best, uncertain in the midst of the growth of giant corporations in agriculture and traditional retail trades. They swelled the ranks of government and corporate bureaucracies, which also doubled in employment in the postwar period, and joined the ranks of technical and scientific employees as well. In retrospect, the GI Bill was, perhaps, the most dramatic material basis in American history for the American ideology of exceptionalism and, particularly, the belief in mobility. It provided palpable evidence that the economic circumstance of birth was not destiny by providing the tools by which occupational if not class mobility could be attained.

At the same time, these were the years during which the discovery of digital logic, first enunciated by Alan Turing in the 1930s, was being developed and applied to the new science of cybernetics by Norbert Weiner and John Von Neumann. In Steve J. Heims's words: "World War II was the watershed in the slow transformation whereby the deepest problems of science changed from the epistemological to the social and political. The war pulled the mask from the illusion that science was apolitical." In the prewar period "basic research in science and mathematics had been relatively aloof from and independent of the U.S. government. . . . The War changed all that very rapidly. Scientists were quickly mobilized by the government to participate in generally costly war-related projects with practical objectives . . . scientists themselves had become workers for the state."[34] After the war had ended and many scientists had left the military projects, the close relation between science and the government forged during wartime persisted. The primary source of research funds was henceforth U.S. government agencies, including in particular military services. Heims argues that this changed relation coincided with "the loss of innocence on the part of the scientific community," in large part a result of its wholehearted participation in the development of nuclear and other weapons "that had not only wrought great destruction but permanently altered the conditions of life."[35] Von Neumann was among the major figures in the march of science from the universities to government service during the arms race of the postwar era.

Von Neumann, a leading mathematician, made a relatively smooth transition from work on problems of nuclear fusion to crucial work in the development of methods of numerical computation that made significant contributions to both the logic that is the core of modern computers and the cognitive science that viewed the human brain in the metaphor of an information processing machine. Von Neumann continued as an important government adviser as the United States moved into permanent war preparedness during the cold war era. His work in game and systems theory became an important adjunct to decision makers in the executive branch who hoped that science could provide a more precise basis for policy formation. Von Neumann's modeling techniques were adopted, with his active participation, by the Pentagon. Von Neumann's work may be seen as an exemplary case of the integration of a significant stratum of scientific and technical intellectuals into the postwar state.

The long period of the European economic "miracle" and U.S. world dominion after World War II inevitably changed the terms of the debate about the emergence of the new intellectual stratum. Some theorists understood the development of intellectuals as a social category to signal a new close relationship of science and technology, business and the state. In effect, for writers such as C. Wright Mills, this stratum was a vital part of a "new middle class," which, however numerous, was fatally dependent on business and government.[36] Mills, who in many ways was among the last of the interwar independent radical intellectuals, still framed the problem of "white collar" in terms of class analysis. In this theoretical perspective, the rise of industrial capitalism not only witnessed the emergence of a modern, differentiated working class but also gave birth to a veritable army of technical, professional, and scientific employees whose crucial role in production, administration, and the enormous sales effort made them, at least potentially, a significant social category. However, like Veblen, whose work was among Mills's more important influences, the technical intellectuals were fated to be allied to the prevailing economic and political powers. The new employed "middle" class, as opposed to the old proprietors and independent professionals, was caught between the two great adversaries of modern industrial capitalism: the "power elite," of which the corporations were perhaps the dominant "order," and "big labor," which to Mills meant the giant mass unions and their constituents in the industrial working class.[37] While he was aware of the emergence of bureaucracy and science as key actors in the postwar period, he could not envisage an independent role for the bearers of these increasingly important institutional orders.

In this book, we shall discuss some of the twists and turns of class theory, particularly in chapter 9. Here we foreshadow this discussion by noting that, although Mills made no close analysis of the role of scientific and technological intellectuals in the fate of the "new" middle classes of salaried employees, Daniel Bell and Seymour Martin Lipset became persuaded that in contemporary democratic society the old ideological and political conflicts were ended and the ethos of this new social category of salaried managers and professionals was bound to displace class conflict. Bell's celebrated forecast of the end of ideology (except in the communist world) is linked to his assessment of the growing importance of professionals and managers in contemporary society. In place of the conflict between capital and labor, he argues, the rise of the scientific and technical intellectual prefigured a new class formation that addresses conflict by reformulating it as a series of problems subject to technical and organizational solutions. The middle class, in the postideological conception, was no longer a residual category, caught in the vise made by the two great antagonists of the industrializing era Now, in the postideological, postindustrial era, science and technology *finally* fulfill, at the level of social relations, their structural centrality to the labor process. Knowledge as the decisive productive force freed the "new" class of professionals, technicians, and managers from dependence.[38]

More than twenty years after the appearance of Bell's announcement of the end of ideology, Alvin Gouldner, working as well in the intersection of Marx and Weber, offered a new class theory. As contradictions are transformed into problems and solving them is shifted from politics to technics, the experts of all kinds, armed, in Gouldner's terms, with a culture of critical discourse, are likely to fulfill the old technocratic dream. Bell welcomes this development, while Gouldner is more dubious. But they essentially agree that the old class forces have given way to new agents of power. The major difference between them is that while Bell is specifically concerned to bury Marxism, Gouldner attempts to adapt class analysis to contemporary conditions. His category of a culture of critical discourse situates the agency of the intellectuals in their own characteristic culture, which, in terms of the old class/power nexus, is oppositional. For if knowledge may speak in its own name, its bearers may not only challenge the old agents of capital and labor, but also constitute social relations in their own image.

Bell and Gouldner, Alain Touraine, Serge Mallet, and Andre Gorz and the Frankfurt school, particularly Max Horkheimer and Herbert Marcuse, showed that social theory was able to come to terms with the significance of science and technology in the contemporary world.[39] To be sure, Bell built on the neo-Marxist effort, initiated in 1909 by Hilferding and made popular in this country by A. A. Berle, to comprehend the significance of the rise of the large corporation to preeminence within capitalism.[40] At the same time, his unique contribution, together with Touraine, was to have attempted to integrate the rise of corporate dominance with the cultural categories of science and technology to suggest a new class theory.

In the late 1950s, on the eve of the veritable global cultural revolution that, twenty years after the war, swept nearly all advanced industrial societies, significant places in the semiperiphery, especially the Caribbean and Latin America, and the already decaying oligarchies of Eastern Europe, Bell heralded the era when social conflicts would be transformed into problems of system maintenance; workers would no longer march under ideological flags but were in the process (at the time he wrote his *End of Ideology*) of being integrated into a new, rational managerial regime. Bell, Seymour Martin Lipset, and the German sociologist Ralf Dahrendorf were key theorists of the triumph of modernity, identifying liberal democracy and managerial capitalism with an early version of the currently popular "end" of history—if by the term *history* is meant monumental class struggles.[41] The "catastrophe" of the end of working-class agency became for Bell a cause for guarded celebration. To be sure, in his influential 1950s essay "Work and Its Discontents" published in the *End of Ideology* collection, Bell demonstrated that even if he was prepared to abandon the radical vision of a new society, he saw some of the remaining warts in the present as future society. One of them was work, that crucial human activity, which appeared with every passing day to become less varied, less fulfilling in human terms, and therefore more alienating.

Certainly, even in the 1950s, the notion of alienated labor was not novel. At about the same time that Bell's essay appeared, Harvey Swados's "The Myth of the Happy Worker" exploded the idea that consumer society was a nirvana for millions chained to mind-deadening, repetitious, and routinized labor.[42] Whereas Swados saw the contrast between consumerism and boring, repetitive work as a *contradiction*, the novelty of Bell's assessment was to relegate the phenomenon of routinization to "discontent," a problem like all other economic and cultural issues subject, in the new era of managed capitalism, to technical rather than historical solutions. According to Bell and others, industrial engineers, human relations experts, and economists could find ways to make work more interesting within the boundaries dictated by profitability. In any case, perhaps the scene of social "meaning" had already shifted from work to culture. In his follow-up work Bell gave theoretical shape to his thesis: the end of ideology was the product of the evolution of capitalism into a new postindustrial society in which technical reason had replaced class conflict, and those who were its agents—the managers—became the new vehicles of progress. (Later on, however, Bell was to rediscover "cultural contradictions"—a formulation that suggested that, perhaps, world history had not terminated.)[43] But, consistent with his generally optimistic vision of a new managerial revolution to accompany technical and cultural changes, he overlooked, as many others have since, the realities of ownership and profit. As corporations, in the name of cost cutting, shed, in the early 1990s, layers of production, technical, and professional employees, including managers, these realities became difficult to ignore.

The Frankfurt school, which agreed with the analysis, offered a different explanation and abhorred the consequences of a horrific technological rationality. Herbert Marcuse's *One Dimensional Man* may be viewed, in this context, as the mirror of technocratic reason: while Marcuse reviles technology as a new global narcotic precisely because it can "deliver the goods" and create a kind of pseudopleasurable "unfreedom" even as it increasingly narrows the scope of individual and collective freedom, his theory of technological domination differs from Bell's position only in its ethical estimate of the new regime. Where Bell celebrates it as a new stage in modernity's inexorable forward march, Marcuse resumes the rhetoric of catastrophe. But *neither Bell nor Marcuse was looking at the horrifying effects on the Third World or the growing gaps in the social safety net in the developed world.* They accepted advanced industrial societies as effective models for the Third World and prefigured the eventual end of poverty and class struggle. In effect, they both accepted, albeit from different standpoints, the postwar era as a definitive sign of "progress" in the traditional modernist significance of the term. Technology produced affluence and this achievement was virtually irreversible for the populations of advanced industrial societies, including the Soviet Union. If Marcuse remained hostile to the fruits of technological development, he shared with his more conservative colleagues the sanguine view that, at least, material want would finally be relegated to historical memory. Nor did either the left or liberal theorists of post- or advanced industrial society foresee the erosion and even the collapse of the welfare state. For them, economic security was an aspect of the *permanent legacy* of the Keynesian reforms of the Great Depression and its aftermath. Whatever the other virtues of Marcuse's analysis, particularly his category of "one-dimensional" thought, which seems to have come to fruition in the 1970s and 1980s, the premise of our work is that the production of scarcity remains a ruling practice of late capitalism, even as the doctrine of progress remains a powerful political aphrodisiac.

Things Fall Apart

However persuasive this analysis seemed to be, it was unable to forecast the sources of instability and even disintegration that beset both East and West in the 1960s. Even as the postwar era of working-class integration into the neocapitalist welfare state reached its apex, emerging forces of not only economic but also cultural and political disarticulation were eroding the conditions of social stability upon which postwar economic expansion and political hegemony depended.

After two decades of dramatic economic growth, Western Europe began to experience the scourge of stagnation. By the late 1960s Great Britain was, measured by the criterion of growth, already in a long-term decline, and Germany and France were not far behind. By the early 1970s it was generally acknowledged that Europe was in the throes of a recession, the most distressing manifestation of which was the nearly permanent two-digit unemployment rates. From the middle of the 1970s to the present, no major European country has succeeded in reducing its joblessness much below 10 percent; frequently, Europe has suffered 15 percent rates. But there were other economic disasters in this decade as well. From 1973 to 1977, the world was beset with an oil crisis, reflecting chronic overproduction and falling prices but also the effort by the leading oil-producing nations of the Third World to limit production in order to rescue profits. The instability of the Middle East, reflected in the breakup of the Metternichian deal between Israel, the Arab world, and the great powers, intensified the crisis. Arab sheiks took their oil to Western banks. The giant banks were, at the same time, beginning the process of reorganizing international money exchanges to reflect the increasing internationalization of capital. In these new arrangements, the leading oil companies were quick to see their advantage: they boosted prices and became the behind-the-scenes partners of the Organization of Petroleum Exporting Countries (OPEC).

Buoyed by high levels of military spending both during and following the Vietnam War as well as world leadership in agricultural production, the United States, despite oil-soaked surging inflation, managed, for a decade after Europe's relative bust, to avoid the outward appearances of stagnation and the concomitant high unemployment rates. But in 1980 it was apparent that the United States was ready to fall into line. During the 1981–83 recession, American workers suffered the highest official jobless rates since the Great Depression, nearly 12 percent. At the same time, millions of workers were on short weeks or could secure only part-time employment.

Even after the recession ended, owing in large measure to the rapid growth of debt and the expansion of financial and other service industries, many remained out of work or found that part-time labor had become a permanent fact of life. When the official unemployment rate dipped to slightly more than 5 percent between 1985 and 1988, the extent of actual underemployment was disguised by the growth of part-time work and the sharp decline of real wages in many sectors. The erosion of living standards corresponded to the relative decline of unions during the 1970s and especially the 1980s, precipitous capital flight that rendered much of the industrial heartland and the Northeast economically prostrate.

The already familiar tales of plant shutdowns in industrial towns and cities was ascribed, at first, by employers and conservative economists for whom blaming the victim is always a ready explanation to noncompetitive high union wages and benefits, and then to the lack of worker productivity. What caused lagging productivity? Were American workers lazy during the booming 1950s and 1960s? How was it that the United States and Britain, during the postwar era the leading developers of computer-mediated technologies—robotics, numerical controls, and lasers—fell behind Europe and Japan?

At the same time, the major developers of the new computer technologies, the United States and Britain, were using computers extensively in military production, introduced them in the clerical workplace, and, especially in the United States, developed a consumer computer market or were exporting software as well as hardware to other countries. Europe was a wonderful market for U.S. and British technological innovation. The effects of the destruction visited upon major European industries by Allied bombings were nothing short of devastating. But with the help of the U.S.-sponsored Marshall Plan and the cooperation of European unions and socialist parties (including the communists) to moderate wage demands, European economies were rebuilt at a time when automation and cybernation, especially computer-mediated technologies, could be applied to production industries. In the late 1950s and 1960s, numerical controls, robotization, and lasers were widely introduced into key industries such as steel and autos in West Germany, Italy, and Japan at a time when U.S. civilian industries were mired in the old mechanical technologies or had barely introduced the transfer machine, the basic technology of automation.

The major employers in the U.S. steel, auto parts, and machine tool industries, among others, did not introduce these technologies until the late 1970s. The largesse of the federal government's defense establishment provided few if any incentives to productive civilian investment; during the war boom of the 1960s even civilian products benefited from this bounty. When the boom was over and some investment in basic technologies finally occurred, the era of global competition had arrived in the late 1970s; U.S. industry was no longer a net exporter of steel and cars, but had become a net importer of these crucial commodities. Consequently, by the early 1980s, it was almost too late to save older plants and smaller companies. Of course, the companies and the plants that remained were, given the globalization of the U.S. economy, obliged to undertake serious modernization, especially to introduce computer mediation into almost every aspect of life, it seemed. Combined with world economic stagnation and a veritable epidemic of mergers and acquisitions, computerization meant that millions of workers were permanently displaced from industrial production. For example, employment in the steel industry declined from 600,000 production and maintenance workers in 1960 to fewer than 200,000 by 1992; most of the reduction occurred in the 1980s. Auto jobs were slashed by half, and machine tools and electrical machinery have lost some 40 percent of their workforces. At the same time, with the exception of autos, industrial production has, despite accelerating imports, maintained and in some sectors increased its levels.

Of course, the introduction of computer-mediated technologies in administrative services—especially banks and insurance companies and retail and wholesale trades—preceded that in goods production. From the early days of office computers in the 1950s, there has been a sometimes acrimonious debate about their effects. Perhaps the Spencer Tracy-Katharine Hepburn comedy *Desk Set* best exemplifies the issues: when a mainframe computer is introduced into the library of a large corporation, its professional and technical staff is at first alarmed, precisely because of their fear of losing their jobs. The film reiterates the prevailing view of the period (and ours?) that, far from posing a threat, computers promise to increase work by expanding needs. Significantly, the film asserts that the nearly inexhaustible desire for information inherent in human affairs will provide a fail-safe against professional and clerical redundancy. In contradistinction to these optimistic prognostications, new information technologies have enabled corporations, large law firms, and local governments to reduce the library labor force, including professional librarians. In turn, several library science schools have closed, including the prestigious library school at Columbia University.

By the 1980s many if not most large and small businesses used electronic telephone devices to replace the live receptionist. A concomitant of these changes has been the virtual extinction of the secretary as an occupational category for all except top executives and department heads, if by that term we mean the individual service provided by a clerical worker to a single manager or a small group of managers. Today, at the levels of line and middle management, the "secretary" is a word-processing clerk; many middle managers have their own answering machines or voice mail and do their own word processing. They may have access to a word-processing pool only for producing extensive reports. Needless to say, after a quarter of a century during which computers displaced nearly all major office machines—especially typewriters, adding machines, and mechanical calculators—and all but eliminated the job category of file clerk, by the 1980s many major corporations took advantage of the information "revolution" to decentralize their facilities away from cities to suburbs and exurbs. Once concentrated in large urban areas, data processing now can be done not only in small rural communities but also in satellite- and wire-linked, underdeveloped offshore sites. This has revived the once scorned practice of working at home. Taken together, new forms of corporate organization, aided by the computer, have successfully arrested and finally reversed the steady expansion of the clerical labor force and have transferred many of its functions from the office to the bedroom.

Visiting a retail food supermarket in 1992, President George Bush was surprised to learn that the inventory label on each item enabled the checkout clerk to record the price by passing it through an electronic device, a feature of retailing that has been in place for at least fifteen years. This innovation has speeded the checkout process but has also relieved the clerk of punching the price on the register, which, in turn, saves time by adding the total bill automatically. The clerk in retail food and department stores works at a checkout counter and has been reduced to handling the product and observing the process, but intervenes only when it fails to function properly. Supermarket employers require fewer employees and, perhaps equally important, fewer workers in warehouses: an operator sits at a computer and identifies the quantity and location of a particular item rather than having to search for its location and count the numbers visually. The goods are loaded onto a vehicle by remote control and a driver operating a forklift takes them to the trucking dock, where they are mechanically loaded again. Whereas once the warehouse worker required a strong back, most of these functions are now performed mechanically and electronically.

Some of the contraction of clerical and industrial employment is of course a result of the general economic decline since the late 1980s. But given the astounding improvements in productivity of the manual industrial and clerical work force attributable to computerization, as we argued earlier, there is no evidence that a general economic recovery would restore most of the lost jobs in office and production sites—which raises the crucial issue of the relationship between measures designed to promote economic growth and job creation in the era of computer-mediated work. Many economists, most notably Robert Solow, who won a Nobel Prize for his views, have argued that technological innovation is the key to economic growth because of its relevance to productivity gains that enable employers to reduce costs and prices and thereby increase profits and expand markets. In the standard measure, growth is defined as an increment in the value of goods and services, in real terms, within a given time frame. The prevailing assumption is that for every unit of investment, particularly in capital goods (machines and raw materials), a multiplier effect produces a concomitant increase in employment for those who operate the machinery, those who produce, distribute, and sell the machines and raw materials, and those who produce, distribute, and sell the product made by the labor that operates the machine.

If the rate of economic growth lags behind the productivity increases, however, technological investment will displace more workers than the jobs produced by its application. For, from the individual employer's perspective, the real purpose of technological innovation is labor displacement as a vital component of reducing costs. This is precisely what occurred in the late 1980s. While the absolute number of jobs appeared to rise, most of the increases were in the service sector, and many of them were offered on a part-time basis. As the steel, auto, steel fabricating, clothing, and machinery industries introduced robots, lasers, and numerical controls on a wide scale, their workforces continued to shrink even as production remained at previous levels or, as in the case of steel, increased.

The corporate, neoliberal program to persuade displaced blue- and white-collar workers to adapt to the post-Fordist regime, which, among other things, proposed to disrupt the social compact, met with considerable success during the Reagan years. Surely, some workers were not resigned to capital flight, union busting, and other signs of the disappearing compact. But many others made a surprisingly rapid adjustment to the new situation, owing in part to the debt-induced boom that masked the decline of real wages and the partial breakup of the old wage-labor system. Consumerism generated a vast expansion of service jobs, especially in finance and retailing. As millions of women entered the labor force to occupy these niches during the 1980s, family income actually rose and many households increased their consumption of homes, cars, and appliances as well as health and education services even as well-paid factory jobs were disappearing forever. The two-paycheck family, for a time, sustained the vitality of the cultural ideal.

But all is not well in the post-Fordist era. The end of the Reagan boom has produced considerable political and economic instability. In 1992 the twelve-year right-wing hold on the White House was broken, even as the new administration indicated that it would retain many of the Reagan-Bush economic policies. At the same time, the Democrats rode to power not on stasis but on the theme of change, which policy parlance translates into the promise to produce more jobs. Needless to say, there is no substantial economic turnaround, nor have the feeble signs of slow recovery in some sectors produced more jobs. In fact, any hope of a substantial recovery may, for this reason, be thwarted. For it is not difficult to understand why investments may rise without job creation if the existing labor force works harder and labor-saving equipment continues to take its toll.

The American cultural ideal is tied not only to consumer society but also to the expectation that, given average abilities, with hard work and a little luck almost anyone can achieve occupational and even social mobility. Professional, technical, and managerial occupations perhaps even more than the older aspiration of entrepreneurial success are identified with faith in American success, and the credentials acquired through postsecondary education have become cultural capital, the necessary precondition of mobility. Put another way, if scientifically based technical knowledge has become the main productive force, schooling becomes the major route to mobility. No longer just places where traditional culture is disseminated to a relatively small elite, universities and colleges have become the key repositories of the cultural and intellectual capital from which professional, technical, and managerial labor is formed.

For the first quarter century after World War II, the expansion of these categories in the labor force was sufficient to absorb almost all of those trained in the professional and technical occupations. In some cases, notably education, the health professions, and engineering, there were chronic shortages of qualified professionals and managers. Now there is growing evidence of permanent redundancy within the new middle class.

As we show in chapters 4 and 5, whole professional occupations are in the process of changing: some are disappearing, and others are being massively restructured. Chapters 4, 5, and 8 are three case studies examining the relationship between the transformations in the nature of the knowledge—and the technologies—requisite for the performance of crucial professions that have been intimately bound up with mobility aspirations as well as the production of legitimate intellectual knowledge in engineering, scientific research, and postsecondary teaching. In each instance, the introduction of new knowledge, new modes of organizing the labor process, and new technologies has posed exciting challenges for some, providing unique research and design opportunities, offered the chance to acquire new knowledge, and suggested visions for making the work more significant and more gratifying. But like *Desk Set*'s librarians in the mid-1950s, many people have perceived the changes as threats to their professional autonomy, their jobs, even to their status as professionals.

In some instances, knowledge inextricably intertwined with traditional professional practice is being transferred to the computer. In others, the organization of the workplace renders obsolete previously acquired cultural capital. In still others, "science" has been definitively transformed into *technoscience*: the work cannot be separated from its mechanical aspects, which, in the light of the drift of the field, seem to dominate all so-called intellectual problems.

While we are concerned to examine these changes and their implications for what it means to be a "professional" or an "intellectual" today and tomorrow, we go beyond description to draw the consequences of these changes for economic, political, and cultural life. For if becoming a professional or a manager has become one of the central elements of the cultural ideal of advanced industrial societies, for millions of working people and their children the earthquake has already occurred and we are living its aftershocks.

In the two decades beginning in the mid-1960s the United States experienced the largest-scale restructuring and reforming of its industrial base in more than a century. Capital flight, which extended beyond U.S. borders, was abetted by technological change in administration and in production. Millions of workers, clerical and industrial, lost their high-paying jobs and were able to find employment only at lower wages. Well-paid union jobs became more scarce, and many, especially women, could find only part-time employment. But the American cultural ideal, buttressed by ideological—indeed, sometimes mythic—journalism and social theory, was barely affected in the wake of the elimination of millions of blue- and white-collar jobs. As C. Wright Mills once remarked in another context, these public issues were experienced as private troubles.

The persistence, if not so much the real and exponential growth, of poverty amid plenty was publicly acknowledged, even by mainstream politicians, but, like alienated labor, it was bracketed as a discrete "racial problem" that left the mainstream white population unaffected. Job creation precluded serious consideration of the old Keynesian solutions; these had been massively defeated by the state-backed, yet ideologically antistatist, free-market ideologies. We were told that deregulation would free up the market and ensure economic growth that eventually would employ the jobless, provided they cleaned up their act. Even in the halcyon days of the Great Society programs of the war-inflated Johnson years, the antipoverty crusade offered the long-term unemployed only literacy and job training and, occasionally, the chance to finish high school and enter college or technical school. The Great Society created few permanent jobs and relied on the vitality of the private sector to employ those trained by its programs.

The concept of government as the employer of last resort was at best sporadically implemented. Even most social welfare liberals refused to acknowledge that the "failure" of the poor to find jobs was not, in the main, a function of their personal deficits but was instead a symptom of a much broader failure of the labor market.

Talcott Parsons grounded his theory of social equilibrium on the nearly certain capacity of the cultural system to socialize the underlying population into the norms necessary to reach social balance.[44] In Parsons's tripartite social system, the personality system, based on the structural conflict between the superego (society) and the id to capture the ego, constituted a source of instability while, on the other side, the cultural system was the social order's reliable ally. Of course, ego psychology and sociological theory sought to explain through concepts of cultural continuity how the system can maintain itself, the world-historical shocks that have characterized the twentieth century notwithstanding. For some, culture remains constant regulation, despite the end of increasingly high levels of military spending and the fierce investment program of many corporations and government agencies in labor-saving technologies. Placed next to the longer-term phenomenon of the globalization of national economies, including the emergence of a truly international production line, labor force, and scientific and technological apparatus, poverty as a public issue has been replaced by symptoms of a catastrophic future for the middle class as well as the working class in nearly all countries of late capitalism. While the economic consequences of this sea change are profound, we want to explore the political and cultural consequences as well.

In the United States, long before the current *perceived* cultural crisis, these traditions have been undermined by the process of citizenship education—"Americanization" or assimilation, which, in the large waves of immigration at the beginning of the twentieth century demanded the replacement of Old World culture by that of the New World. Among the leading elements of the New World cultural system were consumer society and its technological basis, electronically mediated mass communication; the emphasis on productivity, which until recently privileged quantitatively measured production over quality; and the militant assertion of American nationality as an official state culture, an identity that implies, in the hegemonic political discourse, shedding the habits of mind and body that were identified with European and subaltern cultures, and the values associated with them. This forced shedding of the older cultures has been "exported" on a global level and becomes—the decline of the United States notwithstanding—a universal Americanization on more affluent levels while the lower and impoverished levels fragment in a frenzy of cultural and ethnic warfare.

One of the crucial features of the specifically "American" cultural ideal is the idea that we can reinvent ourselves. From the time most immigrants arrive at an entry point and experience an instant name change because some official finds it impossible to pronounce an eastern or southern European, Asian, or Spanish surname, to school and work, when we discover our own *difference* from the model of a typical American, despite the ideology of "family" values that represents an invocation by officially moral authorities to adhere to the cultural system, we are invited to make a personal break from family and ethnic traditions lest we lose a chance to advance in the social and occupational structure.

Paul Willis has shown how British working-class kids remained loyal to their parents' class identity, at least in the 1970s.[45] Until recently, the idea of leaving your class behind was, in many quarters, viewed with suspicion if not downright hostility. Among some immigrant groups, in both Great Britain and the United States, becoming a professional brings with it similar stigmatization; it is a betrayal to leave the family and the neighborhood. These subcultural rules are frequently broken, especially in the wake of the disappearance of good working-class jobs. Yet beyond the circumstances of economic necessity lies the frailty of working-class traditions in the United States, precisely because of the immigrant—ethnic and racial—composition of the industrial working class.

The question now is not only what the consequences of the closing of routes to mobility of a substantial fraction of sons and daughters of manual and clerical workers may be, but also whether the professional and technical middle class can expect to reproduce itself at the same economic and social level under the new, deregulated conditions. For, as we shall show in chapters 7, 8, and 9, the older and most prestigious professions of medicine, university teaching, law, and engineering are in trouble: doctors and lawyers and engineers are becoming like assembly-line clerks . . . proletarians. Although thus far there are only scattered instances of long-term unemployment among them, the historical expectation, especially among doctors and lawyers, that they will own their own practices, has for most of them been permanently shattered. More than half of each profession (and a substantially larger proportion of recent graduates) have become salaried employees of larger firms, hospitals, or group practices; with the subsumption of science and technology under large corporations and the state, engineers have not, typically, been self-employed for over a century.

Similarly, the attainment of a Ph.D. in the humanities or the social or natural sciences no longer ensures an entry-level academic position or a well-paid research or administrative job. Over the past fifteen years a fairly substantial number of Ph.D.'s have entered the academic proletariat of part-time and adjunct faculty. Most full-time teachers have little time and energy for the research they were trained to perform. Of course, the reversal of fortune for American colleges and universities is overdetermined by the stagnation and, in some sectors, decline of some professions; by the long-term recession; by organizational and technological changes; and by twenty years of conservative hegemony, which often takes the cultural form of anti-intellectualism. Since the 1960s, universities have been sites of intellectual as well as political dissent and even opposition. A powerful element in the long-term budget crises many private as well as public institutions have suffered is at least partially linked to the perception among executive authorities that good money should not be thrown after bad.

And, with the steep decline in subprofessional and technical jobs, universities and colleges, especially the two-year community colleges, are reexamining their "mission" to educate virtually all who seek postsecondary education. In the past five years, we have seen the reemergence of the discourse of faculty "productivity," the reimposition of academic "standards," and other indicators that powerful forces are arrayed to impose policies of contraction in public education. We shall discuss the relationship between these developments and academic teaching in chapter 8.

In the subprofessions of elementary and secondary school teaching, social work, nursing, and medical technology, to name only the most numerically important, salaries and working conditions have deteriorated over the past decade so that the distinction, both economically and at the workplace, between the living standards of skilled manual workers and these professionals has sharply narrowed. Increasingly, many in these categories have changed their psychological as well as political relationship to the performance of the job. The work of a classroom teacher, line social worker, or nurse is, despite efforts by unions and professional organizations to shore up their professional status, no longer seen as a "vocation" in the older meaning of the term. Put succinctly, many in these occupations regard their work as does any manual worker: they take the money and run. More and more, practicing professionals look toward management positions to obtain work satisfaction as well as improvements in their living standard since staying "in the trenches" is socially unappreciated and financially appears to be a dead end. Consequently, in addition to a mad race to obtain more credentials in order to qualify for higher positions, we have seen a definite growth in union organization among these groups even as union membership in the private sector, especially as a proportion of the manual labor force, has sharply declined. This is the subject of chapter 6.

Except for purposes of illustration, we have avoided in this book a detailed examination of the relation of science and technology to the transformation of factory and clerical work. We have, instead, presupposed much of the existing research, and in chapters 3 and 10 we engage some aspects of this relationship. Although some writers such as Zuboff and Adler have argued, from differing perspectives, that contemporary technological innovations produce new skills rather than being introduced for the purpose of skill degradation, we argue two alternative positions: the death of skill and, for most people, the eclipse of paid labor as a defining activity for self-formation. It is this eclipse, perhaps more than any other single influence, that we believe accounts for the decline of public education, including schools that serve the middle class.

The economic and technological revolutions of our time notwithstanding, work is of course not disappearing. Nor should it. Rebuilding the cities, providing adequate education and child care, and saving the environment are all labor-intensive activities. The unpaid labor of housekeeping and child rearing remain among the major social scandals of our culture. The question is whether work as a cultural ideal has not already been displaced by its correlates: status and consumption. Except for a small proportion of those who are affected by technological innovation—those responsible for the innovations, those involved in developing their applications, and those who run the factories and offices—most workers, including professionals, are subjugated by labor-saving, work simplification, and other rationalizing features of the context within which technology is introduced. For the subjugated, paid work has already lost its intrinsic meaning. It has become, at best, a means of making a living and a site of social conviviality.

In chapters 10 and 11, we develop these themes and trace the consequences of the universalization of the applications of scientifically based technology through our economic, political, and cultural life. In the first place, we focus on the future of work. But since our argument is that, despite the displacements of consumer society and the nebulous notions surrounding "community," work remains the fulcrum of our cultural aspirations—among them the values of success, well-being, self-worth—the gulf separating actual from ideal form constitutes the source of both catastrophe and the challenge to received wisdom.

Chapter 2
Technoculture and the Future of Work

Until World War II, science was primarily done by men working in small laboratories or classrooms in universities, and the technological development that derived from their discoveries was performed by large corporations. AT&T, General Electric, the various branches of the Rockefeller-dominated Standard Oil companies, and chemical corporations like Du Pont, Allied, and Union Carbide hired physicists and chemists, engineers and technicians whose principal work was to apply "basic" science to the invention and development of new products. Some—notably AT&T and General Electric—participated in some aspects of basic as well as applied research, but on the whole, scientific discovery retained its cottage-industry character. Government support for science was considered taboo by scientists who wanted to protect their intellectual freedom both from bureaucracies that they feared would determine what could be studied, if not how, and from large corporations, whose passion for the bottom line easily outdistanced any sentiments for truth.

Perhaps the most dramatic transformation in the history of the political and social organization of science occurred with the intense competition between the United States and Germany for the development of radar and a practical atomic weapon. After getting a cold shoulder from high British and American government officials, German and Austrian refugee scientists such as Albert Einstein and Leo Szilard as well as their American counterparts James B. Conant, Vannevar Bush, J. Robert Oppenheimer, Ernest O. Lawrence (whom Richard Rhodes describes as the "founder of big-machine physics" in America),[1] and others, finally persuaded President Roosevelt to undertake the atomic bomb project in 1940. Szilard's key discovery of the conditions for a nuclear chain reaction and his experiments—first with beryllium and finally with uranium—made the development of nuclear weapons theoretically possible. But it was not until the Roosevelt administration agreed to commit relatively large financial and administrative resources to the project that what became known as "big science" was launched.

In the same year, Vannevar Bush became the key figure of the newly created National Defense Research Committee, later renamed the Office of Scientific Research and Development (OSRD).[2] The establishment of the OSRD institutionalized government involvement in scientific and technological research. Equally important, it began a long process of incorporation of major universities into what President Dwight D. Eisenhower termed the "military-industrial complex," which by the 1950s had become the most powerful political force in the United States. From the early 1940s to the present, the overwhelming majority of university-based scientists and a considerable number in independent research organizations and private corporations have been dependent on government contracts to perform all types of research. Grants from the National Institutes of Health, the National Science Foundation, the Department of Defense, and other executive agencies such as Agriculture and Interior, which pioneered, albeit on a smaller scale, government involvement in science and technology, became the lifeblood of America's scientists and many engineers. Failure to win grants is, today, tantamount to exclusion from scientific work.

Ernest Lawrence's development of big-machine physics has raised the stakes of research. In crucial areas, most visibly high-energy particle physics, molecular biology, and chemistry, research requires expensive machinery. The costs of producing knowledge are now prohibitive for institutions that are unable to procure sufficient funds. As a result, most scientists who work in nonresearch universities are, with the exception of theorists, relegated to teaching. Moreover, as we shall see in chapter 7, big science is virtually confined to a handful of the more than 1,700 colleges and universities in the United States.

Big science connotes the industrialization of knowledge and the transformation of the university into a knowledge factory. Slightly more than a century after scientific knowledge became the basis of virtually all technological innovation in large-scale industrial production, and fifty years after American science was decisively incorporated as an important but subordinate apparatus of the state, scientific knowledge has itself become big business.[3] The production of scientific instruments, once confined to a very small industrial sector, has now become a vital business of many Fortune 500 companies. The Defense Department is among the chief patrons of basic as well as applied science because it is the most important component of the military establishment.

As scientific research increasingly depends on government contracts, its main organizations—the American Association for the Advancement of Science and the Federation of American Scientists—have been obliged to field substantial lobbying staffs and to provide information in their journals and newsletters about the effects of government policy on the prospects for funding. Although few scientists are public intellectuals, the president of the United States has a science adviser and a battery of agencies and commissions devoted to selecting, funding, and making policy concerning the production and dissemination of scientific knowledge. Even during the Reagan-Bush years, the volume of government funding did not decline, at least in absolute terms, even as knowledge—both scientific and cultural—became politicized. These administrations were either hostile or indifferent to putting money into research on women's health issues such as breast and cervical cancer; they effectively blocked RU486, the important new French-made contraceptive; and they denied scientists the use of tissue extracted from aborted fetuses for use in cancer research. In sum, science has not only become a commodity, it has been politicized and commodified. Its commodification is revealed by the degree to which we may no longer speak of science without instantaneous links to its applications, a concatenation that is mediated by technology.

Technoscience is the mode of existence of all science and technology: Lawrence's model of machine physics is the model of nearly all natural science and is dominant, via the quantification trend, in the social sciences as well; as we have noted, science is rarely done without advance determination of its practical uses; and, especially in molecular biology and in branches of physics such as high-energy particle and solid-state physics, theoretical and practical sides of knowledge are inextricably linked.

The critique of science and technology that reemerged in the 1950s and 1960s hinged on the judgment, shared by many but most forcefully articulated by the Frankfurt school, that reason in "advanced industrial societies" had become identical with instrumental rationality. According to this claim, reason—the Enlightenment's wager for human freedom—had instead been conflated with the domination of nature. This was an unintended consequence of the transformation of the mysticism of the Middle Ages into modern science and technology. While scientifically based technology had the capacity to overcome the age-old scourge of scarcity, in the final analysis its power lay in the ability of technologically developed societies to forge a new social contract with the underlying population. A key element of the compact was that technology could "deliver the goods" to the overwhelming majority in the West (and possibly elsewhere as well). In turn, people surrendered their social and political autonomy to the techno-logical imperative. Accordingly, not only does the economic order rely on technological innovation for spurring growth, but political and social con-flicts are also considered "problems" subject to managerial solution. Thus, technology is upgraded from a crucial tool of production to a sensorium, a mode of life, perhaps *the* mode of life of the modern world. Of course "man" has *always* lived in a technological sensorium, striving to overcome unmanageable and dangerous nature. Thus, the most modern technology becomes the fulfillment of ancient dreams. It promises to deliver us from the thrall of nature, to subsume nature under "man." *But only for some.* For the great mass of humanity, nature holds a multitude of terrors. Even in the advanced industrial societies, the ravages of industrialization, not the least of which is the looming threat of *generalized* cancer, give the built environ-ment the face of an antagonist—and it is.

Within the terms of the surrender of traditional egalitarian values is what John McDermott calls "laisser innover" (loosely translated, this means let us [them] innovate technologically), which has the force of moral, mythical, almost religious law.[4] Scientifically based technological systems have become, with a few exceptions—notably the feminist challenge to certain reproductive technologies, and ecological objections to technologi-cally driven economic growth—almost as sacred as the national security establishment they have served.[5] All sorts of magical wonders are ascribed to them. "Official" doctrine has it that technology, including organization-al structure, stands between catastrophe and happiness, providing solutions to nearly every "natural" and social problem. For example, the conse-quences of untrammeled economic growth for ecosystems are among those catastrophes renamed "problems," and thus considered soluble by means of state or corporate policy.

Echoing Marx's lament that the more machines rendered human labor productive, the less labor was able to control its own product and the labor process and that the more wealth it produced, the more labor was *relatively* poorer,[6] the Frankfurt school argued that this dialectic strengthened the subordination of the great mass of humanity to a social system that, while appearing to be ever more rational, was fundamentally irrational.[7] The fur-ther technology developed, signifying human (but which class of humans?) mastery over nature, the further it also created new mechanisms for human domination. Social relations become the mirror of our relation to nature: just as science has been mobilized to uncover the secrets of nature, which, through technological application, hold the "beast" of natural forces at bay, the human beast is assuaged by the provision of abundance *without requir-ing a fundamental transformation of the social order.*

For Max Horkheimer and Theodor Adorno, the new technology that

60

came into existence with the maturity of industrial capitalism—a technology that provided both relative mass affluence and institutions of massified, electronically mediated culture—signaled nothing less than the "eclipse" of reason as a transcendent principle and, accordingly, "the end of history." The end of history is a consequence of the disappearance of proletarian, if not capitalist, agency, if by that term we mean both the will and the capacity of exploited and otherwise excluded social categories to emancipate themselves from the thrall of capital and other forms of domination.[8] Contrary to Enlightenment expectations, the development of science and technology had substituted a host of spurious pleasures such as profligate consumer goods for a democratic social order.

In technologically driven consumer society, the question of self-management no longer arises as a practical political demand; rather, the whole population is mobilized, according to Marcuse, to demand of the social order that it meet the new, spurious needs created by the immense productive capacity of "late" or "advanced" capitalism. Clearly, Marcuse and other critical theorists had adopted a view of humans under advanced capitalism as nonproducers; their perception of the advent of consumer society presupposed that the human body was a bundle of needs, a "desiring machine" that could be coupled with goods and reduced to political and social narcolepsy.[9] Without detracting from the validity of many of the empirical insights of critical theory, we emphatically challenge their judgment that, without Culture (capitalized here to connote high art and philosophy rather than the texture and practices of everyday life), the masses are doomed to an inert existence.

These powerful suggestions contributed, paradoxically, to the massive student revolts of the 1960s, when millions of young people appeared for a time to reject the blandishments of the technological sensorium. Of course the students came to these realizations themselves in parallel with these mounting critiques, but their capacity for generating what may be termed "spontaneous" critical theory belies the austere view of the "new" consumerism of the senior generation of critics of technological society such as Jacques Ellul and Hans Jonas, both trained in religious philosophy and ethics.[10]

The heady events of spring 1968, when students and youth in Prague, New York, Chicago, Mexico City, and especially Paris confirmed not the passivity of the underlying population but their capacity for self-activity. The ideas of critical theory and its New Left followers seem to have outlived their moment of totalizing power even though they retain much more than a marginal cultural existence; among important sectors of Western intellectuals they retain to this day the force of self-evident historical understanding.[11] In the 1970s, after the ebbing of the movements that were, in various measures, influenced by the gloomy prognostications of Jacques Ellul and Herbert Marcuse—which in the 1960s enjoyed wide dissemination, especially among youth and intellectuals—the critique of technology received considerable impetus from the powerful ecological movements.

These movements have focused on the calamitous effects of industrial technology on the quality of water and air and have mounted fierce and widely supported struggles against the further industrialization of the forests and other wilderness areas. By the 1960s, some trained in natural sciences, such as Rachel Carson, Barry Commoner, and Rene Dubos, began a veritable literary genre of apocalyptic tracts calling attention to the hazards of unregulated industrial production, of materials such as plastics, which, developed during the war as one of its most transferable innovations, had been heralded as a "miracle" material destined to replace many uses for metals, natural fibers, glass, and rubber.[12] In fact, the industrial transformation of the hydrocarbon molecule was among the crucial technological innovations of the postwar era and won acceptance as a hedge against nature's stinginess in yielding "her" fruits. The chemical and oil refining industries—which, in comparison to the relatively labor-intensive textile, steel, and nonferrous metals processes, were already highly automated—constituted, together with electronics, the technical foundation of much of postwar economic growth. But these technologies also brought new dangers to human life: radiation, electromagnetic waves, pollution of various kinds, the thinning of the ozone layer, and the proliferation of toxic wastes.

The critics and the movements that drew knowledge and inspiration from them succeeded, to some degree, in challenging the logic of capital accumulation as well as the technological "imperative" that knows no bounds. Feminist critiques of the health dangers of reproductive technologies, especially the deleterious effects on women of some contraceptives and in vitro fertilization, have become major issues, especially in Western countries. Concomitantly, whereas abortion rights were and remain contested in some European countries and in the United States, the battle to make abortion legal is being slowly but torturously won. And in the 1970s, with the enactment of federal and state legislation regulating occupational health and safety, both feminists and labor unions made significant inroads into corporate management's unilateral control over health and safety conditions at the workplace, some of which were linked to the effects of electronically mediated technologies such as computers. One example is the growing recognition of carpal tunnel syndrome—a muscular disease of the tendons in the hands and arms that is caused by prolonged repetitive motion work such as typing and meat cutting—among clerical workers. Most recently, radiation as a salient factor "causing" cancer, dysfunctions in women's reproductive systems, and a range of other diseases has become a leading public health issue. The link has not yet been made between radiation-producing health hazards and the routine use of computers or, perhaps more startlingly, the universal employment of alternating current as the main energy source. Thus, the critique of science and technology occurs within multiple frameworks: it begins as a critique of the unintended effects of the fear of nature, instrumental rationality, and, indeed, the whole project of the Western Enlightenment—ethical critiques of their constraining effects on prospects for human freedom—and extends to more specifically focused interventions by the labor, feminist, and ecology movements concerning work, reproductive, and environmental hazards.

Raising questions about health and safety challenged the notion of the benefits of scientific and technological citizenship. One of the more crucial but less widely disseminated perspectives has been the threat posed by the development of modern technological systems to democracy. Echoing the older sociological analysis of the effects of bureaucracy on the modern public initiated by Max Weber and Robert Michels, and also the political observations, first written in the 1920s by Walter Lippmann and John Dewey, John McDermott has forcefully argued that technology is characterized not merely by the development of machines of various sorts but perhaps more importantly by the emergence of experts as the controllers of these systems. Consequently, we have witnessed, according to his view, the decline of a public sphere, in which individuals and groups, especially of the lower orders, have a significant voice in decisions affecting the disposition of public goods and those "private goods" that impinge on the public welfare. Perhaps the democratic polis was little more than a myth of a golden age, especially in Dewey's description of the face-to-face interactions said to prevail in the colonial New England town meeting. Nevertheless, when, in the aftermath of World War I, Lippmann boldly called into question the empirical existence of the active democratic public, Dewey responded by invoking the past as a model for reforming an admittedly grim present. No critic of contemporary society could deny that, if the democratic public sphere had once been alive, it had been reduced to little more than a phantom.[13]

Writing within a specifically North American context, McDermott has called technology "the opiate of the intellectuals." The term *opiate* signifies the degree to which the technological "fix" is, in contemporary economics, politics, and social policy, almost indiscriminately prescribed for nearly every problem. For example, policy intellectuals such as Henry Kissinger, Zbigniew Brzezinski, and Herman Kahn as well as most mainstream students of science and technology have claimed that, freed of partisan controlling interests, technology can be employed to solve common problems ranging from diplomacy and production problems to challenges of urbanism such as social welfare, education, and health. Indeed, the quest for nuclear supremacy and its concomitant technological systems, especially over the Soviet Union but also among its postwar allies, undergirded, even guided, much of U.S. foreign policy for nearly a half century.[14]

Today, health care relies heavily on therapies involving the use of pharmaceuticals and is preparing to deepen medicine's dependence on biomedical technologies such as the use of artificially produced genes and organisms to treat diseases and even try to make permanent, prescribed improvements in the human species. Indeed, for some technophiles, the burgeoning technoscience of molecular biology is, in economics Nobel Prize-winner Robert Solow's terms, the technology that can drive industrial growth in the 1990s as well as provide the key to the creation of a eugenically perfectable race. Moreover, the search for the technoscientific fix has extended to psychiatry. In recent years, we have witnessed a powerful and well-financed attack on psychoanalysis and the promotion of chemical medications to alleviate emotional depression. Clearly, drug companies and scientistic psychiatrists have mounted a campaign to make psychotherapy uninsurable.

Since eugenics has, in the wake of the emergence of genetics as the major biological science, waxed and waned and waxed again, ethics has been preoccupied with the question, Can science or any other institution play God? Or, put another way, just as ecologists challenge the notion of unlimited human domination over nature, the question of science's dominion of "human" nature remains hotly debated. But the overt focus of this debate on the idea of using the results of molecular biology to further the "perfectibility" agenda of the eugenicists masks another, perhaps more profound, question: In the era of scientific-technological dominance over much of our economic and social life, can democracy survive without scientific and technological citizenship? Should the "public" intervene in the determination of key aspects of science and technology policy—not merely when their specific interests such as health and safety are potentially threatened by the technological application of scientific discoveries, but also to determine the priorities of public funding for the sciences? But if democracy depends on an "informed" citizenry, how is it possible for the "new" citizens to have the proper technological information, especially with the emergence of the notion of "intellectual property"? Perhaps the acquisition of knowledge works like an intelligence agency; most of it is restricted to those who have demonstrated their ability to crack the secret code or are connected by the invisible threads of power and influence. For the rest, information is parceled out in an eye dropper.

Of course, in light of the history of the past century, especially the intervention of the Soviet and Nazi states in setting priorities for research but also the symbiotic relationship between the state and science forged during the long era of militarization of major sectors of U.S. economic and political life, scientists have reason to jealously guard their "autonomy" from any form of "outside" interference, whether by the state or by the citizenry. Robert Merton, writing in the context of the blatant interference of the Nazi state in the conduct of research in the 1930s, felt obliged to (re)iterate some hallowed ethical precepts of Enlightenment science. He argued that the progress of science depended on the noninterference of the state and, perhaps more to the point, asserted that the scientific community itself provided *internal* norms to protect its integrity. According to Merton, the rules of scientific inquiry, notably "universalism [which] finds immediate expression in the canon that truth claims, whatever their source, are to be subjected to *preestablished impersonal criteria* consonant with observation and with previously confirmed knowledge."[15] In addition, what he calls the "ethos" of science prescribes that knowledge be freely shared among qualified practitioners, that inquiry be disinterested, and that the scientific enterprise subscribe to the norm of "organized skepticism" to guard against antirational prejudices that might be imposed from without.

Merton's discourse is a prescription, through a kind of internal policing by the scientific "community," to maintain the separation of scientific work from the state and other political institutions and from research that might be driven by particular interests. It assures the polity that although science thrives best in democratic societies, it should not be subject, except in the most general way, to state or popular control, the difference between which is often blurred by scientists themselves.[16] Merton claims that science can be counted on to practice "detached scrutiny" of all knowledge claims, but warns that it does not "preserve the cleavage between the sacred and the profane, between that which requires uncritical respect and that which can be objectively analyzed."[17] Consequently, when science extends its research to areas "towards which there are institutionalized attitudes," such extensions engender protest that science is, for example, trying to control spheres that properly belong to religion.

Since Merton is hardly detained by such transgressions but, to the contrary, holds that the logic of scientific inquiry *entails* widening its purview, his most urgent concern is that the institutional interest does not transgress the research priorities that should be determined by scientists themselves. Presumably, science's prerogatives reside in its standing as the leading protagonist of Enlightenment values against mysticism, of which all forms of religion, racial prejudice, and superstition are contemporary representatives. Further, he tacitly argues that science's logic of inquiry entails a kind of imperium: since its purview is "every aspect of nature and society [it] may come into conflict with other attitudes toward this same data which have been ritualized and crystallized by other institutions" such as religion—but not *only* religion, which has already been weakened by the secular state, but also by other interests including the secular state itself.

Merton's work on the normative structure of science may be seen as an attempt to offer an ideology of scientific autonomy. The forward march of science, a necessary element in the quest for human freedom, requires a democratic state that can protect its inquiry from "outside" meddling. Science will flourish best in an environment where civil society enjoys a secure existence. But there is no need for public oversight of the research program of science, since, in its own interest, science privileges self-examination. These sentiments were expressed before the full impact of the "contract" state on science and technology was felt.

Unfortunately, however much the state has crucially intervened to determine scientific priorities, Merton's formulation still describes the ideology of scientists themselves, even though, in the early moments of World War II, it became readily apparent that science was not separate from the state or large private industrial and financial corporations. The state mobilized science and scientists. Through lucrative contracts, science became an arm of both military and, subsequently, corporate-led industrial research. More and more, scientific discovery was closely integrated with technological applications. This has become a characteristic feature of the political economy of scientific and technological discovery. It is not merely that scientists are employed in these institutions; the industrialization of scientific inquiry is no longer confined to private industry or extraordinary circumstances such as wartime requirements. The permanent war economy is part of a larger *contract state* whose tentacles reach far beyond traditional institutional contexts to universities, which after the war were even more interrelated and inextricably involved with both government and private corporations. By the late 1970s these close relationships, which one writer has termed the "university-industrial complex," raised serious questions about the degree to which we may speak of universities as self-governing institutions and of science, which has become their fundamental agency of legitimacy, as an autonomous inquiry.[18]

Today, the high degree of integration of science with industrial technologies, including those of medicine, has led to the felicitous term *technoscience,* which characterizes contemporary research. In turn, the very notion of "basic" or disinterested, shared knowledge is endangered. Funds increasingly are allocated on the criterion of the applicability of the results of inquiry to practical—that is, industrial—uses. The newest and most glamorous projects and emerging disciplines such as molecular biology and its kin, biophysics, high-energy particle physics (HEP) and solid-state physics, are technodisciplines. That is, their core knowledges are intertwined both with the technologies by which objects and relations are putatively apprehended and with applications of the results. For example, as studies have shown, HEP physical inquiry depends on the efficiency of the accelerator as the premier instrument of what is taken as "observation" just as bioengineering procedures (gene splicing) are intrinsic to molecular biology experiments.[19] Similarly, as we will show in chapter 5 (on a series of small molecular biology labs), work in biophysics has been made possible by the development of computer graphics, which more and more comes to constitute the field of empirical observation, the theoretical "laboratory" made concrete, reified, of this subdiscipline. Core knowledge becomes a description of the relationship between reified hypothesis and the limits of technologically and electronically mediated procedures. And computer-aided design engineering provides a parallel instance of technological mediation of what was considered purely conceptual work, so that the distinction between "theoretical" science or design in engineering and "practical" or technological applications tends to disappear.[20] Contrary to common sense, or the observations by the Frankfurt school that "theory" has disappeared, the result of this merger is to emphasize the primacy of the theoretical. In this context, practice appears as a derivation of a series of abstractions that are labeled "theory."

The burden of McDermott's argument is that "modern technology creates its own politics"—not one of the commonweal, in which its applications are subject to public debate, but one of managerial domination of nearly all public institutions: the economy, the state, and local communities. Contrary to the claim of technophiles such as Emmanuel Mesthene, former director of the Harvard Program on Science and Society, that new technology "creates new opportunities" for social improvement and even social change, McDermott points to danger to democracy posed by the widely accepted doctrine of *laisser innover*, according to which technology is an affirmative social force sui generis.[21] Consequently, those with the talent to understand, operate, and control technology should, like the entrepreneurs of the early mercantilist period and the scientific community, be allowed to develop technological applications to virtually all aspects of life without interference from the polity except, of course, the corporations and the private foundations that provide the funds. For the fact is that, more and more, *active* scientific citizens are buyers of technoscience products, intellectual and activist critics, or those locally aggrieved by specific effects of technoscientific applications such as polluted water tables, toxic waste dumps, and dubious technosolutions (incinerators, for example) to overflowing garbage.

McDermott says that "the *laisser innover* school accepts as inevitable and desirable the centralizing tendencies of technology's social organization, and they accept as well the mystification which comes to surround the management process."[22] Their ideologists proclaim the inevitability of managerial power by arguing, as does John Kenneth Galbraith, that "the growing importance of scientific and technical knowledge . . . *implies* an enhanced socio-political power for the people who have such knowledge."[23] What is at issue is the aura of ineluctability surrounding this discourse. While Galbraith's observations concerning the importance of scientific and technical knowledge for selecting economic options is today commonplace, less obvious is his ascription of power to those who possess this knowledge. Does the centrality of knowledge in economic and political systems "imply" an "enhanced power" for possessors of knowledge or is this determinism the product of the revival of managerial and technocratic ideologies in the service of a new knowledge class of which Galbraith himself became a leading theorist? Or, in fact, are the new technological means of production, to use an old terminology, dependent *not* on a managerial class but on the business owners, those who raise the money, the investors, or provide the government funding? (Although it should also be stressed that the scientists and the managers sometimes outwit the owners in order to keep the funding flowing.)

Clearly, as we have already stated, knowledge has become a central productive force in economically developed societies. And as we shall demonstrate elsewhere in this book, science and technology, including organizational systems, have been placed at the very heart of what is signified by the term *economic development*. But the growing power of experts and expert systems, not only in the configuration of work but also in the shape of politics, the control over culture and everyday life——a historically specific phenomenon——requires an examination of concrete struggles in the labor process, in the political arena, and in cultural life.

Consider the difference between McDermott's perspective and that of the Frankfurt school. While the Frankfurt school's Critical Theory shares McDermott's alarm concerning the link between knowledge and power, they diverge precisely in their estimate of the possibility of conceiving political *agency* not only to resist the formation of the new elites, but also to defeat it. Long before Michel Foucault disseminated the idea, Critical Theory argued that the fusion of knowledge with power removes the last impediment to capital's virtually uncontrolled dominion over every aspect of social existence. Because it understands technological domination as a *systemic* feature of late capitalism that incorporates the opposition as a condition of its reproduction, Critical Theory argues that the possibility of the emergence of genuine alternatives has been effectively foreclosed by the ability of the system to incorporate its own opposition, to generate a society of "total administration" where, as we saw in the first chapter, crises are transmuted into soluble problems subject to the procedures devised by expert systems. Even the possibility of resistance, which in the form of the Great

Refusal is, for Marcuse, the only possible act of defiance, is reserved only for those effectively excluded from the system.

As pessimistic as he was concerning the possibilities for recreating an opposition in the wake of the triumph of scientifically based technology, Adorno extolled the possibilities inherent in marginality. Although in his later years he allowed that only art possessed the possibilities for resistance, his earlier work is punctuated by a persistent faith in the transformative power of the proletariat. After this hope was shattered by the rise of fascism and the merger of the state with economic and cultural life, Marcuse took up Walter Benjamin's emblematic aphorism "It is only for the sake of those without hope that hope is given to us."[24] Marcuse invoked the "outcasts and the outsiders, the exploited and persecuted of other races and other colors, the unemployed and the unemployable"[25] to exemplify the possibility for refusal (although the latest version of this "refusal," cultural, ethnic, and a revived religious fanaticism, would have surprised him).

As late as the 1950s, Marcuse had argued that modern technology, by removing the necessity of alienating labor, could free us for creative, even erotic *work* and might make possible its separation from a (sharply reduced) obligation to perform necessary wage labor. Even in his otherwise grim description of the consequences of modern technology, *One Dimensional Man*, he advances the hypothesis that

> within the advanced societies the continued application of scientific rationality would have reached a terminal point with the mechanization of all socially necessary labor. . . . But the stage would also be the end and the limit of the scientific rationality in its established structure and direction. Further progress would mean the *break*, the turn of quantity into quality. It would open the possibility of an essentially new human reality—namely the existence of free time on the basis of fulfilled vital needs.[26]

But technology removes the chance that such work could be introduced in the commodified labor process. Its arrival presupposed the sharp separation of work from labor, an eventuality that can occur only by the relentless pursuit of the technological imperative. In this passage we can see that the human agency that produces domination would, without the intervention of politics, through pursuing technological rationality to excess, turn into its opposite—human freedom. What was perhaps wrong, and even ridiculous, in this asssessment was precisely the persistence of irrational impetuses toward royal or noble status in the midst of technological rationality.

Trained in the pluralist faith of American pragmatism, McDermott, in contrast, relies on the vitality of democratic tradition to contest the growing power of the experts. He is sufficiently optimistic to offer a "class analysis" of the role of modern technological systems. That is, he tacitly rejects the position according to which these systems obliterate all agents except their human embodiments, the experts. For the absence of class in contemporary social theory may be traced to the judgment, shared by techno-optimists and technopessimists, that "postindustrial" societies banish classes to the trash can of history. McDermott's employment of categories such as "lower" and "higher" cultural and political orders implies that there are agents capable of making choices concerning the growing significance of technological systems in the control over social and political affairs, even if they are relatively powerless. The very notion of a lower "order" remains tacitly bound to the vanishing horizon of "class" without which change seems improbable. Whereas in Marcuse's major work on the power of technology in late industrial societies his rhetoric retains the flavor of the relentless inevitability of technological domination and explicitly announces, in its wake, the death of the subject, McDermott's descriptions are angry but by no means resigned. Rather than promoting the aspiration toward a more egalitarian society, which until recently enjoyed the status of semiofficial national doctrine, the managers of modern technological systems have boldly generated a meritocratic system that has tended to widen the gap between the upper and lower orders of society. Far from being a force for more democracy and equality, the rise of technology has been accompanied by a decline of literacy and more generally by the decline of a participating public. We have, accordingly, become a more "highly stratified society" in which "talented young meritocrats . . . will be able to climb into the key 'decision-making' slots and thus generate the success of the new society, and its cohesion against popular tensions and political frustration."[27] McDermott summarizes:

> It seems fundamental to the social organization of modern technology that the quality of the social experience of the lower orders of society declines as the level of technology grows no less than does their literacy. And, of course, this process feeds on itself, for with the consequent decline in the real effectiveness and usefulness of local and other forms of organization open to easy and direct popular influence their vitality declines still further, and the cycle is repeated.[28]

This growing gap between "ruling class and lower-class culture" is, in part, a function of differential access to the control over the most advanced technologies, especially with respect to their language and their applications: "Certainly there has been a decline in popular literacy. That is to say, in the aspects of literacy which bear on an understanding of the political and social character of this new technology."[29] From this follows the crucial issue of whether "one person in a hundred" can understand the speech of organization and technical specialists: "To the extent that technical forms of speech within which the major business of American society is carried on are not understood or are poorly understood, there is a decline in one of the essentials of democracy."[30] Of course, as in the work of John Dewey (McDermott is among our preeminent Dewey scholars), the key word here is *decline*, whose tacit referent is to a time when democracy thrived on the basis of an informed citizenry. But we might ask, When were the jargons of certain disciplines ever really understood on a broad level?—that is to say, the jargons of specialization in *all* areas at *all* times and the hierarchical structure and the access to knowledge and the jargons that express that in terms of ritual advancement and the ability to pay in some way. All of these modern developments, the technological sensorium, constitute a *retreat* to premodern levels of ignorance. Is the technological merely the enfolding of the new technojargonistic expertise into the *outmoded* but ever persistent relations to production, or has the ubiquity of knowledge ushered in a new regime of power in which knowledge producers contend with forces that Alvin Gouldner calls "old money," the industrial and financial capitalists?

Further, "modern technological organization defines the roles of its members, not vice versa. . . . Professionals who seek self-realization through creative and autonomous behavior without regard to the defined goals, needs and channels of their respective departments have no more place in a large corporation or government agency than squeamish soldiers in the Army." While organizations frequently make exceptions for some "very gifted or personable individuals," for the "garden variety employee (or junior faculty member) company sanctions . . . have the force of law."[31] Against the view prevailing among leading scholars of technology—still in force almost a quarter of a century after these lines were written—that modern technological systems based on "smart" machines offer the possibility for creative work for the majority of technical employees, McDermott forecasts what in our observations has become a commonplace: that for the "garden variety employee" corporations have construed new technologies in terms of highly specialized occupational performances even for those trained as professionals while reserving for a relatively small segment of the scientific and technical workforce niches that afford them wider responsibility for the management of technological systems and, in some cases, possibilities for nonroutine work. McDermott's dissent from the pervasive authoritarian logic of modern technological systems—a logic that has smitten many students of science and technology as well—and his analysis of the political and social consequences of the successful dissemination of the ideology of *laissez innover* are framed within a concern for preserving whatever democratic possibilities still remain against the growing power of technocratic elites at all levels. Where Marcuse, for example, has all but lost hope for democratic, much less revolutionary, intervention by the underlying population, McDermott, writing in 1969, at the apex of the civil rights movement and especially the student and youth revolt, understands it as a struggle against the growing centralized power of the technocratic elites, especially, but by no means exclusively, in the universities.

McDermott identifies what he calls a "social contradiction": between the highly specialized workforce required by advanced technological systems, which, except for the few, are systematically made into a new working class and the spread of educational opportunity, which promises occupational and social mobility to this workforce. He ends his essay with a call for "radical reconstruction" of American life on the basis of the revival of a democratic ethos that recognizes *laissez innover* as a "right-wing ideology" that is shared by many leftist and socialist intellectuals.

The underlying theme of many critiques of contemporary technological innovation has to do with scientific citizenship, that is, asking whether and to what extent laypeople—those not part of scientific, technological, or managerial elites—are able to participate in, much less control, the decisions that determine the direction of scientific research, particularly the application of its results to technologies that penetrate all corners of the economic, social, and cultural spheres. As we saw earlier, many in the scientific-technological establishment and their intellectual retainers such as Merton disdain the cultivation of scientific and technological citizenship on the ethical ground (whose premises have been violated in liberal democracies as much as in oligarchies) that science should be free of political interference. Now that science and technology are routinely subsumed under capital and the contemporary democratic state, we have witnessed the insertion of an old, elitist, technocratic argument against popular scientific citizenship. This elitist view holds that given the highly complex nature of modern science and technology, the masses are too ignorant to be entrusted with such important responsibilities. But ignorance should not be equated with stupidity. Most present-day living scientists and engineers come from families that are not scientific or technical intellectuals. And in the past twenty-five years the spread of considerable scientific knowledge among several layers of the general population, albeit predominantly in the middle class, reveals that when people are sufficiently provoked by what they perceive to be calumny, they learn fast and well.

But it is not the scientists who seem to control the priorities of scientific research. Since World War II, the separation of what is termed "peer review" of proposals for funding research and government bureaucracy, which putatively retains final power over these decisions, from government policies has irked scientists. For example, since the 1980s, federal managers have sometimes reversed the judgment of peer panels or, in the wake of budgetary constraints, have funded only those projects that correspond to priorities that are increasingly determined both directly and indirectly by military, industrial, and commercial interests. More and more, profit overrides the conventional scientific ethos, an ethos that has been a mere ideology for decades. Not only does "disinterestedness" virtually disappear, but the invocation to share knowledge throughout the scientific community gives way to practices such as industrial secrecy. In the past decade we have seen the growth of the idea of intellectual property, especially as university-based scientists have agreed to turn over their patent rights to corporations, which, characteristically, guard their secrets against a form of industrial espionage popularized by the Rosenberg case: knowledge theft. As a result, apart from the horrendously poor access that most people have to *public* sources of knowledge, privatization deprives most of us of what we need to know to become responsible scientific citizens. Patents govern the dissemination of scientific information far more than traditional venues such as scientific journals. Consequently, we have witnessed an alarming decline of collegiality among scientists that resembles the competition among inventors that is a pervasive aspect of the history of industrial capitalism. Contrary to Merton's call for "communism" in sharing knowledge, we have definitively entered the era of privatization.

Among the responses to these vast changes, this intrusion of the technological into the everyday, are distrust and criticism based on attempts to gain alternate knowledges, especially in the medical field. Scientists are not pleased by growing instances of popular, secular opposition to some highly valued research programs: ethical questions about the value of genetic research, new birth techniques, genes as property, establishing biomedical laboratories in populated areas; ecological and health concerns with radiation emissions from accelerators; significant popular opposition to the racist implications of attempts by sociobiologically oriented seers to link genetics, in the most narrow, deterministic way, to socially abhorred behaviors such as crime and voluntary unemployment; and criticism of the university-industrial complex that has spread fairly rapidly among leading research institutions. As Merton defined it, scientists recognize that the liberal democracies promote an environment of free inquiry in the spirit of modernity of which science is a crucial component, but are not generally in favor of extending democratic power in matters about which the public has little knowledge and even less capacity. In addition, concern has arisen among religious "scientists" that manifests itself as a sharp fundamentalist critique of evolutionary biology combined with the proposal for a reinstated "creationist science." This is only one of several premier examples of the potential costs of scientific citizenship.

The past twenty-five years are replete with examples of the *absence* of such citizenship: the disastrous effects on communities of the large-scale introduction by chemical companies of carcinogens into waterways such as Love Canal in western New York; the impact of asbestos-clad tubing on U.S. schoolchildren and school employees; the soaring incidence of cancer among employees of the leading producer of this material, the Johns Manville Corporation, and the residents of the New Jersey town where it is located; the defeat of a Cambridge, Massachusetts, citizens' movement to prevent and otherwise control the Massachusetts Institute of Technology (MIT) genetic engineering program; the West Harlem community's fight to thwart Columbia University's plan to convert the historic Audubon Ballroom into a large biomedical laboratory. Both the Cambridge and the Columbia cases are part of the long-range collaboration between the state and the private sector in the development, patenting, and applications of basic and industrial research, much of which involves questionable genetically engineered organisms.

The effective challenge mounted by ACT UP to the priorities, nature, and scope of AIDS research is another example of citizen action.[32] ACT UP, the leading AIDS activist movement, is perhaps the premier instance of scientific and technological citizenship in contemporary America. Originally a nonexpert organization of gay, lesbian, and other activists, it began to arm itself with knowledge. ACT UP and other groups have begun to effectively influence government AIDS policies concerning approvals of new drugs and dissemination of information by public bodies. But perhaps the most impressive aspect of their activity has been to enter the dialogue concerning the direction of AIDS research. The movement has generated its own agenda of priorities through an alliance of laypersons and scientists. While the members of ACT UP are by no means "typical" of the general population, they have not been particularly trained in science or, for that matter, science policy; they are overwhelmingly without scientific or technological credentials.

Plainly, technoscience cannot claim disinterestedness about the sources of funding and the entailments and the consequences of its inquiries. For example, only the most dogmatic defenders of existing science can fail to notice the arrangements that have been made between major research universities and large biomedical and pharmaceutical corporations regarding the ownership of the results of inquiry. At MIT and elsewhere, in return for substantial corporate funding, scientific faculty have agreed to privatize—that is, sell—knowledge as well as share in some of the proceeds of the sale, and recent government policy encourages such arrangements. One can cite other instances of similar relationships: in the development of the superconductor, and in the whole history of the Strategic Defense Initiative, when substantial segments of the scientific community suppressed doubts about the viability of the concept and the science required to develop it in order to receive federal funds, some of which were used to support unrelated research. Moreover, elements of respectable science have been mobilized to emphatically refute evidence adduced by other scientists that industrial pollutants, food additives, and radiation are fatally carcinogenic.

Nor does the proposition that the choice of the objects of scientific inquiry is free from the intervention of interested outsiders retain more than a trace of validity. Consequently, it is no suprise that there is widespread and growing "organized skepticism" from activist communities about the universality of the knowledge generated by established science. As a host of social studies of science demonstrate, scientific knowledge may be compared to ordinary discourses in that it is produced in economically, politically, and culturally specific contexts that influence all aspects of its practice: theoretical, experimental, and validation procedures.[33]

In the past twenty years, undoubtedly conjoined to the ebbing of the most politicized manifestations of the resistance, the technoculture that was once received as a brilliant if dystopian anticipation in social theory of the coming of, variously, postindustrial society, the end of history, or, in the darker ruminations of Marcuse and McDermott, technological domination has become an intrinsic feature of everyday existence. Technoculture is no longer a "tendency," an anticipation, a prediction reserved for "the year 2000," as it was for Daniel Bell and Alain Touraine in the 1960s, or a dystopian futurity in the dark science fiction of writers such as Phillip K. Dick. It has arrived.

Technoculture has become so ubiquitous that resistance to it appears, increasingly, antediluvian. Consequently, among radicals as well as conservatives the technoculture has its prophets as well as its critics. Like all other cultural spheres in the throes of the ideological shifts introduced by postmodernism—education, sexuality, sports, art—technoculture realigns the combat. One may no longer *derive* a position on the new computer-driven culture from familiar ideological premises.[34] For example, within the new discursive boundaries, it is difficult to discern the differences between critics and advocates on the basis of the traditional divisions between critical conservatives and radicals on the one side and optimistic modern liberals on the other, since both argue (from different premises, to be sure) that the task of public policy is to free the new scientifically based technologies from the thrall of arcane social arrangements so that they may be employed to solve urgent social problems. In turn, critics of the scientific and technological revolution of our time, particularly of informatics, have emerged from both conservative and radical camps, and their technophobic, elegiac invocations of past civilization are remarkably similar.

Informatics now influences *all* technology but is, at the same time, a technology with specific characteristics and uses that demarcate it from other technologies. Informatics is the fusion of Computers and Communications. Among the most dramatic effects of informatics is to compress space and time; the effect of this fusion is to speed up the life process. There are no relatively long intervals of communication. E-mail and fax, which use telecommunications but transmit visual rather than oral signs, have utterly transformed experience. Perhaps most important, informatics both creates, via long statistical runs of probable states of being based on a short historic set of observations, an enviroment that becomes *the* environment, and is its own subject. In here, although this is not the place, the whole question of what empirical fact gathering even means is raised, and then one begins to edge into something called cyberspace, not in the way it might reflect the "out there," but in the science-fiction self. Cyberspace tends to reduce temporality to its function, contributing to the "end" of history.

Those who enter cyberspace, which perhaps includes a majority of contemporary intellectuals, have slowly but surely dropped their old objections to technology. And the critique of technology does not retain its global character, that is, critique is confined to the specific effects on a segmented economic and social world. The critiques of modern technology offered by writers like Ellul, Jonas, Marcuse, and McDermott, who, from a libertarian political standpoint, judged modern technological regimes to be forms of domination and called attention to the disastrous consequences of modern technological systems for the fulfillment of freedom and individuality, have slowly given way to critiques, such as those made by feminists and environmentalists, of *specific* technologies and their applications. This particularist critical focus corresponds, broadly, to what Jean-François Lyotard calls the postmodern condition, at least in the contemporary historical conjuncture. That is, under the impact of the collapse of communism—indeed, the passing of all universalist political ideologies (except liberalism)—the moment of general critique has passed, together with a widespread repudiation of the notion of the totality.

Accordingly, knowledge takes the form of local knowledge, just as Foucault has insisted that there are only *specific* intellectuals. Except for an older generation of, say, feminists and ecologists for whom the task was always to achieve freedom from domination, and younger critics who adopt this discourse, most of the younger intellectuals attached ideologically, even if not organizationally, to social movements rather than classes and parties stop their trenchant critiques short of going beyond specific areas of social and cultural life and refuse to generalize their critique to capitalism as a system or even to the whole technological sensorium.

Many students of technology have no trouble accepting these critiques while at the same time embracing a global celebration of contemporary technologies on what may be termed aesthetic grounds. Technology is taken as a toy, as art, as a source of pleasure. For some technophiles, the old critiques are symptoms of a surpassed, unhip culture. We may observe this in the work of Richard Lapham, Constance Penley, Andrew Ross, and others who earlier might have adopted the perspective of critical theory. These writers are, in other spheres, critics of prevailing social and political arrangements. But they have adopted a certain technophilic response—an analog at the level of discursive writing to computer hackers—to the possibilities afforded by contemporary cybernetics to applications for play and creativity. Like Shoshana Zuboff's invocation of new vistas in work made palpable by the introduction of computers and Herbert Simon's faith in the possibilities inherent in artificial intelligence technology for revolutionary discoveries of how the human brain works, an intellectual movement that transgresses older divisions that can only be described as a new technological utopianism has emerged. Of course, Simon's description of humans as information-processing machines has come under heavy criticism. The Dreyfuses have objected to the analogy between the computer—whose judgments are formally limited—and the human "mind," which is capable of deciding on the basis of ambiguity. And John Broughton among others has argued that computers are no panacea for learning, but on the contrary may inhibit critical thought. Since informatics is not merely a vehicle of communication between speakers who are otherwise independent of it but instead creates a different culture, the question to be asked is, What are the presuppositions of learning regimes based on this technology? Both types of critique tacitly refuse recent developments in the technoscience of cognition, which is now among the principal knowledge products of informatics.[35] Rather, the alterative perspective distinguishes knowledge and information, and argues that cognition is crucially mediated by culture. A regime in which there are *always already* procedures for solving problems as well as prescribed answers to them is inimical to achieving the goal of proactive learning.

Many scholars, critics, and journalists have greeted the phenomenal dissemination of cybernetic technologies from the office and factory to the home and recreational sites as the precursor of a new era of community, of perfect communication and even of a new explosion of eros. For some, notably Latham and Ross, the critique of technology smacks of nothing less than emanations from a threatened elite who wish to preserve their power in the academy and other quarters of institutional life. Tom Wolfe taunted the gatekeepers of high culture in the 1960s, a taunt echoed in much the same terms by Ross nearly thirty years later. Discerning the liberating possibilities of the new technology has become part of a more global effort to overturn the ideas of an old intellectual establishment that railed simultaneously at popular culture and at the technological sensorium and despaired at the global embrace of the communications revolution even as McLuhan announced the birth of the new human community that he called the "global village," signifying how small the earth had become in the wake of the technological extensions of "man."

Rather than berate technology and those who develop, promote, and control it, the new utopians, whose eponymous hero is the anonymous hacker, not the ubiquitous critic, have found a crooked path back to the jointly held position of traditional Marxism and modern liberalism: established power constitutes a fetter on the possibilities inherent in the new technologies. Cyberpunks (the hackers, the conferencers for whom E-mail is a primary community), rock groups, postmodern science fiction writers (especially William Gibson), and performance artists such as Laurie Anderson have, it is claimed, given us a glimpse of things to come. If machines are extensions, not enemies, of human creativity, then the blurring of the boundaries between the "real" and the simulacrum is no cause for grief. If the old industrial technophiles extolled machinery that, in Marx's words, multiplied human productive powers, the new technological utopians have argued from the perspective of possible worlds. The power of cybernetic technology to realize the once remote chance for pleasure as well as new knowledge is already upon us.

Heidegger's subtle questioning of the critical view according to which technology is merely purposive activity subsumed under the characterization "means" may be taken as a starting point for investigating the various claims concerning the new technologies.[36] He brings us back to the Greek origin of the meaning of the term *techne* in order to free technology from its subsumption under instrumental reason. Where modern culture views technology as a regime of powerful tools by which human purposes may be served, particularly the domination of nature and the organization, control, and regulation of humans, *techne* signifies, in Heidegger's recollection of its etymology, an uncovering, a way to truth.

In contrast to *techne* as the human activity by which the object, nature, makes itself known to us, modern technology is an "ordering revealing"; technology becomes a process by which the real is enframed, blocking the truth of the world from "shining forth." In this regime, the world stands "in reserve" for human purpose and loses, at least provisionally, its own essence. Of course, Heidegger's project, in concert with one strain of utopian thought, is to restore to nature an autonomous existence from which *techne* receives its genuine signification. He refuses, however, to identify his critique with that of the antitechnological camp. The instrumental view according to which technology is considered a means remains, even if we restore *techne*. The history of technology over the past five thousand years is irreversible. Rather, in this widening of the meaning of technology, instrumental action becomes a local activity subordinate to its meaning as a way of revealing the Being of the natural world.

Heidegger's discourse conceals a thinly veiled attack on the position of Critical Theory and other critics of modern technology. Against the critics for whom technology has *become* culture in late industrial societies to the detriment of the autonomy of thought and being, Heidegger, despite his acknowledgment of the force of the critical view, wants to reconcile modern technology with culture by reviving its status as an end that goes beyond the notion of the domination of nature.

Having permeated everyday life, technology has become a culture; hence the conflation *technoculture*. It is no longer critical. It accepts, even celebrates, computer mediation as liberating. In this postcritical perspective, we are informed of the wonders of cyberspace, where innovations such as virtual reality offer overworked professionals play therapy to counteract stress. In cyberspace the old-fashioned "pressing of the flesh" yields to what Howard Rheingold calls teledildonics. Here is Rheingold:

> The first fully functional teledildonics system will be a communication
> device, not a sex machine. You probably will not use erotic telepresence tech-
> nology in order to have sexual experiences with machines. Thirty years from
> now when portable telediddlers become ubiquitous most people will use
> them to have sexual experiences with other people, at a distance, in combina-
> tions and configurations undreamed of by pre-cybernetics voluptuaries.
> Through a marriage of virtual reality technology and telecommunications
> networks, you will be able to reach out and touch someone—or an entire
> population—in ways humans have never before experienced. Or so the sce-
> nario goes.[37]

Rheingold tells us that "you can reach out your virtual hand, pick up a virtual block and be running your fingers over the object, feel the (virtual) surfaces and edges, by means of the effectors that exert counterforces against your skin."[38] On the other hand, you can do all this by closing your eyes and exercising your hand and your imagination—which, of course, is what science fiction writers have done for nearly a century.

In these sentences we can see the play of the underlying message of cyberpunk: cybernetic technologies embody the blurring of the lines between reality and its representations, whether fiction or "science." The liminality of the cyberspace is the way to giving full range to fantasy. Electronically mediated machines are toys, but like all good toys provide the objects through which human agency is developed. Like Melanie Klein's theory of object relations, in which the child forms its self in and through objects and the notion of selfhood is impossible without its objectification, many theorists of the affirmative cultural implications of cybernetic technologies hesitate before its dark side. Just as we "work through" the dialectic of the oedipal conflicts through play, so electronically mediated adult play is both more than a way to relax and a way to address the frustrations of our highly bureaucratized and regimented lives. While it offers pleasure, it also signifies the merger of the psyche and the machine, revealing the ways in which we are totally "wired," the degree to which we may achieve integration between computer-mediated work—a situation shared by most professionals and managers—and our "leisure." Here the notion of humans as robots is not a pejorative characterization but a proudly worn badge. We *are* our simulacrum.

Those who herald technoculture as the fulfillment of the next and perhaps final frontier of human striving call us not to mourn the passing of the old culture (if, indeed, a culture not intimately linked to technology ever existed) or to a new cornucopia of leisure, but to a new ground of social existence, especially of what we might term "experience." At last, some say, the full development of the individual is possible because we have finally objectified both our physical and our mental capacities in a machine, just as the machine has attained subjectivity through us.

Technoculture plays with the distinction, made first by the Greeks, between work and labor. Some have argued that, in contrast to work in the era of mechanical reproduction, computer-mediated work eliminates most of the repetitive tasks associated with Taylorism and Fordism: a "smart machine" can interact with human intelligence as a playmate. Chess, for instance, becomes a test of the possibilities of artificial intelligence, a mirror for our intelligence. Thus the distinction between work and play that characterized our collective preoccupation with scarcity throughout history is sundered. For the first time, work and play are identical in occupations besides those of artists and scientists. To borrow a term from Whitehead, cybernetic technology not only *ingresses* into events, it has become an event that leaves virtually nothing untouched. It is not a worldview in the traditional sense but is the ingredient without which contemporary culture—of work, art and science, education, indeed, the entire range of institutionally mediated interactions—is unthinkable. So is it futile and even deeply conservative to rail against the technological revolution? The only remaining question, in this view, is how to free technology from the thrall of the organization of labor, the rationalized algorithms of industrial society. In the chapters that follow, we will make some preliminary attempts to answer this question.

Chapter 3
The End of Skill?

Technoculture

Who does the emerging technology empower? Is it the instrument of play for some, the instrument of repression for others? And what of work? Work is not only central to the issues before us but also the most ubiquitous possible application of cybernetic technologies in our lives. This is especially true of those whose work consists primarily, if not exclusively, of interaction with computers. For without a sober assessment of the implications of the computer and its applications for the present and future of work, those who herald the computer's wonders evade one of the central problems for humankind.

As we will argue in chapter 11, there is no warrant for holding the variant of the romantic view of modern technology that holds that the historical passing of traditional skills is a calamity. We do not claim that the remedy is a return to the craft and artisan traditions of production. Nor, however, are we in accord with the technological utopians who, disregarding the specific conditions that form the context of technological applications, declare victory over backbreaking and routinized labor, forecasting the desublimation of (a computer-mediated) eros at work and play. Ignorance, they proclaim, once individualized, computerized learning machines are introduced, if they are equipped with the correct programs, can and will be overcome. Scientific puzzles such as how human intelligence works will be solved. Social problems such as urban blight and personal problems such as infertility and genetic defects can be eliminated.

Cybernetics is the most widely applied current means by which labor is being progressively freed (without pay) from the industrial, commercial, and professional workplace. As a cost-saving device it is wildly successful. Literally millions have been "liberated" from industrial and clerical labor, and, as we are arguing here, a significant fraction of professional laborers are already experiencing freedom from the work for which they have been trained. However, as we shall further explore in the concluding chapter, neither workers nor professionals experience their separation from wage and salary labor as a new burst of freedom, much less of liberated eros. Instead, given the economic and social context, "unwork" appears to its beneficiaries as a financial blow that often devolves into catastrophe and a psychological trauma that affects all aspects of their lives.

Throughout the modern epoch, proponents of the scientific enlightenment have taken the position that the "idyllic" relations of precapitalist societies were impossibly repressive to the human spirit, and the evils of exploitation notwithstanding, capitalism's historically redeeming feature, led by scientific technological discoveries and inventions, has been the will and ability to change the world. However, it is not excessive to argue that the computer, even more than its anterior concomitants—"motive forces" steam power and electricity and labor-saving fabricating devices such as the self-acting mule in textiles and, in the mid-twentieth century, automatic transfer machines—provides a persuasive case for the claim of the newest technophiles that finally, after nearly three centuries of the rationalization of the labor process, we can envision the reintegration not only of work but also of humans with nature and with their own species. For some, these are not merely hyperbolic pronouncements but practical opportunities for ending the estrangement of humans from the multiple worlds they inhabit. For in its most visionary form, computer-driven technoculture claims to fulfill the dream of the "whole" person by healing the rupture between intellectual and manual labor and freeing time for the full development of the individual.[1]

Thirty-five years ago, business and labor leaders proclaimed the "automation" revolution, an event signaled by the introduction of the transfer machine into auto assembly plants. In 1955, during a much-publicized tour of the new General Motors Cleveland engine plant where the machine had recently been installed, GM chief executive officer Charles E. Wilson was said to have asked United Auto Workers president Walter Reuther, "Who's going to pay union dues?" when the fully automatic plant comes into being. To which Reuther shot back, "Who's going to buy your cars?"[2] Now, the transfer machine was merely an electronically controlled assembly line; where thousands of workers formerly installed parts by hand (or with hand-held electric-powered tools), these operations were now performed by the machine itself. Early transfer machines simply applied the feedback mechanism of the thermostat to production. By the mid-1950s it was already in use in oil, chemical, and food-processing industries. These machines bore a weird physical resemblance to human arms and hands; a prototype was exhibited in that famous scene in *Modern Times* in which Charlie Chaplin is hooked up to the automatic feeding device installed by the employer to keep his hands free to work while he eats lunch.

The transfer machine's arms and hands are skeletonlike, bereft of flesh, but electric power multiplies the motive power once provided by muscle. In this regime of production, human labor is made ancillary to the machine and is absolutely needed only to repair it during its infrequent breakdowns.

Although the techniques of early automation were relatively elementary compared to contemporary cybernetic practices, their fundamental purpose in the workplace was crystal clear, while later optimistic discourses about computerization obscured the economic implications of computers. At the dawn of the latest technological revolution, most of those who were either its perpetrators or its objects understood what was at stake: to enlarge profit margins by eliminating or otherwise reducing the most expensive cost of production—manual labor. Recall the high degree of unionization of basic production industries at the end of World War II, a historic development that established high wage and benefits levels and, in the most successful cases, established rigorous shop floor protections that limited the power of management to direct the workforce at will.[3]

On the eve of the Vietnam War, Japanese and European economic recovery with their higher technologies were already challenging U.S. industry to either modernize or close doors. During the 1960s, however, war-generated profits postponed the inevitable. When the United States turned its attention to civilian production in the seventies, whether it was too late or not, U.S. corporations seemed to lack the will to take decisive steps to save the steel, rubber, electrical machinery, and a dozen other industries. The federal government was already in the thrall of an anti-interventionist, free-market ideology that paralyzed it even as dozens of key industrial plants closed down.

Yet dire predictions of the decline and fall of U.S. world economic hegemony proved premature. Global capitalism, still largely U.S.-based, in which national boundaries were no longer (if they ever were) sacrosanct, produced a parallel industrial regime, the global assembly line. Immediately after World War II, banks and U.S. industrial corporations had invested billions of dollars in European recovery; by the 1950s they were extending their stake in Japan, Korea, Malaysia, Puerto Rico, and Latin America. An "American-made" automobile is likely to contain a Japanese fuel pump, a

Mexican-made exhaust system, and Maylasian-produced windshield wipers—or to have been assembled in Japan, Mexico, or Korea. The only thing "American" about many Chrysler cars, for example, is the name. The technical designation for this change is the new international division of labor.

In the seventies, U.S.-based corporations went global and, in the search for higher profits through lower labor costs, literally sacrificed huge chunks of technologically antiquated domestic heavy and light production industries. Employment in the steel industry was reduced by two-thirds. From being a net exporter of this crucial material, the United States became, despite retaining a relatively high level of production, a net importer. Japanese, German, and Italian steel are routinely imported into the United States because the domestic steel corporations resolutely refused to modernize basic steel processes after the 1950s. Specifically, they failed to install computer-driven basic ingot production processes until a decade after the major competitors to the U.S. steel industries had fully computerized their steel production. Similarly, as late as 1980, many remaining U.S. auto parts plants still had 1920s technology even as overseas competitors had already introduced the major computer-mediated technologies of numerical controls, robotics, and lasers, most of which had been developed by U.S. engineers and computer systems developers. These were applied in limited ways to U.S. production industries, notably aircraft engines and other defense-related fabricating industries where the research and development and purchase and installation costs were underwritten, in the main, by the Defense Department.

In the wake of increasing European and especially Japanese imports, the auto industry reduced both total car production and employment. Chrysler, Ford, and General Motors, which had sold cars made elsewhere in the U.S. market, reduced domestic production and invested substantially abroad, especially in the Americas, but also in Europe and Japan. A substantial segment of the mass transportation equipment industry virtually disappeared, except trucks and aircraft, as this country no longer competed successfully in global markets for shipbuilding and rail cars.

In the late 1970s and the 1980s, U.S. corporations, abetted by bountiful defense contracts that did not contain draconian cost-saving provisions, *chose* not to invest in high-level technologies within the industrial sector. Rather than making new investments in research and development in the civilian sectors of industrial production, and in concert with a conservative national government, employers resolved some of their competitiveness problems by launching an unprecedented attack on wages and working conditions and the relatively strong unions, whose efforts had, since the 1930s, guaranteed benefits through collective bargaining. After a decade of continuous union bargaining concessions, disinvestment manifested in plant closings, and bitter strikes that resulted in a shift in the initiative from unions to employers, collective bargaining was in a shambles, but technological innovation suffered as well.

Curiously, in a culture that privileged practical applications over basic research, the United States retained its lead in fundamental computer hardware components, particulary the crucial memory chip, in part as a result of massive government funding of defense-related research and development activities in the fifties and sixties. When Japanese computer-chip producers began to seriously challenge U.S. domination of this field, free-market ideology was quietly shelved: IBM entered an alliance with its principal domestic competitor and secured the support of the federal government to undertake new research and development initiatives. United States entrepreneurs still dominate the computer software field, although Japan and European countries, particuarly Italy, are beginning to catch up.

Decentering the Worker

The new technoculture in the workplace emerges on the ruins of the old mechanical, industrial culture. From the perspective of the worker, whether in the factory or in the office, the second phase of automatic production—computerization—is merely a wrinkle in the long process of disempowerment. No exciting new skills are needed to operate the console that controls the robot on the floor of the assembly plant. In characteristic fashion, management has chosen to maintain the distinction between craft and semi-skilled labor by training machinists as programmers, ensuring that only a small fraction of people get to program the numerical controls box perched atop the lathe. Operating lasers requires no special technical training. In the large steel, machine, and metalworking plants that have been broadly computerized there are fewer workers on the floor, a sometimes eerie experience for anyone of the older generation who remembers the camaraderie of the age of mechanical reproduction.

The old industrial plant resembles more the closed institution of the asylum, where every movement is strictly monitored, than the democratic streets of the large city. Even when many industrial workers gained the benefit of job security and a grievance procedure through union organization, their lives for the eight, ten, or twelve hours on the job were controlled by others. On the assembly line, workers must get permission from the supervisor to use the toilet, and permission is given only if a replacement is available to relieve the worker. Other aspects of time are strictly regulated, for time is money. Within these shackles develops a culture of resistance that has become the occasion for innumerable studies. This is what the old-timers remember about their working lives. But memory is selective and the past is available only through recollection. However, you cannot spend a day in even the most flexible of industrial workplaces without agreeing with the appraisal of Andre Gorz that the term *freedom* has nothing to do with them. The old factory was a prison, and the job dulled the mind and wore out the body. True. But it wasn't lonely, and whether you had a diploma or not, you felt your power because production literally depended on the application, in the most physical sense, of human labor to raw materials and machinery, themselves the product of this three-way interaction.

Perhaps more important was the solidarity of the line—of machines as well as the assembly chain. As arduous and repetitive as the production work was, workers in the mechanical era were never completely subordinated by rationalization. The interdependence of rationalized production always entails a risk to the employer that if even one section of the line, a single department or group, goes down, the flow may be disrupted. And, with the rise of industrial unions in the 1930s, especially those in the automobile and electrical industries with strong shop floor leadership, assembly-line workers and machine operators were able to exercise considerable power over production by limiting their output. Assembly-line workers occasionally pulled "quickie" strikes to slow the line when management violated its agreement to keep the pace moderate. And machine operators negotiated piece rates that allowed them to make their quotas without having to spend every minute of the workday in production. When management unilaterally changed the rates or proposed "tighter" rates on new jobs, the affected group might start a slowdown or refuse to work under the new rates. This action could disrupt the entire production process because of the close intergration implied by rationalization.

Similarly, much clerical labor has been transformed by the computer. The typewriter is still employed in some smaller firms, but the personal computer has all but driven it from the large office. The computer is more versatile as a word processor than its ancestor, and its computational capacities exceed those of any human brain and even the most advanced mechanical accounting devices. But while typing gives way to word processing and the skills menu required by the new machine varies and is, perhaps, marginally richer than the older one, the operators of the most complex electronically driven and computer-controlled technologies follow routine instructions. The initial sense of excitement and fear that accompanies the introduction of any new technology eventually yields to boredom and rote performance.

Moreover, after two decades of the computerized office, in addition to the persistence of routinized labor we have already discovered another dark side. For men and especially for women, sustained work at a computer can be dangerous to health. Now that the routine use of word processing has spanned twenty years, there is a widespread incidence of diseases of muscle and bone such as tendinitis and, more egregiously, carpal tunnel syndrome that cripple arms and hands. In addition, some investigators have found the incidence of infertility among computer operators has far exceeded the statistical average for the general population.

There is no room or time, in a situation where productivity is measured by keystrokes, for a "culture" to arise; it is just a faster assembly line. For office workers the computerized workplace has tended to eliminate the amenities of socialized labor. Back-office workers in large commercial banks interact with their video display terminals more than with other human beings, sitting in individual carrels typing nothing but disembodied numbers all day. Naturally, this numbingly boring work causes enormous turnover, which in the 1980s reached as much as 100 percent annually in the back offices of some large insurance companies and banks.

In the *Grundrisse,* Marx transmutes his earlier concept of alienation by means of a structural category. Among the main consequences of the introduction of large-scale machinery into the industrial workplace is the radical displacement of the worker from the center of the labor process. Referring to early instances of the introduction of an "automatic system of machinery," Marx writes:

> In no way does the machine appear as the individual worker's means of labour. Its distinguishing characteristic is not in the least, as with the means of labour, to transmit the worker's activity to the object; this activity, rather, is posited in such a way that it merely transmits the machine's work, the machine's action, on the raw material—supervises it and guards against interruptions . . . it is the machine which possesses skill and strength in place of the worker, is itself the virtuoso, with a soul of its own in the mechanical laws acting through it. . . .
>
> The full development of capital takes place—or capital has posited the mode of production corresponding to it—only when the means of labor has not only taken the economic form of *fixed capital* but has also been suspended in its immediate form, and when *fixed capital* appears as a machine within the production process, opposite labor; a*nd the entire production process appears as not under the direct skillfullness of the worker but rather as the technological application of science.* [It is] hence the tendency of capital to give production a scientific character; direct labor [is] reduced to a mere moment of this process.[4]

Marx draws the implications of the emergence of scientific knowledge as the dominating force of production: labor time, which is "posited by capital as the sole determinant element . . . is reduced both quantitatively, to a smaller proportion, and qualitatively, as an, of course, indispensible but subordinate moment, compared to general scientific labor, technological application of natural sciences."[5]

Even though the commodity has value in proportion to the amount of socially necessary labor time embodied in its production, at least initially, Marx argues labor is diminished in proportion to the growth of capital in the form of scientifically wrought machinery. Yet even as late as the era of mechanization, human labor power remained at the center of production, because in the century following Marx's descriptions of automatic machine systems, they were still not broadly introduced into the industrial workplace. There were exceptions. Continuous process was introduced in the oil and some branches of the chemical industry before World War II, but direct applications of human activity remained the heart of the labor process in most industries until the late 1950s. When computerized automation was first introduced, it was applied only in a minority of plants and a relatively small number of industrial sectors until the 1970s. While the machine was powered by mechanical or electrical means, the application of cooperative human labor, however "degraded" from the perspective of the old artisanal mode of production, was visible. Huge armies of workers operated machines. There was still considerable manual labor putting raw material in lathes, ingots into furnaces, and cloth, metals, and plastics into sewing and stamping machines; guiding the material to its proper place; and removing finished work and placing it in bins.

With the advent of the computerized workplace many of these lingering aspects of direct worker activity in the labor process associated with the mechanized era disappear or are further marginalized. The worker is no longer the subject of production in the computerized workplace because this technology is not rooted in the grand narratives of skilled and unskilled manual work, even as its function in the workplace (including the clerical) is absolutely necessary. As we have previously argued, the concept of "skill" associated with the craft era has given way to knowledge, which, although it incorporates what we have called tacit knowledge, is in fact the negation of its underlying element, the unity of body and brain. Even if they were purposively sundered by the division of labor, these narratives depended upon the great metaphor of human conquest of nature through the coordination of head and hand. In the regime of industrial capitalism the "whole" producer was divided (presumably in the interest of efficiency) and reunited by the function of management, now legitimated as a property of capital.

The socialist project, before the Bolshevik revolution and the rise of Western social-democratic reform parties, consisted largely in its promise to restore to the producers not so much the full product of their labor (this was always a utopian demand that Marx himself renounced in the name of necessary socialist accumulation, and was never taken up by the modern labor movement) as much as self-management of the labor process. Now, the function of directing the entire labor process would be performed by the direct producers, who might assign the coordinating tasks formerly ascribed to management as to any qualified worker, holding neither special privilege nor institutionalized powers. At the same time, they would no longer be in the thrall of a division of labor in which what was called intellectual was sharply demarcated from what was called manual. So even if craft and its concomitant traditional skills—the concept of which is identical with the unity of design and execution—no longer described the heart of the labor process, the formation of the detail laborer under capitalism, which also signified relations of power and powerlessness, would, under self-managed as opposed to state socialism, become merely a technical description of certain functions within the labor process, not a signifier of social power.

In the past twenty-five years, computer-mediated work, despite its potential for reintegrating design and execution, has been employed, typically but not exclusively, in a manner that reproduces the hierarchies of managerial authority. The division between intellectual and manual labor and the degradation of manual labor that was characteristic of the industrializing era have been simultaneously shifted to the division between the operators and the professional-managerial employees, but also the division between the "lower" operating and the "higher" expert orders broadly reproduces within intellectual labor itself the older gulf separating manual and intellectual labor in the mechanical era. Hierarchy is frequently maintained despite the integrative possibilities of the technology. Under this regime of production, the computer provides the basis for greatly extending the system of discipline and control inherited from nineteenth-century capitalism. Many corporations have used it to extend their Panopticonic worldview; that is, they have deployed the computer as a means of employee surveillance that far exceeds the most imperious dreams of the Panopticon's inventor, Jeremy Bentham, or any nineteenth- or early twentieth-century capitalist.

Some advocates of computer applications as the embodiment of the restoration of true profession or craft argue that its uses must be separated from its intrinsic character; that it is in the interest neither of society nor of corporations to mechanically transpose some analyses of the organization of work appropriate in the industrializing era to the new computerized workplace. On the contrary, they insist that computer-mediated work *could* provide unprecedented opportunities for the full development of the operator's knowledge and authority over the labor process.

Taking issue with Braverman's deskilling argument, Michael J. Piore and Charles F. Sabel, in *The Second Industrial Divide*,[6] forcefully illustrate that the new computer technologies create the possibility for the reemergence of a new regime of craft production. For them this is the result of a flex-ible-specialization, decentralized, community-based, small-batch, skilled-worker production process. For them, Fordist, corporate-controlled, mass production has stagnated and can deliver neither the goods nor the profits. The logic of postindustrial society is that flexible specialization can save the manufacturing competitiveness of U.S. industry by introducing a flexible, continually reorganized, and innovative production process. Further, in-stead of a labor process that degrades skill, flexible specialization creates new skills and requires more skilled workers than the old Fordist production process.

But first, what is the "deskilling" debate about?

Knowledge or Skill?

The debate over skill and knowledge is framed by the work of Harry Braverman on the labor process.[7] Braverman argued that all workers, including knowledge workers, are degraded in the capitalist labor process. In his view, the degradation of work consisted of, first, the fact that technology and labor under conditions of, capital accumulation are directed against skill. The division of labor is not a neutral process but a force of domination. In the past, workers' power was based on their skills. They had historic skills as artisans. Artisanal workers were craftworkers whose skills made them autonomous and protected them from the vicissitudes of the labor market. The workers' skills made it difficult for their work to be degraded.

As capitalism developed to its monopoly stage, it launched a continuous assault on the knowledge of skilled workers by deskilling, by separating them from the knowledge of craft that gave them power. For Braverman this assault is embodied in F. W. Taylor's *Principles of Scientific Management* because it redefines the logic of capital.[8] As in Marx, the move is from the formal subsumption of labor—the control over the product of labor—to the real subsumption of labor that is the complete control over labor in the accumulation process. Braverman argues that Taylor's work codifies this strategy of capitalist domination, which entailed the degradation of work through the process of deskilling: the separation of conception from execution. Skilled workers who did both the manual and the mental aspects of work were to be systematically separated from their special knowledge of the labor process. They were to be transformed into semiskilled or unskilled workers no longer involved in mental work. The workplace was fractured, and a detailed division of labor became its organizational principle. The new worker was a nonautonomous, detailed laborer. With the degradation of the skilled worker, new specialized workers were required to do the mental work: professional engineers and scientists to design the labor process, the machinery, and the new products and managers to command and coordinate the whole process. In the name of science and productivity the worker was dominated.

Secondly, there is the displacement of labor. Deskilling results in the recomposition of the workforce: less-skilled workers replace more-skilled workers, low-wage workers replace high-wage workers. Workers lose their jobs. Displacement is a continuous process, and all levels of workers, from unskilled to engineers and managers, will be deskilled. Machinery and new forms of organization will replace skilled work, creating permanent tendencies of structural unemployment.

With the destruction of skill the labor movement is disempowered as well. The only direction that it can take is to resist new technologies that further degrade the labor process. In Braverman we have the end of Marx's worker revolution. We are left with Weber's "iron cage," with only domination in the present and "no exit" in the future.

Mike Cooley, an industrial designer at Britain's Lucas Aerospace who uses a computer-aided design (CAD) program, is in basic agreement with Braverman's degradation of work thesis. In his book *Architect or Bee?* Cooley writes, "There is already evidence to show that CAD, when introduced on the basis of so called efficiency, gives rise to a deskilling of the design function and a loss of job security—particularly for older people—which leads to structural unemployment."[9]

Cooley sees science and technology as forms of capitalist domination. But the forms of domination themselves can be a contested terrain in which workers, engineers, and scientists can struggle to design a science and technology that is not about the subordination of human beings (beelike behavior) but can lead to a future

in which masses of people, conscious of their skills and abilities in both a political and technical sense, decide that they are going to be the architects of a new form of technological development which will enhance human creativity and mean more freedom of choice and expression rather than less.[10]

Cooley on the one hand agrees with the deskilling and displacement in Braverman's degradation of work thesis. But on the other hand, he disagrees with Braverman and argues that even though workers are in a desperate situation, they can still struggle to redesign their world. For Cooley there are alternatives to the "iron cage."

The critique of Braverman's work is quite extensive.[11] Our purpose here is not so much to add another critique as to relate it specifically to the relation of skill to knowledge. In this task the work of Paul Adler is significant. Adler's work is in direct opposition to Braverman's. Where Braverman theorizes that capitalist technology degrades skill, Adler theorizes that the same technology upgrades skill: "My own research on basic computerization found very similar changes in the types of skills and in the general upward drift in the amount of skill even for low level employees."[12]

Adler views technological change as beneficial to most workers; deskilling affects only a minority of working people. Nor is there a direct relation between deskilling and the creation of low-skilled, low-paid jobs: "There has . . . been some polarization in wages due to the burgeoning of low paid jobs in recent years. But this is the effect not of skill changes, but of a decade of economic turbulence and of the influx of youth and educated women into the labor force."[13]

As old skills are destroyed, new skills are created, and the creation of new skills outweighs deskilling. In general this is true not only of the middle-class sectors of professional and technical workers but of the working class as well.[14] With this upgrading of skill there is an increasing reliance on mental over physical effort, increased responsibility, education, and experience.[15] Adler argues that Braverman's notion of skill is based on a romantic view of the nineteenth-century craftworker and that skill must be redefined in the context of modern technological advances:

> The dimensions of skill I have identified permit us to understand the manner in which automation is shaping class capabilities in the labor process, by increasing technical culture (complexity), by expanding responsibility (socialization), by broadening intellectual horizons and by encouraging recognition of labor *per se* as a historical problem (abstraction), and by forging new forms of group identity that allow the higher forms of individuality (collectivization).[16]

Adler understands that the new technologies of high technology and organization have made notions of skill based on craft inadequate. In their place he has put his own notion of skill framed in the context of the requirements of this new labor process. Adler recognizes the need to develop a new paradigm, but he is wed to an analysis that is structured around the key concept of the old paradigm, skill. We ourselves have criticized Braverman,[17] but here we argue for another notion of deskilling. The history of work in the twentieth century is a history in which machines have increasingly replaced the skills of workers of all collars. In a production process in which science and technology are central, knowledge and not skill defines the process.

The contradictions of Adler's insistence that knowledge is just the current form that skill takes can be seen in the following statement:

The End of Skill?

In job design the knowledge intensity of new technologies dictates a greater problem-solving component to operators jobs than traditional Taylorist approaches would suggest. With automation, the number of operators per unit ouput might fall, but there is typically no net reduction in average operator skill requirements; on the contrary, higher skills of a new type are usually called for.[18]

There is much to be learned from Braverman and Adler. We agree with Adler that Braverman's analysis is the embodiment of a deep romanticism for a pristine craftworker of an artisanal past. But Adler is equally romantic about the inevitability of technological progress.[19] This can first be seen in his arguments against deskilling and displacement in the workforce. The fact is that there has been significant displacement and deskilling of the workforce. In the United States, 26 million jobs were created between 1973 and 1986, the great majority of them both low-paid and low-skilled.[20] Bad jobs with low wages have increased in all sectors of the economy, including jobs for college-trained professionals:

The sharp declines in employment are no longer restricted to old line manufacturing firms or old blue-collar workers. Plant closings, layoffs, and pay cuts have swept high-technology industries as well. In just the first six months of 1985, employment in the computer and semiconductor industries—the core of the new technology—shed more than twenty thousand jobs.

The BLS [Bureau of Labor Statistics] reported that, betweeen 1981 and 1986, more than 780,000 managers and professionals lost their jobs as the result of plant closings and permanent layoffs. And the pace has apparently increased, even as the economy entered its fourth year of recovery. In the drive to make the ranks of management "leaner and meaner," nearly 600,000 middle- and upper-level executives lost their jobs between 1984 and 1986. Such companies as AT&T, United Technologies, Union Carbide, and Ford are leading the management massacre.[21]

This massacre occurred before the Wall Street crash of October 1987. Not only did 50,000 workers, from clerks to investment bankers, lose their jobs because of the crash, they were also displaced by new computer networks, expert systems.[22]

The work of Salzman demonstrates that computer-aided design (see next chapter), for instance, does not necessarily lead to the deskilling of drafters; unintentionally, he seems to have demonstrated that it led to the deskilling of engineers.[23] If the design work once performed by engineers is now being performed by drafters, who are lower skilled and lower paid than engineers, work has been degraded and they are in the process of being displaced from the labor process. As Salzman states:

Since drafting work is directly related to the output of engineers, drafting task productivity increases alone should result in any increases in drafting employment following a trajectory no greater than engineering employment levels, and presumably increasing at a lower rate. If CAD bifurcated drafting tasks into lower level system operations and higher level design tasks, there should be additional displacement of drafting work as lower level tasks are assigned to technician level personnel and higher level tasks are integrated into engineering work.

In contrast to these predictions, employment of drafting level workers increased at a faster rate than that of engineers and other professional technical workers.[24]

In our own studies of engineers, architects, and drafters in New Jersey and New York, we found their experiences with the new design technology to be different. In New Jersey, the work of these knowledge workers was degraded; there was clear displacement and deskilling of both engineers and drafters. In New York, however, the architects', engineers', and drafters' work was not degraded, yet new drafters were not going to be hired. New drafters were to be displaced by the technology. In an extensive training program, the existing drafters were formally trained in computer-aided design and drafting, while they would be involved in doing the design work at their terminals. In New York the goal was to train all the engineers and architects. This was not the goal of public agencies in New Jersey.

Always privileging the creation of new skills as new technologies are developed, Adler would argue here that deskilling and reskilling occur simultaneously. Here is the core of our disagreement with both Braverman and Adler. Our contention is that skill itself is no longer central to the labor process. It has been decentered. There has been a vocabulary shift, and knowledge has become the emergent center of production. Knowledge is not just another form of skill, as Adler contends, but it redefines the world of work. The past, in which skill is central, is incommensurable with the knowledge-based future. This future embodies a whole new set of alternatives.

Adler's goal is to offer a critique of the view, associated with Braverman, that technological change in the capitalist labor process *always already* entails deskilling, but he is not willing to question skill. As long as one speaks the language of skill, it is difficult to avoid Braverman's conclusions. Adler's substitution of knowledge/skill for Braverman's craft/skill was insufficient. Both forms of skill were based on their relative economic values and the length of training required.[25] Knowledge/skill required formal educational training, while craft/skill was based on the experiential knowledge of a trade. Knowledge/skill was an inevitable evolutionary development from craft/skill.

But there is a break between knowledge and skill. Computer-aided design, computer-aided manufacturing, computer-integrated manufacturing, and other computer-mediated technological innovations have taken over the skill component. In this sense there has been deskilling. The twentieth century, the industrial-capitalist era, is marked by the displacement of skill by knowledge. It is the knowledge component—the conceptual, the theoretical—that is now the basis for the scientific, technological, and social relations of production.

The history of skill is embedded in feudalism, and then in handicraft in the era of manufacture. Skill was a concrete property that a person could attain only as a result of a long apprenticeship in a trade. The apprentice was trained by a master, and entry to apprenticeship was very selective, often inherited from father to son, uncle to nephew. Thus skill from its very inception was not open to all; it was a mechanism of exclusion. Even the nobility was dependent on the skilled craftsman, whose special knowledge gave him a modicum of power and autonomy. The craftsman thus guarded his skill, the collective property of all those who had learned the knowledge of the trade. Craftworkers formed organizations of protection, first the guild, then the skilled workers' union, then the profession. Thus it continued into the modern world: skilled men, carefully guarding a territory only the few could enter. Here the case of women is especially significant. Even as women enter the workforce in great numbers and highly educated (in Adler's sense, thus more skilled), they are still denied entry into the most skilled sectors of the labor market. Only 3 percent of the top jobs at Fortune 500 companies are held by women.[26] This can be illustrated historically as well. As John Rule writes:

> In fact definitions of skilled and unskilled work were as much rooted in social and gender distinctions as in technical aptitude. The product of nimble female fingers was often less valued than that produced by men with less dexterity. Josiah Wedgewood in the 1770s paid women flower painters only two thirds of the usual rate for skilled men . . . [and] weaving on Dutch looms was taken on by males and accordingly defined as skilled and as having a justified restriction over entry.[27]

In the same vein, Paul Starr reports on the breakdown of the exclusion of women in the medical profession two hundred years later:

> The most direct consequence of the feminist movement for medicine was a sharp increase in the number of women entering the profession. As late as 1970, only about 9 percent of medical students were women; by the end of

the decade, the proportion had passed 25 percent. But just as striking as the change in numbers was the change in consciousness. The older generation of women physicians had felt obliged to prove they could make it on the terms set by the dominant male physicians. The younger generation of women physicians demanded that male physicians change their attitudes and behavior and modify institutional practices to accommodate their needs as women.[28]

The case of the exclusion of women from skilled work points to the problem of the work of both Braverman and Adler. Women are the disruption of the privileged realm of skill—a male property, guarded closely, excluding outsiders. Women trying to compete in male skill fields were always at a disadvantage, not because they were inferior, incompetent, not strong or intelligent enough but because the skill fields themselves were structured by men. Skill is a male discourse. If women were to succeed they had to change the field of discourse. In these gendered fields they could not win. If women forced their way in, the skill was devalued. In Europe and North America this was true of nonwhites as well. This is not unlike how artisans fought against child labor, not because it was a moral battle but because they were protecting their artisanal territory.[29]

High technology created a new knowledge space that is not burdened with the gendered history of skill. It is a male domain, but because it represents a technological and social break with the past, it offers a terrain for struggle, a terrain where women's knowledge—theoretical, abstract, aesthetic, moral—can intervene in ways that women have not been able to intervene in the past in the realm of skilled work. There is no inevitability; history is on no one's side; it offers a space in which transformation is possible. The reconceptualization of work becomes possible. It is a space in which the discourse of knowledge replaces the discourse of skill. As Aronowitz has stated about the constitution of science-knowledge workers as a class, it is a class with an identity different from that of past classes:

> new classes whose identity springs not from the relation of ownership or even control of the means of production, even if they are crucial to production itself. The concept of the formation of the intellectuals as a class relies heavily on the relatively autonomous mode of discourse as a social object, the knowledge/power axis and the economic/social formation, without establishing in advance the priority of one over the other.[30]

Architects and engineers are increasingly involved in theoretical work, envisioning objects that are not there and making them visible. The machine has the skills; the bearers of the conceptual, theoretical knowledge are the engineers and the architects.

Flexible Specialization or Flexible Domination?

To return to Piore and Sabel's contention: they mount the counterargument to what has been the pervasive view, that the logic of high technology and the new forms of automatic machine production has made skill increasingly superfluous. Piore, Sabel, and Zeitlin argue that craft production historically was and currently is important in industrial production and remains crucial to postindustrial production.[31] To them it is clear that Braverman's deskilling argument is wrong. From their perspective, flexible specialization is merely the extension of craft-based production processes from the nineteenth century. They provide the examples of silk production in Lyon, the New York garment industry, *zaibatsu* federation in Japanese chemical and steel industries, numerical control in the metalworking industry. All of these have in common small-batch, "just in time" production runs (which implies that firms hold virtually no accumulated inventories) and skilled work in decentralized shops. In a world in which Fordist mass-production industries have outlived their usefulness, flexible specialization provides the impetus for an economy of increasing growth and the rebirth of skilled work.

Flexible specialization has four major components. The first is "the capacity continually to reshape the productive process through the rearrangement of components."[32] Second, it provides permanent innovation and thus can supply the specialized products demanded in a growing and diverse global marketplace. Change is thus continuously facilitated. To reorganize mass production is prohibitively costly. Flexible specialization can produce specialized products for constantly changing and particular needs. Of course, this is all made possible by new computer technologies, which Piore and Sabel regard as artisanal tools. The computer constructs a new worker who is like the nineteenth-century artisanal craftsman: "It is an instrument that responds to and extends the productive capacities of the user."[33] The third component is that it is localistic, bound to an often familial community. This facilitates both increased control and a sense of belonging among these new high-tech craftworkers. The fourth component is regulated competition. There is a "tolerance of competition that promotes innovation."[34] But in the real world, competition is concerned only with cost cutting, and especially with wage and labor cuts: "Innovation is fostered by removing wages and labor conditions from competition, and by establishing an ethos of interdependence among producers in the same market, flexible specialization succeeds only by moderating price fluctuations."[35]

Thus Piore and Sabel see this new post-Fordist production regime as the solution not only to the world economic crisis but also to Braverman's reworking of Marx's claims that work has been degraded and that workers become "watchers" of a production process increasingly dominated by technology.

In sum, Piore and Sabel among others claim that flexible specialization negates deskilling and brings nineteenth-century craft communities into the twenty-first century on the basis of a decentralized production process and new artisanal computer systems.[36]

But, contrariwise, is flexible specialization another term for flexible automation, a last-ditch effort to retain skill in new computer and robotic labor processes? Do these new high-tech labor processes degrade most skill categories, upgrading some but, more importantly, displacing most workers? Have unemployment, underemployment, and "out of the work force" been the result of these new labor-destroying technologies? Do the new technologies, which include the organization of the labor process in a global marketplace, mean that the high level of unemployment and underemployment is permanent? Do we have more skilled workers, but with fewer jobs and lower salaries? Piore and Sabel seem more concerned with arguing against Braverman's deskilling thesis and unconcerned with the impact of displacement that is part of that thesis.

According to the *Wall Street Journal,* at a 150-worker Arkansas Carrier plant that makes air conditioners and compressors, the workers "don't have to punch a time clock or prove illness" and, perhaps equally saliently, "have unusual authority. They can, for example, shut down production if they spot a problem and, within limits, they can order their own supplies." But "workers . . . are nonunion and earn $16,000-$17,000 a year," about half the annual salary in comparable union shops. Carrier and other employers who have introduced flexible specialization are careful to locate their plants in "isolated towns," no doubt to avoid paying higher wages to a more contentious labor force.[37]

Many other companies are locating in rural areas. Even though investment in plant and equipment is high, "Miles Inc. has announced plans to build a $140 million facility in Berkeley County, S.C.," to make synthetic fibers; it plans to employ only 150 people. Similarly, many steel companies have built expensive "minimills" that employ fewer than two hundred workers in traditionally nonunion rural areas. While workers learn several jobs in the Carrier plant and have participated in hiring new employees, their shop floor tasks hardly correspond to those of earlier crafts. To be sure, they are required to take a six-week course *without pay* before they are hired by the company, and only one of sixteen applicants actually gets the job. But this hardly constitutes craft; on the whole, they work with computer-mediated machines that, like other typical products of automation, have incorporated the knowledge of scientists and craftspersons.[38]

One of Piore and Sabel's favorite examples of flexible specialization is the garment industry. We interviewed a worker in a factory that uses this technique. Angela is a floor girl on Seventh Avenue in New York's garment center:

> I check the operators' work. I've worked here for thirty-eight years. It's very bad here now. The work is all done outside of the United States and then we put it together here and then we say it's made in the USA. We do very little in the New York area. Everything is done outside. You pick up a blouse and it's made in China. And they pay them terrible wages, fifty cents an hour, how can you live on those wages? And you know who sets them up all over the world? We do, American manufacturers, all of the shops, all over the world. We only assemble them here. They have kids in Chinatown, nine and ten years old, nonunionized, terrible. You can't survive on those wages. You can't be a garment worker and have a decent life. At one time you could but not anymore.
>
> Now more and more of the work is made by machine; the machines do everything, rolling, piping, felling. They put a gadget on the machine like a funnel and it rolls and stitches at the same time. Now piping—we used to call it French piping—now it's done by machine. We used to do bottoms by hand and now the machine does the felling. We used to do everything by hand, but now it's like an assembly line. In the old days it was all custom made, we did everything by hand. Now everything is by machine. They call it progress. They used to have twenty people to do work, all by hand. Now they have three people doing the same work and it's all by machine.
>
> The only new work they create are the sweatshops in Chinatown, Greenpoint [Polish and Mexican immigrants], Brighton Beach [Russian, Jewish immigrants]. I don't know what they're going to do. These people work for peanuts. They can't collect unemployment, they have no benefits, and they work harder now than we did. People need to make a living wage, people got to live. But it's always the same baloney. The immigrants are afraid and everybody else wears blinkers.

Sabel and Piore are wearing blinkers when they argue that flexible specialization's new regime of craftworkers are mastering new skills. As Fred Block rightly points out, flexible specialization has contradictory tendencies on the skill issue.[39] If flexible specialization is the wave of the future, four crucial questions must be answered. First, does flexible specialization increase skill levels in postindustrial production regimes? For Angela the answer clearly is no. Others think the death of the skilled worker has been exaggerated. Tom Forester seems to agree with Angela in his discussion of computer numerical control (CNC):

The computer itself controls the operation of the machine tool and enables the human operator to carry out calculations and make adjustments much easier and quicker.

In the late 1960's, the Japanese started to link sets of NC or CNC machines to central computers which control their operations. This is called *direct numerical control* (DNC), and it has the advantage of dramatically reducing manpower, because only one human operator is required to control maybe a dozen machines in the linked production cell.[40]

Here Forester is making at least two points about CNC: first, that it simplifies many of the skills of both setup and calculation and second, that the operator increasingly becomes a watcher of a machine process in which fewer skilled workers are required. Still others, like Block, Adler, and Jaikumar, find that flexible manufacturing systems do lead to an upgrading of skill and require workers to master the new technologies.[41]

The second question is, Does flexible specialization create more jobs? Almost everyone agrees that the answer is no. Thus we have a situation, even in a best-case scenario, in which flexible manufacturing systems require more skills but ultimately reduce the need for both skilled and semiskilled workers. As Block succinctly puts it, "It seems probable that there will be continuing shrinkage in the number of operatives, and that the general tendency in manufacturing will be toward higher skill levels."[42] Block sees this as the result of what he calls "flexible automation": fewer operatives, but the remaining operatives have to be multivalenced (we deal with this question in more detail in our case study of architects and engineers). For the most part, the job categories that require more skill are those responsible for maintaining and repairing the new technologies.[43]

David Harvey views the post-Fordist flexible-specialization fix as more problematic than Piore and Sabel or Block do.[44] He believes that flexible specialization has potential for increased worker control but that so far it has been introduced into the workplace almost completely in the terms of the capitalist as "flexible accumulation." Here competition has not increased innovation, but flexible manufacturing has cut costs by replacing workers with automated computer technologies. The control over workers has been privileged over skill-based autonomy and worker control. This new post-Fordist flexible regime has sbeen a power move in which small-batch, just-in-time production runs do not empower skilled workers:

I also want to insist that flexibility has little or nothing to do with decentralizing either political or economic power and everything to do with maintaining highly centralized control *through decentralizing tactics.* The last decades have witnessed an increase in the concentration of multinational capital; the difference is that this power is now increasingly organized through networks of seemingly autonomous firms and activities.[45]

Third, do we want to bring back a nineteenth-century craft tradition? Were all workers more autonomous in the past? Piore, Sabel, and Zeitlin all hold up the nineteenth century as the golden age of craft traditions and worker autonomy.[46] These traditions were built on the mastery of skills and the domination of nature. Women and non-European men were seen as close to nature and thus ripe for domination. They were excluded from the craft unions and guilds. Most European men were excluded from these groups as well. In the United States this was translated into the exclusion of southern and eastern Europeans from skilled work and in the union movement to elitist, corporatist, Gompersism. In Europe and Japan, flexible guilds or federations such as the *Système Motte* and *zaibatsu* were hierarchical forms of skill exclusion; workers actually suffered a reduction in rights and were put in competition with each other. Exclusion was enhanced over innovation and quality. Many groups—women, for example—were automatically excluded.

What Piore and Sabel present is a *metaphysics of skill,* a romantic vision of a past that is preferable to a frightening technological future in which workers are subsumed under machines. Thus, they reconstitute twenty-first-century work regimes as regimes of high-tech artisanal workers. But even if they were right, why wouldn't they reflect the nineteenth-century forms of domination? Would we now have "flexible domination," in which unions would have even less power and white men, women, black, and Latino workers would be pitted against each other for fewer and fewer jobs? Wouldn't this struggle between workers lead to lower and lower wages and to ever worsening working and living conditions? Piore and Sabel seem unconcerned about this. Like *laisser innover,* "skill" as the means of innovation becomes the legitimator of new technological regimes of domination.

The fourth question: Is it skill that is being increased in these flexible specialization regimes? Our contention is that skill is being decentered in all high-tech production regimes, and that production regimes based on conceptual, theoretical knowledge are coming to the center. When skill is measured in terms of the length of time required to learn a procedure, increasingly we are talking about formal education and university degrees. Skill training through apprenticeships is disappearing. The new technologies perform not only the physical, manual work but increasingly can perform the tacit knowledge component of the skilled worker. As Fred Block says of the components of skill:

Skill depth refers to the time it takes to learn a particular task, such as machining a complex job. Skill breadth refers to the range of different types of knowledge that employees must have to carry out their jobs. Flexible technologies tend to reduce skill depth precisely because the relationship of worker to materials is now mediated by machinery. This kind of hands-on knowledge that production workers often accumulated over a long period of time can become obsolete.[47]

Block seems unaware that skill breadth is increasingly obsolete as well. The tacit knowledge of the skilled worker is being incorporated into the new technologies by scientists and engineers. As a result, the remaining labor process is increasingly intellectual, abstract (head separated from hand), theoretical, and scientific. Shoshana Zuboff struggles with the same problems. She is not ready to acknowledge that theoretical and conceptual knowledge is different from both skill breadth and skill depth. Thus she differentiates between "action centered skill," which is "implicit in action," "context dependent," and "personal," and "intellective skill."[48] Intellective skill, the wave of the future, consists of "high order analysis and conceptualization."[49] It is our contention that intellective skill is not skill but knowledge. *There has been a rupture with the past.* We agree with Sabel and Piore that skill is based on artisanal craft guilds from preindustrial production regimes that continued into industrial production regimes. Mass-production regimes did begin the process of deskilling, and high-tech production regimes are not skill-dependent but knowledge-dependent. Yes, skilled work still exists, but it is increasing only at the margins of the new production process. "The logic of the tendency" of future production regimes is a technoscience-based labor process, with knowledge at the center.[50]

Democratic Workplace or Corporate Panopticon?

Zuboff argues that the computer gives a new dimension to communication by creating a "richer social text," enabling people to communicate for a variety of purposes, especially sharing knowledge, to a degree impossible in either face-to-face interaction or print. But the ethic of knowledge sharing—celebrated in the stories sociologists and philosophers tell when they have been mobilized as publicists for the scientific "community"—has been seriously undercut by rampant privatization; knowledge is organized on a need-to-know, ability-to-pay basis.

Informatics may, in time, permit the emergence of a countertendency, at least for those who can afford the hardware, the software, the telephone bills, and the user fees—in short, a wider circle of professionals. Conferencing, a technique made possible by modems, permits people in widely separate places to talk to each other by creating computer-mediated texts. But this is merely a kind of collective privatization, not a genuine democratic development. The newspaper, the magazine, and the book occupy a different temporal order from that of the ordinary workplace, in which minutes and hours rather than days, weeks, and months are the primary units.

Zuboff's most important claim for the computer, that it can facilitate communication and shared decision making between the higher managerial echelons and subordinates, remains a hypothesis that, in any case, presupposes a different system of control. Zuboff's own studies of current industrial uses of the computer demonstrate exactly the reverse: that Bentham's Panopticon, combined with the traditional imperium of the early nineteenth-century entrepreneur, a figure made vivid by computer innovator Ross Perot, for the most part still characterizes the worldview of the managers. The opportunities for improved communication among peers are, indeed, greatly enhanced by the computer because, logically enough, peers do not take the electronic text as a command; on the other hand, Zuboff reports that one manager was mortified by the rapidity with which a suggestion communicated to subordinates on a computer was received as an order, a response conditioned by the hierarchical regime of power in the corporate workplace.

It may be concluded that the computer is Janus-faced and mirrors its masters, the plural indicating a fissure among those who developed it. Charles Babbage, credited with inventing an early pre-Turing version in the mid-nineteenth century, was already aware of its industrial potential. Although the idea of automatic production was already part of the lore of high British capitalism, the systematic use of science as a basis for technological development was still in its infancy. Some latter-day Babbages, despite being employed by corporations who own the rights to dispose of their inventions, retain the utopian, even anarchist, impulse that infuses every effort to integrate design with execution, to finally abolish the historically and socially constructed gulf between intellectual and manual labor.

Whether the computer is employed as a powerful instrument in the corporate Panopticon or as a way to facilitate the emergence of a democratic workplace in which *techne* signifies the unity of humans and nature rather than their split is indeterminate from the perspective of the internal constitution of computer technology. In what follows, we want to suggest that many of the widely cited instances in which the introduction of the computer into professional work formerly performed either by hand, as in the case of engineering, or by conventional oral and printed texts, as in the case of teaching, manage to elide the Panopticon of power only by "forgetting" the context within which computer-mediated work is done.

Chapter 4

The Computerized Engineer and Architect

Computer and Profession

The cornerstone of the ideology of professionalism is that the distinction between true professionals and other categories of labor lies in their autonomy from most forms of managerial authority.[1] This claim presupposes that professionals possess specialized knowledge that requires specific training and that even managers are unable to grasp it without undergoing the same regimen. Further, as the narrative goes, even when, say, physicians or engineers occupy high administrative positions, their ability to direct professional work is limited by the rapidly changing knowledge needed by scientific, technical, and service professionals. Typically, administrators have little time to follow professional literature or consult with colleagues regarding innovative practices. Thus, rooted in the distinction between intellectual and manual labor, the model of professional work is self-management made possible both by specialized training certified by credentials and by the organization of work that accords professionals, uniquely, a high degree of autonomy.

Of course, the relatively rapid post-World War II unionization of professionals in nearly all technologically advanced capitalist countries—not only in teaching and social services but also among "high" occupations such as medicine and engineering—is a sign that the gap between ideology and practice grew in proportion to the emergence of large-scale corporations and state bureaucracies as the characteristic sites of knowledge-based labor (for further discussion of this point, see chapter 5). While it would be a mis-

104

take to overemphasize the degree of collegiality in the older institutions of knowledge production, the appearance of the knowledge factory, whose characteristic form is the laboratory, at the turn of the century demarcated the history of science and technology. The famous examples—Edison's later laboratory at Menlo Park, General Electric's research facility in Schenectady, New York, Bell Labs' huge laboratories in western New Jersey, the Manhattan Project during the war, the University of California's Livermore labs, and the Brookhaven, Long Island, radar and nuclear labs—prefigured the dissemination of this development throughout industrial society during the past forty years.

Moreover, while the birth of welfare-state capitalism can be traced to Bismarck's initiatives—under the impact of pressure exerted by a mass workers' and socialist movement in the fourth quarter of the nineteenth century—the consolidation of huge state bureaucracies in the West and in the countries of state socialism in which technical professionals and managers played a pivotal role is chiefly a post–World War II development. Within the regime of the corporate/state bureaucratic organization, in which the professional manager rather than the specialized practitioner takes charge, the traditional professions have been decisively subordinated to manager-dominated organizations, if not with respect to income, certainly in the area that defines them: their right to control the conditions of their own labor, especially to make autonomous decisions based on judgments linked to their credentials and knowledge.

The computerization of whole categories of industrial work has attracted intellectual labor in the first place because it seems to provide a scientific and technical, rather than political, basis for reintegrating design and execution. As we showed in chapter 1, the preference for technical rather than political innovation is a major component of contemporary professional ideology. Social conflict is eschewed in favor of negotiated solutions in which interests are mediated by policy compromises, or issues are resolved by procedural—that is, organizational—means.

The widespread introduction of computers into industrial production, administration, and the services reduces toward a zero point, at least in tendency, what Antonio Negri has called the "mass worker."[2] The mass worker was produced in the midst of the mechanizing era, and especially during the prolonged period of capitalist regulation that followed World War I. This emergence was the result of the triumph of Fordism, the industrial regime within which production and consumption were integrated even as workers were deprived of even the small vestiges of control within the labor process that they had enjoyed before capital's full implementation of techniques of mass, mechanical industrial production.

Now, the heart of the new post-Fordist labor process is no longer marked by the separation of intellectual and manual labor, but by the imperative for its integration on the basis of the dominance of scientific knowledge over the worker. At the General Electric engine plant in Cincinnati, for example, the logical and temporal sequence between design and production and within design itself has been dramatically foreshortened by computer-aided design and manufacturing (CAD/CAM). Close communication between the once-distinct sectors of a plant means that both sides both adapt to and assimilate each other's specific characteristics and problems. At the minimum, the frequent bottlenecks between the blueprint and production problems in traditional industry are more rapidly solved without disruptions. At best, CAD/CAM blurs the lines of demarcation between design and execution, so that the shop floor "participates" in the design process to a degree that depends on management's ability to broaden and even democratize power over decision making to include "manual" labor. In fact, even in the most advanced plants, the distinctions are blurred but not entirely obliterated. After all, the selection of *what* is to be produced rests with owners and managers, not workers, even if, according to neoclassical economics, the ultimate choice lies with the market or, to be more exact, the consumer. In either case, workers' participation is confined, in the most democratic cases, to issues of production. Everything else is out of bounds.

In this regard, corporate-sponsored "quality circles" that provide an opportunity for nearly the entire workforce to give their views on the labor process, conditions of work, characteristics of management, and other shop issues carry two burdens: cooptation where an enlightened management attempts to head off union-sponsored protest, and training a new workforce under conditions of vast changes that require their participation and, to a certain extent, their independent judgment of production decisions. This is possible where CAD/CAM has become typical of the industrial workplace and is materialized in both individuals and groups who have global knowledge of the labor process by virtue of their grasp of the new integrated computer-mediated systems. But in the main, especially in the United States and Britain, corporations attempt to impose the old managerial regime on the new "socialized" worker who is fully capable of managing the labor process.

CAD/CAM in a Private-Sector Company

We wanted to find out the effects of this innovation in a group of professional and technical workers and others who, even if they are not conventionally credentialed, perform professional and technical labor. Specifically, we wanted to know how the character of the work changed, whether power relations between managers and employees were significantly altered. Finally, we were interested to find out whether, in contrast to industrial and clerical labor, which, in the main, had experienced computerization as an extension of the old industrial culture, new work relationships were forged among this highly qualified group.

Over seven years, we studied architects and engineers in New York and New Jersey, in the public and the private sector. We have observed the transformation of the work of architects and engineers who design with computer-aided design and drafting (CADD). The physical skill of drawing has become secondary to the knowledge component. Conceptualization has become primary. Engineers and architects using CADD have significantly changed the object that is produced, though there is much evidence that the departments are doing more work, with greater accuracy and more creativity. The notion of productivity itself is changing. How can you talk about productivity when the product itself is changing and different from what it was in the past?

During a visit to the General Electric (GE) aircraft engine plant in Cincinnati, we observed the ambiguous uses of CAD/CAM. This plant employs some two thousand design engineers, divided into several groups. A fairly large section is employed in CAD/CAM, a smaller section works exclusively on designs of engine parts and employs CAD in what may be described as routine tasks, similar to those done by civil engineers in infrastructural agencies. And an even smaller group, clearly the elite corps, is engaged in basic and applied research on CAD itself (GE produces CAD software for its own operations and commercially).

The employment relations manager of the plant told us that much of the research work was dedicated to finding ways to further cut the labor force, not only manual labor in the factory but engineers as well. He acknowledged that the productivity savings in drafting made possible by CAD would reduce the number of engineers, unless defense and commercial orders increased by the amount of the savings, estimated to be about 600 percent in time. When we visited the plant in 1986, half the engines were slated for civilian aircraft and the other half for the military. Since the pressure on military costs mounting in Congress and the Department of Defense had already been felt by the company, they anticipated considerable eventual reductions among drafters and engineers. It was clear that the proportion of manual workers to professionally and technically credentialed employees had already been severely reduced. The computer-driven numerical controls sitting inconspicuously atop most of the machine tools, the impressive computer-mediated laser technologies used for boring holes, and the extensive presence of robots on the shop floor told the story: millions of square feet of space stood relatively empty of living labor (although not of its materialization in the machines).

In the old regime of industrial production the floor would have been crammed with people; the introduction of the computer had given the plant a ghostlike quality. The live workers were not only scarce, they appeared to be marginal to the production process, although this appearance is partially deceiving. Nevertheless, these were knowledgeable workers who were not merely "watchers" in the sense Marx used the term in the *Grundrisse*.[3] They were in constant communication with the CAD/CAM engineers and really had input into design changes based on their production know-how. And they constantly made adjustments in the programs. The machinists had lost many of their older skills but were now programmers. They missed some of the features of the old industrial regime but enjoyed considerable range in the performance of their jobs. Some machinists and toolmakers said that they had always interacted with engineers but that the quality of the interaction had been improved by CAD/CAM because they shared a knowledge base. The production workers were less subordinate than in the old system. The socially constructed distinction between intellectual and manual labor was preserved, however, in part by corporate labor relations policies and union contracts. The managerial and engineering staffs shared only a limited quantity of power with the machinists, corresponding to the company's commitment to preserving knowledge hierarchy. Although the integration of design and execution had become the official discourse of company managers with whom we talked, hierarchical relations remained (ultimately) in force, especially when a disagreement arose. There was no real collective mechanism for ironing out differences. The machinists were invariably obliged to back down despite their considerable experience in the technical aspects of design, their newly won familiarity with computer programming, and their understanding of production problems.

The engineering design work is performed in a building separate from the production plant. In the engineering department a small elite of about sixty people is experimenting to improve the three-dimensional capabilities of the CAD software that the company sells as a separate line to other industrial companies and engineering firms. The environment in this part of the building is animated, almost exuberant. "Supervisors" sit next to engineers and interact with them as peers. The exchanges are often heated, but there are few signs of overt use of managerial authority to settle arguments about issues linked to the common objective of developing a versatile three-dimensional CAD program. The group meets every week and discussions are intense; the group leader chairs but also participates. Most of the debates concern technical issues of how best to design the prospective program.

There seems to be no pressure at all for "results," and, of course, there are no immediate production norms to fulfill.

The second group is much larger, comprising perhaps three hundred engineers. They are working on various designs to improve and otherwise alter the basic GE engine(s) in order to meet military specifications and those of private aircraft manufacturers. Subgroups are working on specific problems of the overall design. They are mostly mechanical and electrical engineers, but some are specialists in aeronautical engineering. In group meetings, the technical discussions are somewhat wide-ranging, but group leaders exercise much more authority of knowledge. There is a noticeable difference between the experimental group of peers and this somewhat more hierarchically organized and limited section. Yet most of the people seem relaxed and the groups function semiautonomously.

The third section is composed of about fifteen hundred engineers who, for the most part, do not work in subgroups, but sit at individual drafting tables and computers in rows in several large spaces much as clerical workers in the back offices of banks and insurance companies do, or they work on problems that arise on the production line. The section head has an enclosed office, and the engineers interact with him by knocking on his door. These engineers are working in three broad subareas: very specific design tasks (flanges, subelectric systems, switches, castings) handed "down" from the second group; production problems that cannot be solved on the shop floor; and quality control and work flow issues that entail frequent interaction with the shop floor.

The bulk of the engineers are the detail workers of the design function. Their work tends to be routinized and repetitive. Not all of them work on computers; perhaps a third of the engineers still perform drafting functions by hand. This situation was often dictated by the requirements of the specific assignment, but also reflected a hierarchical arrangement in the workplace. And, most important, they were subject to informal but strong production norms. There was a definite factorylike environment in this section, and they were required to undergo frequent evaluations of their productivity, their attendance and promptness, their congeniality to authority, and the quality of their work. Rules were far more rigid than for the other groups, and poor evaluations could result in the withholding of merit increases, downgraded work assignments, or discharge.

This plant is one among several we have visited in which the Janus-faced character of the CAD/CAM technology is displayed. The engine plant management employed computer-mediated processes to upgrade the qualifications of some workers to enable them to *participate* in some decisions, but also used the technology to cut costs by reducing labor forces at all levels and, above all, retained not only ultimate authority at the level of top management, but retained operational authority as well. The Panopticon is alive and well at GE.

CAD in a Public-Sector Engineering Workplace

In 1984 we first started our studies of CAD use in the New York City Department of Environmental Protection (DEP) and the Transit Authority. Civil, electrical, and mechanical engineers employed by public agencies design a variety of infrastructure installations such as water systems, pumping stations, switch signals for subways, and rail and auto bridges. Traditionally, nearly all of the design work was performed by hand and brain, employing little or no mechanical or electronic equipment except drafting tools, most of which were as old as engineering itself. Computer-aided design and its counterpart, computer-aided manufacturing, mark a work revolution. Now almost all of the painstaking labor of making multiple stages of blueprints, which by hand could take weeks, is reduced to several hours. The machine electronically performs the drafting under the guidance of the operator. After the basic model has been put on the screen, adjustments and variations can be made without making a new print.

CAD had just been introduced and only a few engineers and architects were working on the machines. We observed that the architects, engineers, and drafters who were doing manual design work always looked busy. They were always at their boards, continuously drawing. Those few engineers and architects who were working on CAD never looked like they were working. They were sitting, using a mouse, staring at the computer screen. They weren't drawing, they were thinking—a major change: from continuous drawing to continuous conceptualization.

Mickey, a DEP architect, says "With CADD we can build a functional and beautiful city." For him any notion of productivity would have to include not only cost and function but aesthetic considerations as well. On the possibilities of CAD, he said in 1984:

> At first I thought they would be just drafting machines. You would have a drawing produced electrically . . . but I am really excited by it, it is like I can manipulate and take apart any drawing. I can stretch it, move it, take a piece out of it. They call it object intelligence. It is like an infinite layering system. . . . Whereas, let's say on your jacket I can sew the gold buttons on [he points to my sleeve], those gold buttons always exist in my memory. To an architect that is unbelievable. It is like having everything you have ever seen or worked on at your disposal to use again. Now we can say, "Didn't we use that detail

110

somewhere?" and you can't remember where it was. With CADD you can scan and ask for details and it will list them immediately.

After the prototype after the transition [to CADD] we will shoot another [similar] project out and cut down the time on that. We expect to cut down the time by 70 percent. That's what the expectation is now.[4]

Where in the old design process mathematical calculations took considerable time, the CAD program provides a menu of mathematical options from which the engineer can select the appropriate one for standardized jobs. Since much of the design work of civil engineering is based on standard dimensions, only a minority of jobs present special problems of calculation. They still require that the engineer know the mathematics that has been transferred to the machine, but the time required for these calculations has been drastically reduced.

In the old work regime drafting and math occupied nearly 90 percent of the engineer's time. These were more or less routine activities that were necessary preconditions for whatever innovations were needed in the design of, say, a waste disposal system to accommodate limitations of space or terrain. The specificity of the design function consists largely of taking account of these limiting conditions to fit otherwise standard sizes. CAD sharply reduces the time necessary to acheive these routine preconditions. But it can do much more.

Some CAD programs developed in the past decade are equipped with three dimensions, a feature that reduces the degree to which the designer must be able to visualize the end product. This ability to visualize has been among the crucial components of engineering design. Now the (relative) literalization of the image removes a major obstacle to the elapsed time of design. In work on standardized designs, visualization is still a part of the job but is, to a significant degree, taken over by the computer.

The engineers at one of the largest agencies were absolutely delighted to have CAD because it made the job infinitely less tedious by removing many routine, manual aspects of the work, notably technical drawing, and made designs more precise without a seemingly endless succession of drawings. The introduction of computer terminals for design purposes throughout the agency enhanced their solidarity by providing them with knowledge that managers did not necessarily possess. And they learned CAD together and made designing, unlike drafting, a collective effort. The management and the engineers were eager to expand CAD so that everyone had the opportunity to work on a terminal. Line managers and engineers hoped the widespread use of the technology would support their long-standing contention that contracting out design work was unnecessary on either professional or labor-power grounds. The new technology was understood as a means to make them more versatile, and because of productivity gains they were able to handle more work.

The dream of merging work with play seemed to be not just within these employees' grasp, but in the process of realization. In the new technological regime, the actual work process resembled a basic science laboratory in that they were now easily able to experiment with different approaches to a given problem; intellectual work dominated the portion of the work that had until now been considered routine, manual labor within this profession. For even if drafting required considerable skill, it was often experienced as boring. At least in its early phases, CAD represented liberation from tedium for most of the engineers.

We want to emphasize that the actual technical problems associated with design in public infrastructures are complex but are made less daunting by CAD, which reduces the lengthy processes needed to set up the design problem. We observed engineers who delighted in engaging the process regardless of the intellectual challenges posed by the design problem itself. They were learning something new, and it was the learning that excited them as much as the possibilities for gaining greater autonomy from managerial bureaucracies whose obligation to higher authority often created conditions that restrained creative work.

Of course, it is still too early to determine whether these hopes can be realized in hierarchical organizations, even for qualified professionals and under conditions where a particular management encourages the general upgrading of the skills of engineers without subordinating them to excessive limiting conditions. The work process is, after all, subject to the economic and political exigencies of larger organizations and struggles. For even if these engineers are being transformed from relatively highly paid qualified workers to "hackers" (the term denotes people who are dedicated to exploring the full potential of and the intellectual issues entailed by computers), they are not free to determine the product of their labor, only to exercise some power over the process.

When we talked to Mickey again in 1990 he told us about a new fence design for the Coney Island sewage plant, a wave fence that CAD made possible. When CAD is combined with computer-aided manufacturing, a whole new industry of artistic fencing can be created. As the hardware and software capabilities increase, as three-dimensional capabilities expand and the database increases, the design possibilities increase exponentially. As the machines are upgraded, the architects and engineers must increase their knowledge as well. As Mickey says, "We can digitize the city." Professional architects and engineers using CAD seem to have an unlimited horizon.

Yet this vision of CAD is being constrained. Mickey said in 1989:

I feel that my profession, designers at the higher levels, are resistant to change. The field is stagnating. We're falling behind. We aren't using CADD to its potential. You always have designers knocking CADD in the journals. Maybe that's just sour grapes. For me CADD is very powerful.

Robinson, a mechanical engineer, said, "Right now we're just using the machines as a drafting tool"—their transformational powers were not being developed. CAD increased the productivity, accuracy, and creativity of the work of the engineers and architects that we studied over seven years, but the work basically remains the same. For most architects and engineers, CAD is just a better way to draw; it has not expanded the intellectual side of the profession.

In the discourse on professions, the debate is about whether the professions are increasing their knowledge/power and thus becoming mandarins or whether their knowledge base is decreasing and they are becoming redundant and being replaced by machinery or degraded to technicians—in a word, whether they are being proletarianized. The engineers and architects that we studied are neither mandarins nor proletariat. The only cases of displacement we saw were at the semiprofessional level, the drafters. In New Jersey some were replaced by CADD and in New York they have been reduced to correcting design drawings, but many of them do their work on the computer.

Many of the engineers and architects that we have studied believe that the failure to customize CAD and use it to its full potential is a result of organizational problems. The organizational model has been insufficient to the task of change; it is outmoded. It is also our belief that because CAD is expensive, it has been increasingly difficult to finance the capital expenditure. The New York City DEP was able to buy forty-five machines—at a cost of $2 million. A planned move from Manhattan to Lefrak City in Queens would bring the whole department together, increasing coordination, but at the same time, city and state budget cuts were likely to mean less work.

It is evident that the budgetary constraints imposed on local and state governments during the Reagan-Bush era have an impact on the development of the public sector. This has a direct effect on public-sector professionals. It is in the public sector that professionals can have autonomy and power because they are not servants of private enterprise. To be more specific, a well-funded public sector would enable the free development and experimentation required for engineers and architects to develop CADD to its full potential. CADD can only fulfill its promise if they can, as Lopez, an architect who is very proficient in using CADD's three-dimensional power, says, "play with the machines so that we can customize them to our needs."

Budget cuts and a continuous fiscal crisis do not allow public-sector professionals to develop to their full potential. The public sector offers professionals a sphere in which they can have more autonomous power than in any other sphere in society. Thus privatization is a form of class war against new class professionals. These themes will be expanded in the rest of this chapter.

The DEP CADD Room at 40 Worth Street

The DEP has spent $2 million on the expansion of its CADD facility. It has increased the number of its CADD machines from eight to forty-five. Sixty-one architects and engineers have been trained. After the move to Lefrak City the training will resume. The CADD room at 40 Worth Street in Manhattan is no longer the place where most of DEP's CADD work is done. Now the other bureaus have their own CADD capability, and the CADD room is used by the Bureau of Heavy Construction and the Architectural Design unit.

The CADD room on the thirteenth floor now has ten Tektronix 4225 graphics processors with nineteen-inch color display screens that have a fifty-six-color capacity and are fully three-dimensional. These ten CADD machines have been added to the three Tektronix 4109s that had already been in use and were traded for Genescos, which had neither color display nor three-dimensional capabilities. They also have a CalComp 5735 thirty-six-inch electrostatic plotter with a resolution of 400 dots per inch with a 25 megabyte disk and paper takeup. They already had a CalComp pen plotter. Pen plotters use liquid ink pens; they are slower than electrostatic plotters but the work looks better. The pen plotter produces drawings that look like manual drawings because it draws with pens. The electrostatic plotter works like a photocopier and the ink tends to smudge. Both are capable of two-dimensional and three-dimensional drawings.

The plotters are very noisy when they are printing. Until recently, they were in the main part of the CADD room, and their noise was a constant factor there. With the upgrading they were put in a small room that is part of the CADD room to protect them from moisture, and the noise has been reduced.

When an architect, an engineer, or a drafter wants to print from either the 4225 or the 4109, the drawing is sent to the central processing unit (CPU, the mainframe), the Prime 6350, at 1250 Broadway and then back to the plotter at 40 Worth Street, where it is printed. Robinson, a mechanical engineer, describes the process:

Right now we go from here to a T1 [telephone company modem] downstairs on the eighth floor and then to another T1 at 1250 [Broadway] and then finally to the computer [CPU] and then the same way back.

The actual computers are all Tektronix and CalComp. The software itself that runs on the Tektronix machines is McDonnell Douglas [GDS, General Drafting System, though Robinson often refers to it as Graphic Decision System]. So we have three companies involved, Tektronix, CalComp, and Prime. These are the hardware companies. And one software company, McDonnell Douglas.

Right now we have ten 4225s for just our bureau, three 4109s, and one Prime PC 400. That's really the terminals we have. The Prime PC is not really a terminal. It's a screen. You can't do graphics on it. All the other ones can do graphics. The 4225s when we bought them were state of the art . . . [but they are] now outdated. They have two meg local memory and nineteen-inch screens.

In March 1989 we watch Lopez, an architect, do three-dimensional designs on the CAD. He is designing the circular roof on the Oakwood Beach pumping station on Staten Island. He tells us that the circular pumping station would not have been possible if it was designed manually; it has been made possible by CADD's three-dimensional capabilities. The screen is clear and colorful, with appropriate shadows and trees and shrubbery. Lopez is constantly changing the screen to get different perspectives on the pumping station.

The deadline for sending the design to the state for approval is approaching. All the terminals are busy. The noise of the plotters can be heard. Everyone is working and talking. There is a high-pitched nervousness in the room. They will finish on time and even win an award for the design. They are working hard; they are a good team. Robinson is sitting with Pedro, working out a difficult design problem.

This is a difficult task: in-house design of a pumping station in the midst of a major DEP systems upgrade. CADD has meant many changes, including a need for fewer consultants, in the way design work is accomplished. At the same time, the department is getting new contracts: $1 billion for Newtown Creek and another contract for the Coney Island sewage plant, both in Brooklyn.

Reggie, an engineer, is working on changing the arrow because the lines are too thin. He wants thicker lines on the arrowhead so it will look better for the presentation to the state:

I'm trying to make this line thicker than it is in the program. You can do it, but you just have to fool around until you get it to be the right measurement. Once I get this line right I can pull it out anytime even though it wasn't originally in the program. I can build a library that can be standard. That's what's good about this software is that you can create new line styles and characters.

GDS is a design and drafting software that allows for the customizing of line sizes, the unlimited layering of drawings. The GDS training manual overview says:

GDS (plus its library) is under continual refinement to improve the usefulness and keep it abreast of hardware development and user's requirements. GDS is the central graphics "hub" of an increasing number of compatible systems (marketed by ARC and others) such as: electronic land surveying, bills of quantity, road alignment, 3D modeling, reinforced concrete detailing, perspective viewing, etc.

Jerry, a drafter, and Nino, a civil engineer, are brainstorming. They have been working for hours and are still unable to solve a problem on the Staten Island carousel pumping station. They take a short break. Jerry talks about the problems he is having with his car. The secretary is scolding one of the sewer engineers, who is talking to Robinson about a computer problem. "The door must always be closed," she reprimands. The tension is high in the CADD room. The noise of the plotters adds a heavy beat to the operation. There is continuous teamwork; everyone is helping everyone else. The engineers, architects, and drafters are very skilled in different disciplines, but the CADD machines and the other computers, though frightening to many of them, are a unifying force. Their physical skills are now less important; for the most part, the machines have replaced those skills. CADD draws better. When the DEP purchased the machines they were to be used as drafting tools, to reduce the backlog of drawings and to replace the drafters because the machines draw better. Even senior drafters agreed that the machines' drawings were as good as if not better than their own work. The machines have powerful memory; they are flexible and can be upgraded; they have both three- and two-dimensional capability, almost unlimited color, light and shading, and multiple layering capabilities—but they cannot conceptualize. They *can* make the engineers' and architects' ideas, thoughts, and innovations visual in a way that manual drawings cannot. Thus architects and engineers as knowledge workers are empowered by the CADD machines.

Most of the architects and engineers are excited by these possibilities; others are resistant. Whole bureaus are resistant. We chose the Architectural Design section because they are a best-case scenario of the promise of computer-aided design and drafting. Among the CADD users in the department it is thought that the application of CADD is the most beneficial for the Structural and Mechanical Bureaus and the least beneficial for Architectural Design. But the engineers and architects in the Architectural Bureau are committed to the possibilities and promise of CADD, and thus we studied them in 1984 and 1985 (along with the Transit Authority, which has an Intergraph system and does almost all of its work in-house) and again between 1989 and 1991. In general, the knowledge workers in this bureau are also committed to big design projects and to the advantages the public sector offers design work. As Jerry, a senior drafter, whose profession is endangered by the technology, says, "A year ago. That's the last time I was at the drawing board." He has been making corrections on the pumping station roof. He continues, "If we had an endless amount of money everyone would have a CADD terminal. I love this machine. It makes pretty pictures and I'm learning all the time. Of course sometimes it gets tedious. After a lot of hours you get bonkers like everything else."

Nino breaks in: "We're always learning, and when we get stumped we ask Robinson."

In this free-floating teamwork the specialized skills and knowledge that each was trained in break down. The different disciplines are increasingly flexible and blending. Robinson is a key example. He was trained as a mechanical engineer but is so knowledgeable about computers that Mickey, the CADD coodinator who is also the chief architect, appropriated him from the Mechanical Bureau and made him his systems analyst. Mickey now continually fights to keep Robinson in architecture. The blending of disciplines is limited by a bureaucracy stuck in an old paradigm.

Pedro, an architect who has just learned how to use 3-D, is working on the carousel roof of the pumping station. He is constantly aligning and realigning the structure. His screen is covered with orange and white lines. The orange lines designate what is completed. Pedro seems worried and says to Lopez, "What's going to hold these bricks? We need insets in the drawings, otherwise we have bricks all over the place."

Lopez replies, "The contractors usually leave them out."

"We want them in the drawings now. That's why I'm looking for Nino [the deputy to the chief architect]," says a structural engineer.

"I'll put them in," says Pedro. "I'll put them in after, then. Something's wrong with this mouse [he is continually moving it over the menu tablet but nothing is happening]. The yellow button is not working." He types in the command. Lopez helps him locate the required drawing and he calls it up. At the same time, at another workstation Robinson is helping an engineer at another computer. He is demonstrating to her the capabilities of the printer.

Pedro looks at me and says:

> I did everything by hand for many years, since I was a kid. But this is so much fun. I was scared at the beginning, but once I got the hang of it, they couldn't get me off. I had to retrain myself. Sure, Lopez and Robinson helped. But I had to retrain myself. It's not like drawing on a drawing board and no one can really teach you. You have to teach yourself.

Jerry is finished with his work and he walks over to the pen plotter, the CalComp 1077, which prints his drawings of the pumping station in ten minutes and twenty-seven seconds: lines representing 287 meters, 14,810 lines of plot code on the thirty- by forty-inch standard sheet.

The next day Mickey walks in the room and says, "Remember, the commissioner challenged us on this project. He didn't think we could do it quickly on CADD. I want to surprise him. I want to bring it in early. I want to show him that we can scan a plant and build a 3-D model and do it quickly."

While he is talking, the architects, engineers, and drafters are handing in their overtime sheets to the secretary. Frankie and Robinson are planning the number of trees and shrubs at the CADD machine. They match the legend to the landscape plan. Frankie says, "We don't have an architectural landscape person. We make the decisions on the plants and the pumping station. We do everything ourselves."

The power of the CADD program allows them to make landscape decisions independent of hiring an expert engineer. Two years later, in 1991, the same logic is in use in another project. In this project, also on Staten Island, there is a flooding problem. They want to do as little damage to the area as possible because of ecological considerations, and they plan to use the natural aquifers for drainage as much as possible, reinforcing them where it is necessary. Budgetary constraints prohibit hiring engineers with expertise in hydrology. Instead they plan to purchase geographical information system software that will allow their current architects and engineers to design the project. Whether this works or not, this is the direction in which these knowledge workers are moving. The continuous expansion of software combined with the permanent budgetary constraints of the Reagan-Bush era increasingly has forced these architects and engineers to become multivalent knowledge workers.

Robinson, looking at the drawing of the shrubs and trees around the pumping station, says:

This is what the old-timers are afraid of when we do these drawings on CADD. They're so perfect that an engineer doesn't check the calculations. They take them for granted and if there is a mistake, it just keeps growing. And everyone bases their calculations on the mistaken calculation and it grows and grows.

When you do things manually everyone comes with their calculations and such mistakes aren't made. I've told them that this isn't true. Even on CADD everyone comes with their own calculations, so the checks exists. It's really not a problem.

They're almost finished with this job. Robinson says, "we seem to have everything under control, but I'm not sure. There will probably be some people working late tonight. I was here until 11:30 last night."

"It will be sent to Albany by courier," says Lopez.

Jerry adds, "This is how you know we're in good shape because everyone is joking around. When we're in trouble we're intense. Everybody is glued to the terminals. So you can see we're in good shape." Everyone in the CADD room is joking and laughing or talking about the television show "A Current Affair" or about the Knicks. Jerry sees this as a definite indicator of a successful job, almost completed.

Frankie comes into the CADD room with two consultant engineers, giving them a tour of the facilities. He introduces them to Robinson, Lopez, and Bill (the CADD room's own sociologist). Nino asks, "Are these Tektronix 4115Bs going to be the new computers?"

Lopez: "No, it's a 4225. Smaller, more capacity, faster. This kind of work is very difficult to do by hand. The machine can do twenty-two-and-one-half-inch panels for the whole roof. This is true even for this carousel building. Once you have a good basic drawing you can play." Lopez blows up the panel on the screen. The detail is sharp and seems to me to be quite complicated. "You need the 3-D detail for the contractor," he says, and begins to explain the job in great detail for the consultants.

Nino: "Is it cost effective even for the structural detail? The detail that pertains for each structure?"

Robinson: "Yes, even for that. We can do the detail real close and we have to do nothing twice."

Nino: "What about a job when you do all different angles?"

Robinson: "Even then. You do it, you have it, and then others can use it."

Lopez explains the upgrading of the department, the capacity of the machines, and the McDonnell Douglas software.

Nino, a little overwhelmed, says, "Do we need all of this stuff?"

Robinson: "Yes, because if it can't be done on the machine it can't be done. And this helped us show what can be done. It's helped us convince the Art Commission what can be done."

Lopez: "We used this for the Art Commission and it's because of the 3-D drawings that they bought it."

One of the consultant engineers: "We have release 10, though we haven't hooked it up yet."

Lopez: "That's got 3-D and everything."

Consultant engineer: "I'm looking forward to playing with that."

Robinson: "We have overkill with GDS. We have 600-line styles."

Lopez: "We tend to use fine lines." He proudly shows the original drawings for the circular pumping station that is also called the carousel building.

Nino laughs and says, "I can see the chief architect's name all over the changes in that building." It is because the chief architect views aesthetics as important, or as part of the definition of the functionality of that building, that Nino laughs his friendly laugh.

The project is finished and sent off to Albany for approval. A year and a half later, in a Thai restaurant on Bayard Street in lower Manhattan near 40 Worth Street, William DiFazio is eating with the engineers and architects from the CADD room. It is the last time that they are going to eat together in this restaurant now that the whole department is moving to Lefrak City. The chief architect tells them that they have won an award from the Concrete Institute for the carousel pumping station. In a constantly changing computer environment that develops faster than these engineers and architects can adapt, relearning and uprooting of knowledge recently learned are required. In this environment, continually being reorganized, these architects and engineers want to play with the CADD program with the freedom of children.

Productivity

The competitive advantage of high technology is debatable. Many claim that it has not lived up to its promises. Immense productivity gains were the sales pitch of vendors, who in the name of "state of the art" got both government agencies and the private sector to invest billions of dollars in the new technologies. As Lester Thurow, dean of MIT'S Sloan School of Management, asked: "Why do we have all these wondrous things and productivity gains aren't showing up?"[5]

In general, analyzing productivity gains in nonmanufacturing sectors has been complex and elusive. If anything, it has been a slow process with many economic and organizational barriers to overcome:

White collar productivity is not amenable to precise measurement like manufacturing productivity and therefore assessing the possible gains has been extremely difficult. The claims made by vendors have often proved illusory, with the result that improved office productivity has become a kind of elusive Holy Grail.[6]

Productivity gains resulting from computer-aided design are as difficult to measure as those from office technologies in general. Complications and "bugs" in both hardware and software have led many experts to emphasize the negative effects of CAD. A Swedish National Union report concluded:

A general conclusion was that CAD has the attraction of the novel, but can have negative long range effects on the work of the designer in many areas.

Work at terminals can cause ergonomic problems in terms of strain on the back, neck and eyes.

Social communication is replaced by technical as attention is absorbed by the display.

CAD creates an information flow not adjusted to human logic and unnatural decision-making routines can inhibit creativity.

An assumed division of labor into routine work to be done with CAD and creative tasks for the humans could possibly be founded on an unrealistic view of the not fully understood creative process.

An increased proportion of variant designing, where ready-made part solutions are being combined into new designs, can lead to dequalification of the designer.

There is a risk of increasing polarization into "routine designers" and "creators."

The decreased amount of drafting can cause competition between different professions for remaining tasks.

Negative effects of bureaucratic forms of organization are not automatically abolished by the introduction of CAD.[7]

Paul Adler, in a study of CAD/CAM integration in the electronics industry and the aircraft industry, finds that they are struggling to capitalize on the potential of this new technology. He includes new product development in his criteria for productivity in the design of printed circuit boards (PCBs) in the electronics industry:

When I asked firms to identify boards of comparable complexity designed in 1980 and 1987, only three of the nine organizations could show any evidence of overall improvement in time, cost, or quality over [the] whole cycle of design, fabrication, assembly, and test. The weak impact of CAD/CAM in

PCB efficiency is further evidenced by the fact that in cases on which I could collect the data bare fabricated boards (profiled, printed, drilled) of identical complexity have become more, not less, expensive, suggesting that productivity increases have probably been slower than the relatively modest inflation rate over the period.[8]

One impact of increases in CADD productivity has been the prediction that drafters will become obsolete. These conclusions have been supported by the work of Leontieff and Duchin, Shaiken, and Cooley.[9] In our own work in the New Jersey Department of Transportation (DOT), we found that the department purchased CADD with the expectation of using it as a drafting tool to improve productivity by a factor of ten and to facilitate staffing reductions by eliminating the need for many drafters.[10]

In Shaiken's explanation of CADD, drafters become redundant because the engineer's design automatically becomes the drawing, eliminating the drafting function:

CAD lays the basis for a vertical integration of operations from the designer's concept of a part to the point at which it is made. With conventional methods, an engineer would design the part, a drafter would draw it, and a machinist would build it. Now CAD is capable of translating the design directly into a part program that guides the cutting tool on an NC [numerical control] machine, eliminating all intervening steps between design and production.[11]

Salzman's research contradicts these findings. In fact, even with the increased investment in CAD by the automobile and aerospace industries— "from $50 million in 1977 to over $300 million in 1986 in both industries"—drafters' jobs increased:

According to projections based on CAD's productivity impact this dramatic increase in investment should result in a dramatic decrease in employment of drafters. Employment of drafters during this same period, however nearly doubled in the aircraft manufacturing industry, from 3600 drafters in 1977 to 7100 in 1986 and nearly tripled in automobile manufacturing from 2600 drafters in 1977 to 7400 in 1986.[12]

The studies of Adler and Salzman seem to add weight to the argument that CAD has not significantly increased productivity and competitive advantage in the design process. Though everyone agrees that CAD still has great potential for productivity gains in the future, these gains cannot now be demonstrated. The future gains seem to be coming very slowly even with significant capital expenditures.[13]

Our studies, though they may be specific to architectural and engineering design in the public sector, contradict these findings. In the New York Department of Environmental Protection (DEP) and the Transit Authority, productivity increases were demonstrated. As early as 1984, when we began our CADD studies in New York City and the Transit Authority was just beginning to use its Intergraph system and the DEP was using McAuto, productivity gains were already apparent:

> Certain facts, however, are clear concerning design benefits. In both public and private sectors, there is general agreement that the quality of their product is higher because of the demands and capabilities of the machine. Managers report that there will be little need for artists' renditions of drawings now done by hand because the machine does them as well. The work is improved with regard to precision and the versatility of the machine broadens the range of capability of the users; where data bases are established for calculations, a considerable amount of time can be saved in mathematical operations.[14]

Though productivity gains were already apparent by 1984, when these agencies' CADD facilities were in a primitive stage, the gains were based on little data and limited experience. The projections of 200 percent to 400 percent increases in productivity and resulting savings of over $6 million may have been overly optimistic.[15] In the second phase of our New York City studies (1989 to 1991) there is better evidence of productivity gains.

At a conference in Cambridge, England, in 1988, Michael T. Cetera, chief architect and CAD coordinator for the Bureau of Heavy Construction, New York City DEP, summarized the productivity gains that architects, engineers, and drafters using CADD have achieved:

> Bureau of Sewers: Demonstrated productivity gains by creating a library of standard details which are used in the development of contract documents, supplemented by the customization of their character and line styles to improve drafting operations. The engineers are working toward automatic transfer of survey to design contracts, using computerized survey methods, and have developed an X-BASIC program which allows for the automatic generation of sewer regulator contract documents.
>
> The Bureau by creating 247 CAD drawings, saving a total of 3,250 hours of design time (when compared with the output of manual drafting), has justified the purchase of additional 10 work stations, and ensured the use of GDS software for all future sewer mapping and contract documents.
>
> The Division of Plant Design has developed techniques to train during

production, utilizing the accurate geometric capabilities of GDS software to assist in design and during construction, and is working toward the creation of a database by digitizing our existing infrastructure.

Organization of our data to provide a library of standard details, automatic material take-offs and cost estimates, in order to reduce current design costs and exceed our present production capabilities, is our goal.

CAD has enabled an increased work load as predicted and quality has greatly increased.

CAD software has enabled designers to input accurate graphic data and automatically transform design sketches into contract documents, saving design time.

The integration of training costs. Trainees contribute to library creation and digitize existing structures, while more experienced users create 3D models to aid project visualization and concept development. The byproducts are two-dimensional preliminary background drawings, which are utilized by our designers to develop contract documents. Plans, sections and evaluations are generated to maintain dimensional accuracy; limited access prevents the inputting of details, but standards are digitized and re-used from CAD libraries.[16]

In December of 1988 the Computer Aided Drafting and Design Committee (CADD Committee) did a pilot project evaluation in the New York City Department of Transportation. The purpose of the project was (a) to facilitate the automation of drafting, design, and engineering work, (b) to achieve economies of scale in drafting and design, and (c) to reduce drafting time, create standardization of symbol and line styles, and thus to facilitate better coordination both within and between bureaus. This was an attempt to establish standardized educational qualifications and training procedures. They summarize their findings:

Productivity Savings during Pilot Project

Project	No. of drawings	Manual creation	CADD	Benefit
Lenox Ave.	20	80 hrs/dwg	70 hrs	200 hrs
Herald Sq.	5	70 hrs/dwg	21 hrs	245 hrs
Flatbush Ave.	10	4 hrs/dwg	2 hrs	20 hrs
Muiry Sq.	5	21 hrs/dwg	10 hrs	55 hrs
Totals	40	175 hrs/dwg	103 hrs	520 hrs

Performance by end users on designated projects demonstrated that significant savings in labor hours can be achieved with the introduction of CADD. Although the total hours saved is reduced when training time and costs are factored in, future productivity increases will be considerable once a firm foundation of training is established and true economies of scale are realized. The jobs selected for CADD evaluation represent a typical sampling of highway engineering projects. Quantification of productivity gains "on average" is difficult to estimate given the diversity and uniqueness of projects.

Elimination of redundant survey efforts by DEP and DOT (Flatbush Avenue) represents a significant reduction in project cost and completion time and illustrates the successful coordination of interagency efforts within the context of computer-aided drafting and design.[17]

Robinson confirmed many of the productivity gains from an experiential perspective:

> One thing I always said, if there's a change that has to be made there is no way you can beat this machine. When it comes to the input it varies on the user itself. Some users are faster than others; they can approach design differently. But once you have their design on the system, there's no way . . . any kind of modification, no one can beat this machine. You can't beat this machine with modifications; it's practically impossible. The hard part is getting design on the system. From that modifications just flow.

CADD with 3-D capabilities increases the capabilities and the creativity of the engineer and the architect in the design process, according to Robinson. CADD allows for the quick comparison of alternate designs of the same project, in both two dimensions and three. Its speed makes possible multiple design schemes and increases time for experimentation in design. As our engineers and architects often say, they have "time to play." Robinson continues:

> It allows a lot more creativity, because you can start from one design and copy it and modify it slightly while the design is still intact. If you are trying to do it manually we copy the whole thing and then modify it. That's the way you have the time saving because you have a lot more schemes a lot faster. You know that you can overlay each and every scheme and then see how well they fit your design. Manually it's not as simple as that.
>
> And then 3-D increases it further. . . . . [On] one job that we were doing, the structural section was having problems visualizing the design and . . . the machine . . . showed that it can be done. They saw it. And they went ahead and did the calculations. When they couldn't visualize it they couldn't do the calculations.

Ultimately, CADD increases the traditional conception of productivity. It creates economies of scale and competitive advantage where it is fully utilized. There are inevitably "bugs" in the system, as Salzman has correctly pointed out, but they are eliminated over time and are mostly the result of a technology that has exploded the traditional occupations of engineers, architects, and drafters. The transformation of the work of these knowledge workers has created organizational problems, physical problems, stress, even electrical wiring problems as the result of increasingly complicated networks that require immense amounts of cable and electrical and telephone resources. The solution of these problems is an ongoing project, as are the productivity gains that have occurred. The most important transformations have been in the reconceptualization of the fields themselves—and of the notion of productivity. Increasingly, the knowledge component of design work, in the sense of the importance of abstract knowledge, has increasingly come to the center. The skill component, the actual drawing, is increasingly done by the machine. It is still important—there is no design without the drawing—but it is stored in the memory of the machines in libraries that can be called up instantly; the skill of drawing is increasingly secondary.

Douglas F. Stoker, president of AESI, a system integration and software development company, says of CAD's impact on this reconceptualization of productivity:

> Design is the discovery and sythesis of information. The process of design is the process of describing something that doesn't exist . . . the product of design is information.
> Information handling is one of the things that computers are supposed to be good at. With the obvious trends of lower cost, higher power and greater functionality, CAD hardware and software will continue to be applied to an expanding list of information handling tasks in the design process.[18]

For Stoker, the product of design is information and CAD greatly facilitates the increase of information. For Raoul, productivity is the increase of creativity and the time that CAD saves the designer. The engineer or architect can play with the design, creating further possibilities. An important contribution by Adler is that he adds learning to the criteria of productivity. This requires constitutive changes in the organization of the work of engineers in design and manufacturing: "CAD/CAM calls for a subtle change in the whole fabric of the organization, away from the conception of the organization as a production system with a dual objective of both production and learning."[19]

For Mike Cetera, CADD allows for the reintegration of aesthetics and functionality; its redefinition of productivity also allows for the merging of art, scientific technology, and competitive advantage. What we are witnessing is the reconceptualization of productivity. Productivity as an economic calculus—hours of labor divided by output—is now insufficient. In all of these cases, productivity is being redefined, and knowledge is becoming the central factor.

Privatization and the Assault on the Knowledge Worker

In chapter 2 we discussed Gouldner's work on the "new class" in the context of the privatization of the public sector. In that discussion we argued that the Reagan-Bush policy of privatization was a form of class war against the poor through the defunding of social welfare programs; it is an attack on the "new class" as well. In short, we agreed with Gouldner's argument that a new class whose basis is the culture of critical discourse became increasingly independent of the private sector. Cultural capital and not economic capital is the basis of its power. Potentially, public sector knowledge workers become at least semi-autonomous. Gouldner wrote this during a period of expansion of the public sector. He did not live to see the Reagan-Bush era of privatization and the downsizing of the public sector. He did not see the policies of privatization as the assault on the culture of critical discourse and its bearers. Privatization and the creation of an austere public sector are strategies for controlling "new class" public sector workers.[20] The redefinition of productivity in which knowledge is a central component plays directly into the possibilities of a more powerful new class of knowledge worker. Specifically in terms of CAD, it becomes possible to all design and fabrication work in the public sector (in-house). Private sector consultants are displaced. CAD enables the public sector to become more than competitive; it can do the work at lower cost and better quality. Public versus private sector should be a significant question, but in this age in which the ideologies of entrepreneurialism and privatization rule, this debate is muted.

Where the public sector has been allowed to develop a concept of professionalism in which the notion of public accountibility is crucial, the concern with community, the public good, and quality work can flourish.[21] Specifically in terms of CAD, a funded public sector would allow the architects and engineers that we studied in the New York Department of Environmental Protection and Transit Authority and the New Jersey Department of Transportation to have the luxury of time, to truly customize the technologies to their fullest capacities. They could discover the new design possibilities that are the promise of the interaction between the engineers, the architects, and the machines. They could create and use a database in which the whole history of design is just a command away. In this sense the library of design in the computer memory becomes history as a living project, not history in the reified academic sense of "dead facts" stored on library shelves. The architects and engineers in the New York City DEP are like the "scholars and intellectuals" that we discuss in chapter 8, on college and university work. They are visionaries who are redefining the parameters of design in *the public interest.* CAD is the tool that makes this possible. Their vision is one of community enhancement, public participation. They design to meet community needs, both aesthetic and functional.

Thus, Mickey describes the process that was involved in designing the fence for the Coney Island sewage treatment plant:

> We said why not just take the elaborate sculptural pieces and use them on the
> ends, the head-blown flames, the imagery and in between put a chain-link
> fence covered with vines and that would make it look sought of like a set
> from . . . *Sleeping Beauty*, where the vines are growing over civilization,
> where the vines are growing over the imagery. So . . . the artist liked the idea,
> [and] as he did that, he started coming up with another idea. And from that a
> new idea started to evolve, which . . . he calls "wave wall in green." He elimi-
> nated all the old imagery and he made the imagery out of chain-link fence
> that would be covered with vines and give the appearance of a topiary.
> Where now he's forming waves with the shape of the fence. . . .
>
> He decided he would do it by hand. We kept telling him to do it by
> computer. Lopez is here. He said, it's not necessary, I can visualize it. We
> went to the planning board, the community board; they found it very hard to
> visualize—see, this is a straight-on view. So we told him to build a model. He
> said no, it's too expensive. Lopez modeled systems, modeled his design from
> his contract drawings, built an actual representation of it, which we're now
> going to use to show the botanical gardens, to verify that the vines can grow
> in this configuration. With the different lighting effects, with the CAD we can
> make computer shadows on this to show different lighting effects. Also to
> show the community board laypeople so they can better understand the pro-
> ject they're going to pass on. It only took thirty man hours. You couldn't
> have built the base for a model, and for the model to read, it would have had
> to be so huge, it would cost more than thirty man hours to transport it. On
> top of that, . . . the three-dimensional computer model gives us automatically
> two dimensions [and] orthographic projections (geometric), which can be
> transposed into contract documents. And it can be changed

This is a design process in which two architects and an artist can collaborate through the medium of computer-aided design. They can create an aesthetic conception that they can render into a three-dimensional model that can be displayed. They are consciously involved in a paradigm shift in the field of design.

But there are limits. The public sector is under constant attack. Mickey says, "We have [a] budget crisis and everything is flying around, budget cuts, layoffs, projects dying, social services cut."

Privatization as class struggle: public sector architects and engineers versus entrepreneurial private sector engineering and architectural firms. Privatization is private-sector accumulation with public-sector funds. Public services are, increasingly, restructured for the needs of private profit. In this struggle there are different strategies. In bridge and highway engineering we found three strategies for the use of consultant engineers. In the California model, design is done almost 100 percent in-house by public-employee engineers; California is recognized as the country's performance leader in design work. In the New York model, 50 percent of the design work is done in-house.[22] New Jersey offers the perfect example of the destruction of quality public-sector design through the process of privatization. The New Jersey engineers and drafters are like "proletarians." They are being degraded and displaced in their design work. Even with a tremendous increase in design funds as a result of the $3.3 billion transportation and trust fund for highways, bridges, and transit systems, the proletarianization of design work is not reversible. The lion's share of the increase in design funds from $150 million to $800 million is going to benefit consultant engineers. Public-sector engineers are only involved in the design work that the consultants find unprofitable, or they correct consultant errors and monitor consultant contracts. In New Jersey, CAD is being used to further degrade the work of and displace engineers and drafters:

Contracting out is preferred in the state sector because it is believed that private sector consultants are cost efficient and that they perform higher quality work than engineers employed by the state. Public sector engineers deny that this is so. They argue against DOT's use of consultants with the claim that public sector engineers are more cost efficient, that they do high quality work, and that it's only because of managerially produced incapacity that they fail to successfully compete with consultants. DOT engineers claim that if they were adequately staffed and designated for important projects they could out perform consultant engineering firms. However, management maintains permanent understaffing which makes it impossible for them to compete with the private sector. Permanent understaffing also means that they cannot be assigned projects that require a significant amount of design work.

DOT engineers can neither maintain their skill levels nor develop the requisite skills because they are never assigned important complicated projects. The engineers refer to the process of managerially produced incapacity as a "Catch-22." The engineers' Catch-22 takes the form of a self-fulfilling prophecy. The management and the state believe that consultants do cheaper and better work than in-house design engineers. As a result, they assign projects in such a way that it is impossible for in-house engineers to be good as outside consultants. They have not built into the system an internal capacity for in-house engineers to do design work. Most engineers who are confronted with this structural bias against in-house design feel overwhelmed. Most of the men interviewed were angry with a system that is structured against them [and] as a result they feel that their work has been degraded.[23]

Why can California engineers who work in the public sector do the best design work in the country, while New Jersey engineers are proletarianized? The answer is ideology, not technical efficiency. In fact, in New Jersey, even with all of the managerially produced incapacities, the public sector still outperforms the private sector. Plus, because the public-sector engineers are continually correcting consultant errors, the productivity of the consultants is artificially inflated:

> In the Evaluations Bureau, where it can be demonstrated that in-house work is cheaper and of higher quality than the work of consultants, the state's commitment to contracting out is fiscally unsound. The research staff of Local 1032 of the Communications Workers of America (CWA) found in 1984, consultants did 59 percent of the total bridge inspection and received 75.5 percent of the money spent on inspections. DOT bridge evaluation engineers did 40.3 percent of the inspections and received 24.5 percent of the budgeted money. Researchers compared costs on 14 completed projects and found that consultant inspections cost $3032 on average, while in-house inspections cost $1453 on average.[24]

The situation in New Jersey exemplifies the destruction of public-sector capabilities during the Reagan-Bush era. Privatization has been a potent ideological weapon that has successfully degraded public-sector design capacities. Austerity as class war has permeated all levels—federal, state, local—of the public sector. It has successsfully structured design incapacities into the New Jersey DOT. Big entrepreneurial consultant engineering firms are guaranteed big projects and big profits. The New Jersey engineer has been proletarianized, but not in the revolutionary sense, only in the sense of being degraded. In New Jersey there appears to be no future; in New York there are still possibilities. But the engineers, architects, and drafters must be willing to struggle to get the luxury of time to learn and use CADD, the opportunity to experiment and to produce new designs that embody their knowledge, their aesthetic sensibility, and the public good.[25]

In his study of hackers, Stephen Levy found systems analysts and programmers in areas of advanced "computer science" completely oblivious of or indifferent to the context that framed their enthusiastic, almost fanatical dedication to computers in MIT's defense operations.[26] In this respect, they reproduced the complicity of those scientists who, without reservation, worked for the defense establishment because the Department of Defense was virtually the only source of funds for basic research. In the case of civil engineering, employees are relatively free of considerations such as these, freeing them to exploit the opportunity to turn their relatively routine labor into creative work.

The two other agencies we studied provide a fruitful contrast to the relatively optimal case of New York's Department of Environmental Protection. In New Jersey's Department of Transportation and New York's Transit Authority, a quasi-governmental agency, management chose CAD programs that were far less versatile. In these agencies, the programs did not have three-dimensional capability, nor was training afforded to the entire engineering workforce. Management saw the introduction of CAD as a productivity gain: its chief function was to reduce and eventually eliminate drafting and to facilitate designs of a completely routine nature. The engineers were unable to respond enthusiastically to the introduction of this limited CAD program. There was little hope assigned to the introduction of computer technologies. Instead, they were concerned with issues of job security and suspicious of management's exclusive preoccupation with productivity gains at their expense. In both of these agencies, engineers and drafters had long been engaged in a dispute with management over the issue of contracting out. They charged that management had reserved the more interesting and complex work for outside contractors who, management alleged, were more highly qualified than inside employees. Over the years, the amount of work that the agency farmed out grew steadily; besides routine design assignments, inside employees were performing inspections on the finished jobs submitted by contractors rather than enlarging the scope and variety of their own responsibility. The engineers complained that they were really repairing the mistakes of the contractors but were prevented, by political and bureaucratic authorities, from attaining their full potential or, indeed, making full use of the possibilities of CAD.

Our studies showed that the Panopticon is not easily dislodged, that even the most revolutionary technology can be recruited in the interest of reproducing power, in this case to further degrade the labor of engineers, and even to be used as an instrument of proletarianization.

Artificial Expectations: Computers in Education

One of us (Aronowitz) was a volunteer at a small alternative elementary school in Brooklyn. The school is using several instructional programs developed by Apple to help sixth graders learn to type and improve their language skills. My job was to insert and boot the programs on the computers, help with any problems the kids may have, and observe that the programs are being properly used. They come to the computer room, a small atticlike space on the top floor of this old building.

The typing program is a multicolored version of my seventh-grade curriculum on the old Underwood manual typewriter provided to us in Junior High School 118 by the New York City Board of Education forty-five years ago. Although the typing is very quiet, the routine is identical: FRF space, SWS space, JUJ space . . . and so on. The "communications" from the program are cuter than my typing teacher's watchful eye and disapproving words, but no more exciting. The programs are indexed to the kids' improvement, but the new levels—copying business letters and disembodied, abstracted texts—are geared to training typists and secretaries.

Nor are the language-skills programs examples of alternative ways of learning. The same old "supply the missing word" and practice drills for vocabulary building that I experienced in my time were replicated by the programs. Most of the kids were appropriately bored; they dutifully followed the commands that appeared on the screen with more or less difficulty and seemed anxious to finish their assignments and get out of there.

Of course, this example may be criticized as an extreme case in which the computer takes on the function of rote teaching and learning. It can be used to help students learn physics and math. Students can learn at their own pace in decontextualized disciplines such as math and physics that require, at least at first, step by step, rote learning. And the sophisticated programs developed by computer scientist-educators, notably Seymour Papert, by combining play with rigorous, rule-driven algorithms, make learning a challenging and enjoyable activity that helps develop the capacities of the student to master traditional analytic procedures.

Proponents of computer-aided education have advanced several central arguments: Computers can individualize the learning needs of the student in a way that the collective modes of instruction cannot. They are versatile. They can be tutors (as in the example just mentioned); they can be tools—for example, when they make my writing easier by providing editing features; or they can be tutees, that is, can be programmed by the student, who, as Papert suggests, becomes an epistemologist.[27] The activity of programming helps the student learn a metaskill, but since, following a long tradition in biological and psychological thought, Papert believes we are information-processing creatures whose thinking parallels that of a computer, he argues that when we learn how the computer works we also learn the rules governing thinking.

Hubert and Stuart Dreyfus, John Broughton, and Edmund Sullivan have provided powerful critiques of this bold but mistaken proposition.[28] The Dreyfuses acknowledge that for subjects that require rote learning—

that is, acquisition of knowledge by following rules embodied in step-by-step procedures—the computer as tutor has a valuable but limited place in the classroom. But these critics challenge the underlying assumption that the mind can be defined as an information-processing machine that observes rules analogous to those governing the computer. Consequently, they challenge the applicability of the computer to context-bound thinking, where knowledge of the terrain must be obtained more by intuition, memory, and specific knowledge of the actors or the geography than by mastering logical rules. As Jeremy Campbell states:

> In sum, learning to program and to use the computer may provide the student with reflexive understanding of the rules governing thinking that are employed by sciences that work according to Aristotle's laws of thought and its contemporary modifications, but cannot embrace processes of thinking entwined with indeterminate situations in which the rules are, likely as not, invented by actors themselves, modified to specific situations, even if established in advance. Since our relationships are forms of negotiation about the rules of the game and any everyday interaction, even those that are routine, like shopping, are frequently indeterminate a priori rules are routinely violated in order to accommodate the specificity of the terrain.[29]

Whatever the physiological and biological presuppositions of thinking, it develops in concrete situations, profoundly shaped and frequently altered by many determinations, including the choices made by people themselves. This perspective does not deny that the physiological dimension establishes *boundary conditions* for the functioning of human organs: despite tremendous progress made possible by affective and logical means, some individuals reach a plateau of learning (which continually expands as we learn more about learning) because of the limitations imposed by the physical constitution of their brains and nervous systems. And even if psychological research has established that the brain is actually a binary organ, and that each side is the repository of certain functions, these are only general preconditions of thinking. We can design a computer that helps us address logical, analytic problems that are, surely, an important goal of education. But the computer is not designed to anticipate or respond to problems that lie outside its logical ordering. So the computer may be used to uncover the epistemological foundations of technological thinking that is rule-driven rather than context-driven, but it cannot be taken as a model for thought. It is an *interested* machine beautifully designed for some purposes and not for others.

The elementary school that uses programs that replicate traditional pedagogies is, consciously or not, wasting the instructional potential of computers. Better to employ peer tutoring in cases where some students need additional assistance to master the rote features of a discipline. The interaction with another person is more pleasant than facing a machine for routine exercises, and the insight of the tutor into the nonlogical problems the tutee may be experiencing is brought into play.

Papert is convincing in his advocacy of computer programs that contain games to teach problem solving in math and science. This creative employment of some of the discoveries of artificial intelligence has successfully been introduced into schools all over the country. Yet the epistemological and educational claims for these programs demonstrate the severe limitations of computer-aided instruction. They presuppose a theory of thinking according to which the brain is a rationally calculating instrument whose characteristics may be described and measured with the same precision as a bridge or a water distribution system. The most recent expression of the mechanistic worldview is molecular biology, which has drawn a picture of all life in the image of the computer program. DNA, which constitutes the genetic material of life, is structured like a microchip: the program of the course of an individual's life is inscribed in the material. This determinism carries on the great tradition of genetic science and challenges the largely indeterminate character of the modern theory of evolution.

The image of the body as a type of machine has dominated biological thought since the sixteenth century. Vitalism arose in the nineteenth century to oppose this reductionism and proposed instead a teleonomic model of the organism according to which, from the cell to the body, *purpose* is seen as part of the process of reproduction and decay. Most important, while the mechanists obliterated the distinction between mind and body by considering mind to be identical to the brain and the nervous system, simply in terms of variations and the combinations of atoms and molecules and a specific organization of cells, ganglia, and receptors, the vitalists insisted on a place for spirit, which in biological parlance was reinterpreted in teleonomic language in order to avoid the idealistic baggage the term implied.

Between mechanism and vitalism, a third group has insisted on that thinking developed as a dialectically complex process involving the internal relations of the organism, its relationship to its environment—the specific conditions that provide a range of situations and issues to be addressed—and the larger economic and political and cultural milieu that is expressed largely within the specific purview of the actor but is nevertheless part of the biophysiological situation of the actor. Needless to say, this view holds that the life process in all of its aspects bears on the configuration of thinking. The actor is shaped by, but also shapes, the multiple aspects of her or his thought—modes of communication, discourse, and ideology, all of which constitute knowledge of the life-world within which the individual is ensconced.

Thus, learning does not exclusively, or even principally, consist in acquiring logically constructed, decontextualized systems of knowledge but consists in the capacity to selectively test acquired, fixed knowledge in current situations, the reflection upon which constitutes new knowledge. Whether the student is the tutor or the tutee, what is essentially *a regime of truth* transmitted by the computer either as the metaskill of programming or in a traditional content does not necessarily contribute to education. As Broughton has shown, students are obliged to surrender control over their education to the computer because of the deterministic algorithms imposed by its programs. Students submit to an externally determined order of knowledge, regardless of how playfully it is taught.

Even the language of computer technology, particuarly the terms *command* and *menu,* indicates the degree to which the self-management of the learning process is constrained from without. The computer's binary structure (yes/no) provides few opportunities for "maybe" yes and no, and even if these commands were available, Turing's logic would inevitably reduce choices to the the Aristotelian imperative.

To be sure, the playfulness of computer-mediated learning is best exemplified by the development of the computerized hypertext. Logging printed texts of all kinds on the computer allows the reader to literally reconstruct the text without observing its linear ordering. What may be revealed by this intervention is the degree to which texts are produced, not merely imbibed, by readers. While reader response theory in literary criticism has alleged this to be the case even in quite conventional reading, the indeterminacy of the hypertext gives this supposition a material reality.

Of course, the implications are quite revolutionary. Armed with a bountiful hard disk upon which has been transcribed a "classic" novel, poem, or play, the reader is free to become the producer of a new text by juxtaposing words, sentences, and paragraphs and creating new narratives and images. Scrambling the symbolic order of the "original" author might reveal hidden dimensions that had been otherwise enframed.

The hypertext brought into being by computer technology may become a form of *techne* in this case uncovering the text's subterranean significations, but it cannot explain the will to playfulness that makes this text possible. Clearly, we are entering the era when irony and its comrade, skepticism, no longer occupy an oppositional crevice in the repertoire of linguistic and art forms, but are ubiquitous. In a time when we generally admit that the past can be recaptured only from the perspective of a present that doubts itself as well as every other time, the invocation of the sacred texts of Western civilization on the computer monitor no longer enframes our culture; its enframing consists in revealing that we can now transform the permanent into the contingent, that high culture as much as popular culture is ineluctably entwined with technology. And we now suspect that the statement that the reader is the author and the text is a system of signs whose meaning possesses no fixed referent is more than a pretty formulation of the literary critic but has become, at least in tendency, a reasonable description of the new situation.

Some claim that when reading becomes nearly identical with the game of writing—that is, when it is no longer a putatively passive activity but an intervention, an act of transformation—the problem of literacy may no longer possess existential significance. For just as work becomes play for those empowered to remain at the controls, art is no longer, properly speaking, the work of specialists, but may be broadly dispersed through the technoculture. This applies not only to verbal speech but also to images that no longer require the painstaking preparation of art school or its equivalent in endless practice. Just as hand drafting is being relegated to memory, the visual hypertext, computerized or not, bears the same relation to painting that the mass-produced automobile bears to the hand-built car.

In this respect, Walter Benjamin's meditation on the fate of the artwork after lithography and photography applies fully to computer texts.[30] The auratic has by no means disappeared; throngs of people crowd into museums to view authentic paintings in the latest Picasso or Van Gogh retrospective, and these objects bring millions to dealers and collectors. The significance of going to the museum is encapsulated in the function of the witness, of "being there" rather than viewing as an activity that attempts to discern the intrinsic *meaning* of the work of art. Museum statistics are a sociological category and have to do with little more than the aesthetics of presence.

So, too, with technoculture's appropriation of the literary work. Its transmutation from the finality of the printed page to the "soft" word images on a screen enables the reader to deconstruct the work and otherwise play with it. Some would recoil from this idea, not only on grounds of bourgeois taste but also as an educational and political issue. After all, is it not the case that those who ignore history, literary or otherwise, are doomed to relive it—or worse, are just plain doomed? Yet hypertextualization provides, even if it is to the detriment of canon, the possibility that the reader may become a producer through demystifying the authority of knowledge. Here we are not confining this statement to the literary sense of the consumer as producer suggested by cultural and literary theory, which assumes that every reading is, at the same time, a rewriting through contextual interpretation. Or, employing the thesis that signs have no fixed signification, the text may be considered open, not only with respect to its intratexual dialogue, but also in relation to the act of reading as itself a dialogue. What we mean here is that through technological means, the text is now subject to physical revision in which the authorial voice becomes indeterminate.

Under these circumstances, some believe books will become objets d'art, museum pieces whose survival is a measure only of the reluctance of a conservative publishing industry to develop new markets. For books are repositories of fixed knowledge; even if they are subject to interpretations and variations, they also lend themselves to being seen as cultural gate-keepers. The unfulfilled promise of the hypertext is that it abolishes the authority of all but the computer's capacity to reveal the authoritarian character of taste. What the hypertext reveals is that standards are socially produced, usually on behalf of the claims of the powerful to be the legatees of culture.

Thus, unless the process described by McDermott forecloses its salience, the hypertext is potentially a weapon of the powerless in the struggle for control over the signifiers of culture. For the hypertext is at once a playful reading of the past and a production of the future in the present. The question is, by whom and for what? For those who object to Benjamin's withering critique of high culture, it is enough to recall that he was among its most dazzling products. The mechanical reproduction of art was to him an important element in the effort to transform power relations since he understood culture as a discourse of power. In effect, capitalism's development of the means of mass reproduction of art, which attested to its compulsion to revolutionize the means of production, also signaled its loss of control.

For technoculture, its appropriation by capital as the condition of its growth notwithstanding, is, in all of its forms, inherently destabilizing to the regime that gave it birth. This state of affairs comes about because the dispersal of knowledge by the technology separates it, at least in part, from power. Or, to be more optimistic, power/knowledge shifts away from the centers where the new technologies can be easily assimilated into the old cultural as well as industrial order. Contrary to the earlier judgment of writers such as Andre Gorz that technical intellectuals are merely an adjunct to the corporate capitalist order because they have been deprived of critical functions and capacities, it may be argued that the reverse is the case. Although traditional intellectuals had a wide ideological influence because they dominated ethical and cultural discourses even when they were politically at the margins, the manager remained the key intellectual of the industrial order. To the extent that scientific and technical intelligence was merged with management through its final separation from manual labor in the nineteenth and early twentieth centuries, capital could regard workplace resistance as a large problem within the social order rather than working against it from the outside, at least in Western countries.

But technoculture has transformed the nature of information and knowledge beyond the workplace. The growing skepticism of the underlying populations of all countries concerning their economic and political systems is not the result of ignorance but of the countersurveillance effects of mass communication and computerization. That is, one of the hallmarks of the Panopticon—the one-way mirror—has, at least putatively, been smashed. Surely the countersurveillance function of the established media, especially in the United States, is unintended, but it can be devastating. Even though the Iran-contra investigation in the mid-1980s was botched by a political system lacking political will, even though the Reagan and Bush administrations succeeded in concealing the enormous savings and loan scandal for months, the discoveries of these events have the majority's tendency toward antipolitics. The eventual dissemination of information about these and other scandals has produced a quantum leap in the political skepticism of perhaps a majority of the citizens of Western democracies. For to understand electoral politics profoundly is to despise and shun them, even when there are no alternatives.

Despite the truly herculean efforts to quarantine the fruits of technoculture in commodified, trivial pursuits or to confine its applications to entirely conventional areas, its corrosive effects persist. We believe there are no grounds for the technological determinist view that industrial societies can no longer contain these effects. On the other hand, there are strong reasons to believe that one of the inherent features of the computer chip is the one presaged by Marshall McLuhan's nimble phrase "global village." For one of the true treasures of all prevailing powers is the evocation and reproduction of the exotic, the esoteric, and, above all, secrecy. In the mediated world, even the subaltern have lost the aura of otherness, prompting the invention of dialogics and the philosophy of difference. The terms promise that there are no barriers that communication cannot overcome even if the spurious quest for universal truth seems a more distant goal than ever.

Chapter 5

The Professionalized Scientist

Throughout this book we contend that knowledge is central to the production process, that knowledge is stratified, and that scientific knowledge is valued most highly. Thus our case studies of scientists are of central importance. For this chapter we studied biomedical scientists and the stratified professional labor process at a major research facility in the New York metropolitan area. We talked with the scientists in their workplace, focusing on their own accounts of how they "make science" in their laboratories.

Professional Technoscientists

A cell physiologist describes his work on toad bladders:

It deals with the general issue of water balance. How a person maintains a water balance. A large part of the story is how this hormone vasopressin that is released from the brain when we become dehydrated works. This hormone goes to the kidney, then the kidney opens up membrane permeability to water, and water kind of pulls back so that you retrieve water from the urine and it concentrates the urine and most of the water goes back to the blood. And that's how we save water. The question then is how a hormone traveling through the bloodstream, when it hits a cell in the kidney, how can it tell it to open up and then close down and only let water through and nothing else, just once. They are very fine holes. The question is, Does it make new holes and then tell the membrane to close it up? Or is there sort of a new door that opens and closes? Or does it sort of insert whole new doors, are they always open or if they open can they be closed? One way to get at that is to use the toad as an experimental animal because in the toad these cells are not bunched together in the kidney, which is sort of a big piece of meat. In the toad bladder it's just one cell as a sack. And so you can take this sack and you can put something on the inside and something on the outside and you can see what makes the bladder permeable to water. And then you can weigh this and you can put a tracer and you can measure this facet. All you have to do is observe when the door opens and when the door closes.

Well, these kind of experiments I started out in the late 1960s and there was a lot of interest in measuring what inhibits them and so forth. How does a hormone adapt to its receptor, what interacts with the receptor. Then it became known that arginine and cyclic AMP and that somehow is included in triggering water permeability, but now all that is pretty well known. And this has been applied back to the whole kidney and so forth and has been integrated in the general field of kidney physiology. Now the next stage, at least for my interests, is to find out what the molecular nature of the receptor is, so I'm trying to isolate the receptor and that gets into the biochemical. How do you get the receptor out of the membrane . . . [and] how do you pull it out in such a way that it is still active? Because you can't see it, it's sort of a soup, and the only way that you can detect what you have is by putting radioactive hormone in there and seeing if it binds in the way that you're accustomed to the hormone finding its place. So and this then requires different techniques. You have to deal with other fields I'm not familiar with. And then we get into biochemistry, and synthesizing different probes that are linked to the receptor and by the receptor. And these probes can be radioactive or they can be fluorescent, like a lightbulb. In fact, by using a microscope we can see where the light goes, as a receptor, and that requires different techniques, electron microscopy, fluorescent microscopy, and you find a field in which established people have been doing that for many, many years and for others you have to learn again those techniques. So I'm a little bit behind. You're ahead of the game so far as receptors are concerned, but you're sort of trailing behind in really knowing how to use that microscope and identifying the right techniques. . . . The latest technique is that instead of trying to isolate the receptor chemically, now it's to find the gene that makes the receptor.

The cell physiologist talks in simplified terms, avoiding the full gamut of technical scientific terms that he normally uses. He uses only the technical terms that are absolutely necessary and does not explain scientific techniques or technology. He does not explain thin-layer chromatography, or fluorography, epifluorescence microscopy, differential interference-contrast microscopy, the silicon-intensified target video camera, silica gel 60-F-254 plates, gel filtration, or a Beckman liquid scintillation counter. Nor is he going to explain the binding of the vasopression analog, or the effect of dhydrin and flu-hydrin on the urinary bladder of *Bufo marinus* (Dominican female toads), or of the spatial pattern of rhoda-LVP binding.

He is right to assume that he should talk in a simplified language because we lack his level of expert scientific knowledge. All of the biomedical research scientists at the research institute speak this way to us, and of course their assumption is correct. They also talk in this simplified way to the students in the medical school. They can speak to them with a higher level of scientific sophistication, but in general the researchers assume that medical students and even practicing physicians do not have the required scientific training to really understand the "cutting-edge research" done by the molecular biologists, the neuroscientists, and the theoretical biophysicists at the research institute. In a sense the medical practitioners and I are in the same boat, even though we are sociologists; our knowledge of biomedical science is very limited. The nonarticulation of knowledge between working scientists and professional practitioners has led to a knowledge deficit for the practitioners. The medical practitioners are increasingly in a subordinate position, dependent on biomedical research that they only minimally understand. They know the practice and they have social status, but they are trained only in the basics of laboratory research and theory. Biomedical science is moving beyond physicians' knowledge, yet they are dependent upon it as a knowledge base.

But as biology moves to the level of genes, biophysical structures, even the quantum level, the cell physiologist realizes that he is not at the "cutting edge of research" either. He tells us that organ and even cell physiology are basically dead sciences. Cell physiology now takes a back seat to molecular endocrinology and theoretical biophysics. Though he is still engaged in laboratory research, his major functions now are teachering and curriculum design. He prides himself on his teaching abilities—medical students need to learn organ and cell physiology—but he is aware of the changes in biomedical research. The whole discipline has been redefined by molecular biologists, molecular biochemists, and study of the biophysics of molecular biological structures. No one escapes this paradigm change. In this structural molecular world, he is still working on physiological functions and is now being forced to learn the techniques of gene manipulation to continue to get grant funding. He is aware that his primary purpose in the medical school is as a teacher and not as a scientist engaged in research.

A molecular endocrinologist describes his work:

> OK, the endocrine system that we concentrate on is the pituitary gland. [The] two hormones out of the six or seven that we concentrate on [are] prolactin

and growth hormone, and for each of those there is a gene. We're concentrating really on two questions. One is why those genes are activated only in the pituitary gland even though every cell in the body has [them] and that's kind of a developmental question: What is the development that makes it so specialized? And we're asking what makes the cell . . . make prolactin and growth hormone and not make other things like collagen. The pituitary is . . . a very central gland and it makes these hormones but it also responds to the hormones. So we're trying to figure out how the cells that make prolactin and growth hormone integrate . . . a battery of hormonal signals that they get from the rest of the body and in addition signal neurotransmitter signals that they get from the brain and how all this gets integrated. So the pituitary cell decides whether or not to make prolactin and growth hormone and how much to make. There are positive and negative signals.

What we're specifically working on . . . is how other hormones regulate these genes. Like for prolactin there [are] a number of different kinds of hormones and there are two things that we'd like to know. We'd like to know how they work in general on gene expression, how these hormones turn genes on and off in general. And we'd also like to know specifically how they work on growth hormone and prolactin. Those are two closely related questions. It turns out that prolactin is a particularly good gene to ask that question about because there are a lot of different hormones that regulate it. We're looking at positive peptide hormones made by the hypothalamus, actually a neurohormone. We're actually concentrating on two neurohormones; they both come from the brain, if you consider the hypothalamus part of the brain. One of them is a positive signal for prolactin and the other is a negative signal. We're trying to figure out how they work individually and . . . together. And to make prolactin synthesis go up or down. . . .

We use recombinant DNA techniques. Everything that we're doing today has been made possible by recombinant DNA. It really is the most amazing technological revolution in biology. . . . You can identify DNA with it, . . . you can transfer it, you can modify it. So what we really do is take these genes out of the pituitary cell that they're normally in and we stick them back into the same cell but we modify them first and see what happens. Or we take little bits and pieces of the gene and stick that back in and see what happens. . . . The main thing we do is that we try to figure out which parts of the gene have important regulatory functions by using recombinant DNA to take bits and pieces of the gene and putting it back . . . into the pituitary cell and then asking things like which part do you need for a hormone to regulate, which part do you need to do without. Or sticking it back into a nonpituitary cell where it's actually silent. And then asking what kinds of things do you have to add back in order to get expression. So the first question has to do with how hormones work and the second really has to do with the question of why the gene is silent in a nonpituitary cell. . . .

One big advance in the past years besides recombinant DNA has been the ability to determine exactly the sequence of bases on DNA. You know

DNA has four bases and it's the order of them that decides everything. So there are ways by which you can chop up DNA up into various size pieces. And then you take the DNA which is negatively charged and put it on top of something that is like old Jell-O and you apply an electric field and the pieces all move in there but the ones that are big move slowly because they're big and the ones that are little aren't held back as much by the gel, they move fast. Then when you're all through, . . . you expose this [radioactive mesh] to a sheet of film and from the pattern that you get it's sort of like a ladder. So you can determine the sequence of the DNA. So we do that to be sure that we know, to be sure that we got the constructs that we felt we had.

But the molecular endocrinologist's work is not all in the lab. He teaches in the medical school. He must write grant proposals to continue his funding. He must manage the laboratory and its postdocs, students, and technicians. He must write papers. As a senior researcher he also has an important place in the institute and must attend various administrative meetings. He must attend meetings and present papers in his field. This is a world that is not for everyone. The professional discourse of biomedical scientists is not just elitist in Magali Larson's sense;[1] there is a gap between practitioners and knowledge producers, between doctors and scientists. This gap is not just the result of differences in schooling and in the work but reflects the constitution of the knowledge itself. The esoteric content of the production of scientific knowledge is framed within an experimental methodology. This occurs in laboratories, where machines are used to do the experiments, data is produced and measured with other machines, and the whole process of producing knowledge is summarized in a professional scientific paper. The process of being trained as a scientist also occurs in the laboratory. You are professionalized as a student in someone else's lab, doing someone else's experiment. Eventually students learn to define their own problems, do their own experiments, write their own grants, and write their own articles. (Of course these articles are collaborations, and grants are dependent on the world outside the laboratory, and the technologies are the result of years of study, designed by scientists and engineers and produced by factory workers and machines for a corporation.)

Central to the work of the molecular endocrinologist are the issues raised earlier concerning professions within the context of knowledge production and the use of a dominant technology, recombinant DNA. Knowledge production is relatively nondistinguishable from the technologies used in its production. This is so not only in the lab and in raising funds to continue the research, but in the writing of papers as well. Bruno Latour calls this technoscience and includes all of the activities listed here and the extensive activities outside of the laboratory as part of the whole network of scientific practices. In *Science in Action*, Latour describes the world of technoscience, and one way to read this book is as a treatise on scientific professionalization in a professional labor process.[2] For Latour this is a process in which science is always "in the making" and science as certainty, as a completed task, occurs only after the fact. Scientists are "fact builders" and machines are presented as "black boxes." Latour's black boxes consist of a series of negotiations in which facts and theories become "well-established facts"; they are objects that measure or process data with certainty, in which groups of supporters of science have been enrolled and recruited, and all of this functions as an automaton—"that is when the many pieces act as one."[3]

In the molecular endocrinologist's description of his work, recombinant DNA is a black box within which biology as a discipline has been reconceptualized; it structures the careers of professional scientists. This is the revolution that he talks about. Grants aren't funded, labs aren't built, institutes aren't run unless they are framed by recombinant DNA and molecular biology. In the following example, a cell biologist explains how her work has been transformed through the black box of recombinant DNA:

> I study how proteins are directed into cells. How to get a protein across a membrane is totally unknown. Protein topogenesis, a secretory pathway, protein is transported across the membrane and then transported into a different compartment. I do recombinant DNA. I'm not a gene jock but I do it all. I've had to learn genetics but I'm not a geneticist, but I make mutants, I use mutants. . . . There are no biochemists anymore; everyone is a cloner. Working with DNA is so much easier than working with proteins. Recombinant DNA is easier and sexier. It means you feel oh so powerful. It's a powerful technology.
>
> With proteins they're all different and you can transpose DNA. You can clone something in a week. The first time I made a mutant, I felt like GOD. But now anybody can do it. But it's a powerful tool. It allows you to ask questions that you couldn't ask ten years ago. But there are problems. People are going into biotechnology for the money. What I really object to in the biotechnology industry is that I feel it is important for a scientist to be an academic and a scholar. Another problem is that biotech people don't share their work because everything is proprietary. That is a problem for a lot of scientists. But not all conditions are optimal and people are coming back.

Professional scientists are continually creating new methods. In the biomedical sciences grants are not approved by funding foundations unless the most up-to-date methods are used. On our first day in the field the department chair tells us:

The transformations in biomedical research are so rapid that the technologies are changing every three or four months. Even the best people are having trouble adapting to the current changes in their fields. . . . This creates problems in funding; funding has been destabilized because of the continuous transformation in methods. I can't recruit the best people and the people that are here cannot get grants unless we have the latest technologies. And that's a very expensive undertaking.

These professional scientists are engaged in writing grants, producing research findings with the current methodologies in their laboratories, writing papers describing their research findings, with the appropriate citations, and locating themselves in their fields. Latour sees all of this as part of the continuous process of "fact building," in which scientists mobilize their resources in the "paper game" through which they present their findings as indisputable. That is, "dissenters" are forced to build a bigger laboratory (very expensive), give up, or agree. In this reading of Latour, he is describing both how scientific knowledge is constructed and how the labor process of professional scientists is constructed as well. It is clear that laboratory experiments are a fraction of the work of these scientists. The laboratory is not only a place where scientists work, teach, conduct experiments, write papers, and produce the scientific knowledge base for medical practitioners, it is also the place where instruments are located. As Latour states:

> I will call an instrument (or inscription device) any set-up, no matter what its size, nature and cost, that provides a visual display of any sort in a scientific text. . . . The instrument, whatever its nature, is what leads you from the paper to what supports the paper, from the many resources mobilized in the text, to the many more resources mobilized to create the visual displays of the texts. With this definition of an instrument, we are able to ask many questions and to make comparisons: How expensive they are, how old they are, how many intermediate readings compose one instrument, how long it takes to get one reading, how many people are mobilized to activate them, how many authors are using the inscriptions they provide in their papers, how controversial are those readings. Using this notion we can define more precisely . . . the laboratory as any place that gathers one or several instruments together.
>
> What is behind the scientific text? Inscriptions. How are these inscriptions obtained? By setting up instruments. This other world just beneath the text is invisible as long as there is no controversy.[4]

Latour's work continues to show how scientific knowledge and the scientific professions are structured by silencing controversies, how even when controversies are resolved in terms of the counterposition it then becomes a source for mobilizing articles, instruments, and laboratories for the purpose of producing and reproducing its black boxes, its certainties, and its "less disputable claims." Controversies always threaten to disassemble black boxes, but this is only possible with the mobilization of more resources, that is, bigger laboratories. A black box can be disassembled only to the extent that a new and better black box can be constructed. Thus professional scientists are engaged in the labor process of constructing knowledge as mobilizing more papers, more instruments, and bigger laboratories so that their black boxes cannot be deconstructed. This is a continuous accomplishment that enables scientists to continue to define reality, to define nature with their technoscience. Technoscience constructs scientific knowledge by building facts that are incontrovertible because controversy has been silenced.

For Latour, science and technology become increasingly difficult to distinguish. They become interchangeable. As a result, the laboratories, the instruments, the articles that constitute technoscience become increasingly expensive. Thus, the world of the laboratory, because of the immense costs of science, becomes more and more dependent on the world outside of the laboratory. It must mobilize financial support. It must mobilize the scientifically uninformed, either for their tax monies or as a profitable market. In the United States, where there are 240 million citizens and only 485,000 scientists and engineers with doctorates and only 64,000 scientists with doctorates in basic research, this is a serious problem.[5]

It is here most clearly that we can see Friedson's point about professionals attached to institutions, what Latour calls "centers of calculation."[6] The biomedical scientists we talked with are part of a medical school and a research institute. They are dependent on funding institutions to continue to produce scientific results. We chose to observe successful scientists, that is, scientists who routinely get funding, because only funded scientists can function as professional scientists. It is not that we wanted to make our arguments about scientific work from a best-case scenario, but that scientific work is dependent on producing fundable science—on writing the type of grants and doing the type of research that public and private granting institutions will approve.

A neuroscientist tells us:

> I would say that I devote about a month a year to writing grants, which is pretty good. It means that I am fortunate to be successful in writing them. If you're not you have to keep writing them and you have to resubmit them. . . . I've written them once and I've gotten them funded, thank goodness, and I haven't had to think of them again.

A cell biologist talks about grant writing and funding for basic research:

> I believe in big government when it comes to funding science. Science has to be overfunded. You have no way of knowing what's going to be important scientifically in the future. The breakthroughs of the last fifteen years, oncogenes, genetic diseases, etc. All of that work came out of basic research that originally seemed esoteric, seemed like research that had no practical use. Viruses in mouse cancer, people worked for years and years with all of these esoteric viruses and genes and now we see that this is so in humans. This research would not now be funded. The work of the early seventies, which the whole genetic revolution is based on, scientists sitting around studying esoteric enzymes, viruses, and genes, would not now be funded. This was all based on what is now considered wasteful. You need wasteful funding because no one can predict what will work out. Maybe only 10 percent works out.

Science is expensive. In this time of fiscal conservatism, there is insufficient funding for basic scientific research. Privatization would not be the answer here because corporate labs are mostly interested in research and development (R&D). But now even research and development is being defunded. Thus the 1992 National Science Board report "The Competitive Strength of U.S. Industrial Science and Technology: Strategic Issues" finds that:

> The real rate of growth in U.S. industrial R&D spending has declined since the late 1970s and early 1980s. In addition the nation's position has deteriorated relative to that of its major international competitors whose investment in nondefense R&D has been growing at a faster pace than U.S. nondefense R&D since the mid 1980s.
>
> Domestic industrial R&D expenditures slowed from an average growth rate of 7.5 percent (constant dollars) during 1980-85 to only 0.4 percent during 1985–91. The federally supported portion of these expenditures dropped from a growth rate of 8.1 percent to -1.7 percent over these two periods; industry's own support dropped from 7.3 percent to 1.3 percent. Almost all major R&D-performing industries contributed to this reduced growth rate.[7]

This is so even though R&D may result in marketable, profitable products. Basic research is not immediately profitable, and may not be profitable even over the long run. Basic research is primarily funded by the government; in 1991 the federal government funded 61 percent of all basic research.[8] Basic research funding is only a fraction of R&D funding. Even though basic scientific research is the basis for most R&D, it is insufficiently funded, and R&D is increasingly central to the economic future of the most powerful countries. It is increasingly at the center of global economic competition. In this context the molecular endocrinologist states:

> Well, I don't know about overfunding. It seems to me that a country that spends, what's the NIH [National Institutes of Health] . . . around ten billion dollars. It spends such a proportionately small amount of the federal budget on science . . . and it doesn't cost that much to fund science in this country. It wouldn't cost that much to double the funding for science. . . . The success of a country in the world is going to depend a lot on technology and I think that this country has clearly declined technologically relative to Japan. And I think if we continue to discourage young people from going into science we are going to be in big trouble. I don't think it should be overfunded but it's underfunded right now. I'm not surprised that people aren't going into science right now. I know that 20 percent of the grants that are applied for are funded. How can you even suggest to a student that he go into sciences and it would be a crap shoot that he would ever do what he's trained to do.

The cell biologist feels the same way. Her American-born postdoc has both an M.D. and a Ph.D. "If Karl asks me if he should continue in research or go into medical practice, I would have to be honest with him," she says. "I've told him that he would make a lot more money as a practicing physician than in research. He's very talented but with so little funding there's probably no future in science for him."

All of the scientists we talked to saw lack of funding as a major factor in the shortage of American-born Ph.D students and postdocs in basic science; one told us that "our postdocs are almost all from Eastern Europe, mainland China, and Holland." They have American-born medical students, but relatively few of them are going into basic research. This leads to a crucial problem for the United States in the biomedical professions: the increasing break between scientific knowledge and the skills of practicing medicine. The M.D./Ph.D.—trained both in the laboratory in basic research and in the practical skills of medicine—would be the optimal solution. Though some trends point in this direction and this may be the future, in general it is not happening now. The chair of the Department of Physiology and Biophysics says this problem has been exacerbated by the dominance of molecular biology in the biomedical sciences. At one level, to get biological research funded you must be doing research at the molecular level. Thus cell physiology is only marginally funded, and cell physiologists are even in the process of cloning their first genes. These trends in funding are a form of control over what is acceptable biomedical research. On another level, biomedical research is becoming too esoteric for the practicing physician. Medical students are only minimally trained in molecular biology, which is now at the "cutting edge" of biomedical research. There is a widening chasm between the fundable research of their professors and what they are taught in medical school courses. The medical students do not have sufficient training in the new disciplines of biomedical science. Biomedical science is dependent on a broad and detailed knowledge of the sciences from physics to molecular biology and molecular chemistry—a knowledge that neither medical students nor practicing physicians have.

The head of the Department of Biophysics tells us:

> One hundred years ago you had dying sciences, now you have dying disciplines. Everything has changed because it has become molecular. In science you used to be your own boss, but now they pressure you and the pressure is on change. Thus all the disciplines are combining and everything in biology is using the facilities and theories from other fields. We have a Department of Biomathematics. This is a biophysics department.
>
> You can't do the old stuff because you can't get the money. But you teach the old stuff, physiology. What we teach the medical students is physiology and you can't get a penny for that. The old physiologists are the teachers. The young turks do research and teach at the graduate level. There is an incommensurability between research and teaching, and there is a clear gap. This is new, it's the last five or ten years. What do doctors need to know about research? How much molecular biology does the average medical student need? They need to know some of the literature. As the new disciplines emerge they will need more knowledge. But in general the medical students are the last to know. The medical students are technically graduate students but they're really undergraduates. In order to know what to do, to be a doctor you need the old knowledge. And we teach it to them. But the new stuff we can only teach in a limited way—the totally new approaches required to do the work that we're doing now. We can't teach it to them.

The Historicity of Professional Scientists

By reading Latour's *Science in Action* as a description of the professional labor process of research scientists, a labor process that is about the production of scientific knowledge, we have developed some interesting insights into the relation of knowledge to professions. The first is that as the labor process of scientists develops, they produce not only new knowledge but also new disciplines. These disciplines produce new institutions of knowledge that organize the theories and practices and production of newer and better black boxes—that is, knowledge that is increasingly difficult to contest, that produces theories and experiments and instruments and results and papers that seem indisputable. They produce centers of calculation in which the disciplines are based and in which all data will be processed and explained. All of this is technoscience, and all knowledge is increasingly processed through it or becomes the model for all knowledge production in the natural sciences, the social sciences, and the humanities. This is the knowledge machine of technoscience.

Second, there is a disarticulation between scientists and practitioners. Practitioners in the biomedical fields increasingly depend on scientists doing basic research at the same time that scientific knowledge itself is becoming noncommunicable to the practitioners. They rely on scientific knowledge but they must take it on almost religious faith because they do not have the background to understand it. This is the same contradiction we observed among engineers and architects who use computer-aided design: theoretical knowledge increasingly replaces the skilled knowledge of the old engineers and architects. In this sense, doctors are like craftworkers; though they are still highly rewarded, they are increasingly being displaced by theoretical scientific knowledge and by technoscience.

Bruno Latour's analysis has been profoundly helpful in our analysis of the labor process of professional scientists. But Latour never terms the work of scientists and their collaborative efforts in constructing science—from doing laboratory work to writing papers to getting funding to creating centers of calculation to mobilizing the world outside science to benefit science—as a labor process. We see all of the production of scientific knowledge as part of the labor process of professional scientists. He acts as if his discourse on science, and all of his painstaking analysis of how "science in the making" becomes "ready made science," indisputable black boxes, are independent of the historicity of the scientists: that is to say, in terms of the increasing centrality of scientists and the power of their instruments and discourse. Scientists have moved to the center of professions and to knowledge production. If he talked specifically about the subjectivity of scientists and the normative representations that operate on a societal level, he would have to note, in Alain Touraine's words, "the pervading power of apparatuses for the management, production and dissemination of not only material but also symbolic goods, languages and information."[9]

The centers of calculation are already in existence and more are continually being created. The market principle and profit are increasingly important in the production of scientific knowledge, and this presupposes a power relation. Latour's insight that knowledge is mobilized back to these centers of calculation also presupposes a power relation, a social structure, and a labor process. The past, present, and future lay heavy on Latour's analysis, as does the historicity of its subjects.

In terms of the historicity of professional scientists, the issue of whether they are "the new brahmins"[10] or the new proletariat is important. These scientists are not a new class, not of the proletariat. For Marx the proletariat was "in but not of" capitalist society. These scientists are both in and of society. Social scientists use the term *proletarianization* to indicate that a worker's labor and product are owned by another. In this sense, scientists are degraded workers, they are like the proletariat. They do not own their laboratories. They are salaried employees in basic research laboratories at universities, or at federal labs, or in corporate R&D laboratories. Though this is true, these are not degraded workers who are "not of" society. These scientists are middle-class professionals who are relatively well paid as compared to most other middle-class and working-class occupational groups, with the exception of physicians, lawyers, and corporate managers. They are accorded high prestige, especially in terms of being the guardians of scientific truth. They possess the knowledge that is most highly valued in society in general. These are not proletarians doing degraded and dequalified work. They are not easily replaced by new technologies but are at the cutting edge of producing the knowledge that is central to creation of the new technologies. It is in the sense of their centrality to new production processes that they have power in the sense that Marx's proletariat had.

According to Marx, the power of the proletariat came from two sources. The first was that even though the proletariat was propertyless, in the sense that it did not own productive property, it had the potential to transcend itself as a "class in itself." That is, as an inert class, as a purely political-economic category, it could become more—a "class for itself," a class that could become conscious of its own power and form a culture in which it could politically represent itself. The second was that even though the proletariat consisted of dequalified workers, "human labor in the abstract," it was central to the capitalist production process. Even though all of the workers' labor could be reduced to "simple labor" and all workers were replaceable, only they could produce the "surplus value" that made the continuously growing process of capital accumulation possible. In this sense today's professional scientists are like proletarians. They are becoming more powerful because they produce new scientific knowledge that is increasingly important to the postindustrial production process. It is only in the sense of their central role in the postindustrial production process that they are like Marx's proletariat; in all other senses they are not. The historicity of these professional scientists is different. To talk of them as a "new working class" only confuses them with the potential agents of another era.

Though, because they are the knowledge workers and "fact builders" of the new production-of-knowledge processes, they are increasingly central to new power matrices. At the same time these scientists are not a new class of brahmins or mandarins.[11] Of course there are scientists who fit these descriptions, but they are exceptions; in general, scientists are not an autonomous class. They are dependent on corporate and state funding, and thus corporations and the state largely determine the science that they do. They are based in corporate, federal, and university labs and are dependent on those institutions and subordinate to their administrations. Their subordination to the institutions of grant allocation forces them to frame their science, their research problems, their research designs, their methodologies and instruments in terms of the current funding "fashions." These professional scientists, neither proletarians nor mandarins, still occupy a special place in the new scientific production processes and as a result have potential for constituting their own political agency. It is at this conjuncture, where power and knowledge meet, that we must talk about professional scientists as intellectuals.

Professional Scientists as Technical or Public Intellectuals

A theoretical biophysicist describes his research:

> I used to have an experimental component to my research but that has closed off. I have left this to colleagues, and my research is entirely theoretical. It is molecular biophysics if you want. It is an exploration of the molecular basis of biological mechanisms. And it is an exploration that is based on physics, mathematics, and theoretical chemistry and it is performed with computer simulations. What I'm interested in are the most basic mechanisms of life . . . or biological structure. I work on the structure of DNA, the regulation of gene expression that is the interaction between DNA and protein, the structure of protein, the structure and function of protein. . . . My oldest roots are in research on neurotransmitter receptors, again from the point of view of the structures of the neurotransmitters that allows [sic] them to perform the signaling process. And more recently, because it's become available from experiments, is the structure of the receptors that recognize those neurotransmitters and transduce the signals. And so it covers in fact an enormous amount of cellular and mollecular biology. It covers most of the physiological mechanisms that are understood at the molecular level. But it deals with all of this computationally through simulation, and theoretically there is a difference between theory and computation. . . . I call theory things that are based on laws which then are extrapolated either numerically or analytically to specific formulations. Simulation is taking specific formulations, equations that govern a certain process, and propagating them as if you were the system and accumulating the knowledge of the behavior of the system within a given formulation.

To come up with a formulation you do theory. To make this formulation work and give you data on how the system evolves you do computational simulation. And one is purely theoretical and is close to theoretical physics, and theoretical chemistry, and theoretical geology, and theoretical astrophysics, all of those things. And the other is closer to experiment, in fact, but it's a computational experiment. Because you design a system, although it's theoretical, and then you make it work and observe it and try to measure things on it.

We ask him to show us his lab:

I would bring you over to the graphics lab and I would show you the computer. That is new . . . in biology. It is certainly not new to math or to physics. And in fact the people who are sort of credited with being the beginners of the new era of molecular biology, Watson and Crick, had no lab. They were in principle crystallographers; they took some measurements of low-angle scatterings of these fibers, these DNA fibers. But after that their work was entirely theoretical. And they tried to interpret other people's crystallography experiments. So even the founders of molecular biology in that sense, not in the sense of manipulating DNA but of understanding DNA, were not really heavy experimental scientists.

We asked him about his computer requirements:

The bottleneck is on one hand computational power. As fantastic and gigantic as it has become, it is still a bottleneck. And the other bottleneck is not conceptual but pragmatic in the sense of embedding what we already understand about molecular behavior into formulas in which we can work on computers in a reasonable amount of time. But it has not always been so in this field. . . . When I began you had to write your own programs for everything, even if you wanted to calculate the energy of an atom. Now you can calculate energies with advanced quantum mechanic methods, the energies of a symbol of large assemblies of atoms, with essentially canned programs. You still have to know what you are doing, but the programs themselves are canned. You don't have to write the program. . . .

I compute on . . . most supercomputer facilities in this country—the ones that are established by the NSF [National Science Foundation] to the ones that are established by the Department of Energy. We even had a peek into the Department of Defense, when I once had a grant from them. Wherever and whatever I can get my hands on. . . .

It's all from this computer here [he points at the computer in his laboratory], which in fact could be a terminal. No small terminal would suffice, but . . . this is a Mac II. See I now will exchange it for an FX. The order is in because I need a little bit more power and certain other options. But anyway this is an entry to every one of the supercomputers.

The theoretical biophysicist has definite hierarchies both within the profession and in terms of the construction of knowledge. Ultimately they are the same. For him, basic research is more highly valued than applied research, which is more highly valued than biotechnology. University-based research is always privileged over both government and corporate research. Government research would be more valued if the government had both funded and managed it more competently. In terms of the structure of scientific knowledge, his hierarchy is the conventional positioning of the sciences: physics highest, followed by chemistry, followed by biology. Within biology, the molecular level is privileged over the organ level and the cell level—thus explaining the relatively low status of the cell physiologist in the department. This is all part of a process that values structure over function. The cell physiologist is involved in explaining organ function, while the theoretical biophysicist is involved in explaining the molecular structure of biophysical systems. The increase in the value of biology is not its connection with curing and preventing diseases; it is that the development of molecular biology has opened the discipline of biology to structural explanations relying on physics and chemistry. Thus biology has achieved greater scientific status as a result of its becoming more like physics.

Within the natural sciences, he also has the conventional hierarchy of privileging the theoretical over the applied, which he extends to the biomedical sciences. It is in terms of these hierarchies that one should understand the professionalization of scientists. To be a professional scientist, to do technoscience, you must be affiliated with a research institution, and you must have control over the use of time in that research institute. You must have time to think, to do your research (which includes having access to the appropriate technologies), conduct experiments, measure your results, give seminars, write grants, write articles, and so forth. You must produce knowledge that has the certainty of fact and that is defined as fact within your discipline and among your colleagues. Essentially, it is the production of this kind of knowledge that defines a successful career in technoscience. Scientists are increasingly rewarded with privilege and prestige to the extent that their work is more abstractly theoretical than applied and practical. To use Alfred Sohn-Rethel's categories, the professional scientist's achievements are more valued to the extent that head work becomes independent of hand work.[12]

We see this same professional structure in Andrew Pickering's social history of particle physics. Quarks are not observable in terms of any kind of test; they are inferred from theoretical frameworks that are viewed by particle physicists as the most appealing. Thus, Pickering describes Richard

Feynman's agnostic approach to the structure of protons based on his theoretical frame of reference, independent of evidence:

> To have to deal with a proton as a particle-cloud rather than as a single particle made the field theorist's task that much more difficult, and constituted another reason for the abandonment of the traditional field theory approach to strong interaction physics. Feynman, however, was not so easily deterred. He took the view that protons (and all hadrons) were indeed clouds of an indefinite number of particles. In a field theory of mesons, nucleons, and antinucleons; in a quark theory, they would contain quarks and anti-quarks. Feynman was agnostic: he simply assumed that the swarms contained entities of unspecified quantum numbers, and christened them "partons."[13]

Though the theoretical biophysicist views his colleagues as good scientists, he also views them as insufficiently theoretical. From his standpoint, the working molecular biologists are doing basic research, but they are more technicians than theorists. For him, a true molecular biologist conceptualizes about the world as a molecular system. He says of his colleagues that "they say molecules, but they don't really think molecules. And their explanations are not molecular and they don't talk about forces and they don't talk about size." The increase of computer power, the massively parallel systems, will enable him to do theoretical simulations of a much higher order of complexity, if not of certainty. From his point of view, this will increase the power of his knowledge and will transform the work of the scientist. With the development of more powerful computer systems, it has become increasingly difficult for him to understand the workings of the computer itself in the process of doing his research. In exchange for this increase in the power of his knowledge he must sacrifice his control over his labor process—a necessary sacrifice for increasing his privileged theoretical simulations, that is, the creation of a new black box. As he says, "I may not be able, anymore, really to understand the process of my program within the tool that I will be accessing. That is an alienation, an extraordinary alienation. I often ask my students, how do you feel not knowing how this program works?"

In the theoretical biophysicist's hierarchy, theory and physics occupy the highest positions. But this is not a static hierarchy; it is in continuous development, and for the theoretical biophysicist its trajectory is toward a more theoretical and physics-based biology. As this occurs, the production of knowledge will become more certain. As he says:

> The Schrodinger equation, Schrodinger's cat, is the famous paradigm for uncertainty. . . . Now Schrodinger was the father of feasible quantum

mechanics. He described the wave equation which governs the solution to the problems of quantum mechanics. . . . We can't solve this even now, we cannot solve this exactly. It's not an analytical solution for more than one system . . . so every other solution is approximate, but the approximations grow, until where they are extraordinarily reliable.

Scientific fact has been reconceptualized as extraordinarily reliable approximations. This is the certainty of knowledge that the mathematical simulations of the theoretical biophysicist produces.

The hierarchies of technoscience will continue to develop and as they do, the theoretical biophysicist believes the structure of the university laboratory will change, as will the scientists themselves. He sees this as the future of work in biomedical research:

> One of the most important of the senior molecular biologists on the floor is MB. He majored in physics but now is a molecular biologist and he never thinks of a molecule. . . . This group of people I believe will evolve into something else, which will include at least what they understand and the molecular understanding involved. For this a lot of insight, a lot of understanding has to be achieved. . . . That's the impact of what I do and what the crystallographers do on the molecular biology and vice versa; of course these two are combined. But in my field the major development will be the penetration of what the quantum people do into what the dynamics people do, into what the classical people do in order to represent truly the properties of these systems. Otherwise we will not be able to represent those properties that make for the movement of electrons, the movement of single protons, the things that actually provide the signaling of biological systems. Till then we won't have a true description of biological systems. So that is a development that will occur, absolutely, but it will require an enormous increase in computational power.

His hierarchies are important to understanding not only the professional scientist doing technoscience but also the structure of knowledge in technoscience itself. Quantum theory exposed the limits of certainty in scientific knowledge, but this problem was solved by redefining knowledge as mathematical representations. This was at one level a break with the Newtonian tradition in which the object is real and can be empirically observed and described. To the quantum theorists it was clear that there are no pristine objects independent of the observer and the instruments of observation. The quantum theorists, trying to maintain objectivity, had to solve the problem of the interactional nature of science in which objects are always mediated by the observers and the instruments.[14] Their solution was "measurability." Thus Schrodinger's equation and Einstein's equation become crucial. Reality and nature are reconceived as numerical representations. In technoscience, we have "reliable approximations" with increasing degrees of certainty.

On another level, science's break with metaphysics, dating from the Enlightenment, is patched up. We seem to have returned to the Plato of the *Timaeus* and his notion of pure mathematical form as the purest level of knowledge, or to Aristotle's pure forms independent of situation-bound content, that is, experience.[15] As Aronowitz states in *Science as Power*:

> Scientific theory describes the relation of humans to the object of knowledge, not the objects themselves, taken at a distance. Further, our knowledge [of] objects is always mediated by the logic of scientific discovery, e.g., by its concepts of causality as part of the apparatus of discovery.[16]

It is the self-referentiality of technoscience's mathematical forms that allows theoretical, mathematical representations to become both experiment and theory independent of experience, and thus its mathematical representations are said to be objective. The power and the appeal of mathematical simulations to scientists is that they enable the technoscientist to observe the unobservable and measure the unmeasurable with precision and increasing certainty. Thus the theoretical biophysicist will inevitably be able with his theoretical simulations to provide a "true description of biological systems."

Though this self-referentiality in technoscience is problematic for "dissenters," it is not for professional technoscientists. The social sciences attempt to mimic this process, and have the positivist unity of sciences. Here, there is a unity from Aristotle to Newton to quantum physics. This is the model for all sciences, and it expands its range to include everything from biophysics to sociology. It is hegemonic in its influence, dominating all fields of knowledge and all professional knowledge workers. Its mathematical representations can explain, predict, and control more and more phenomena in more and more realms. Technoscience in its "quest for certainty," seeks out the essential genes, quarks, and molecules of scientific truth.[17] The real power of technoscience is seen not in its uses and abuses but in its ability to produce reality and nature. It reifies its own categories, its mathematical representations, and makes them real. We can see this type of power in the following statement made by the MIT-based molecular biologist Robert Weinberg:

> Right now our society runs on the premise that everyone has a biologically equal chance to be anything he or she wants. . . . But what will happen when,

157

in fact, the scientists find strong evidence that everyone's fate is greatly affected by the inheritance of a group of very specific and identifiable genes[?] There are many who scoff at such a possibility. But the trajectory of scientific advances makes me believe it's inevitable. How we handle this new information is something we as a society ought to begin thinking about and discussing as soon as possible.[18]

For the most part the biomedical scientists we studied are not as reflexive as Weinberg, though they see possible dangers in unregulated, entrepreneurial biotechnology. As the department chair says, "They're playing with a loaded gun." For them, the real danger of biotechnology is that it endangers the economic viability of the interest-free basic research that they do. They are in the process of objectively answering the key questions in their field. They are attempting to develop questions, experiments, and technologies that will enable their science to explain more phenomena. From their standpoint, they are very much like Mannheim's "free-floating intellectual," unattached to class interest, not bound by situation-based politics.[19] They are involved in explaining reality for its own sake. The genetic code and the molecular structure of biological processes leads to the possibility of factual accounts of how the body works. The accounts are incomplete and the descriptions are approximations, but eventually there will be a better theoretical system that will enable even better approximations of the objects of scientific study.

As the molecular endocrinologist says about the production of truth in technoscience:

I think that in my field of endocrinology we're working out some of the basic ground rules of how hormones can act. Different hormones work in different ways. For some kinds of hormones there are kind of unified field theories where they all work in similar ways and for other types of hormones it is much more complicated. The only way we're going to find out what the possibilities are [is] by getting a lot of examples from different systems, my system and other people's systems. We are in the process of doing that, we are . . . collecting basic data now. The future is going to be that once we know what the characters are and what the possible ground rules are—that is, what kinds of roots can a cell use to have a hormone from the outside, end up making a gene be expressed either more or less, or maybe even turning it off or turning it on. Once we know what the possibilities are, the future is going to be to really work it out in some detail. It doesn't sound too interesting but it is. That is, to find out what all the steps are that go from the top, from the cell to the gene. What the actual molecules are that are doing things. How they interact with each other. And then also how a gene ultimately knows what to do. If it has one hormone telling it to go on and one hormone telling it to go off, or another hormone telling it to go on. How this all gets integrated into the network. So I think the field's future is going to be to continue to work out the possibilities, what can happen in a cell. And the future is going to be to work it out in more detail, for individual pathways to find out how pathways converge with each other and how signals are integrated. . . . Another part of the future is going to be to work our way back, to try to understand how you cannot have an infinite regression. That is, we have a gene that is regulated by a regulatory protein, that's made by a gene, that's regulated by a regulatory protein. Well, that cannot go on forever because a cell doesn't have an infinite amount of genes. So the question is how is that delimited. And that I think is a question for the future.

From the standpoint of technoscientists, this is their task: seemingly endless, but delimited by the molecular structure itself of the production of knowledge. The task is made more difficult now by external factors such as the expense of research and of new technologies, a government that is not willing to fund science appropriately, and corporations willing to fund only profitable science. Technoscientists question the external factors but they do not question the validity of their work. For them, internally "good" science produces good knowledge. "Bad" science is the result of incompetence or of the scientist actively intervening in the process of knowledge production. The molecular endocrinologist told us that it is important to teach his students the difference between good and bad science. We asked him to define the difference, and he responded:

> Well, you know, when you're a scientist you're a reductionist and you believe that there is a truth out there to be discovered. . . . And so a straight definition would be that good science is science that gets you the right answers, that finds out what's really going on. Bad science, I would say, is science that is—well, you could say that there are many more ways to do bad science than good science. Bad science would be where techniques are sloppy, where interpretation of data is sloppy, or where somebody does experiments and tries to clean them up much more than they imply. So in a sense bad science is people experimenting or conceptually going off in the wrong direction unnecessarily because of not being very good at what they do. I differentiate that from people who have theories that turn out to be wrong, but not because there was anything wrong with them as scientist but because that was what was known at the time and that's how science goes. Their theories were just wrong.

The molecular endocrinologist identifies three reasons for bad science: insufficient knowledge leading to wrong conclusions, incompetent scientists, and scientists whose interest distorts research (i.e., "clean them up much more than they imply"). The third reason is most interesting because it posits that the "subjectivity" of the scientist is a key element in bad science. Subjectivity is defined here as any factor through which interest is expressed. History, cultural meanings, social relations, power, and personal interests all make the objectivity of the science suspect. All of these would be included in the molecular endocrinologist's example of bad science. Though he would see "common sense" as outside the world of science, it is common sense that politics and science do not mix. Scientists and politicians hold to the idea of the objectivity of science, even though political decisions of government are important to the development of policy relating to and funding of scientific research. Thus a cabinet member in the Reagan administration in 1988 banned fetal-tissue research and later reversed his position; Dr. Otis R. Owen said, "For the past few years I have been out of public life. However, I feel compelled to come back into the political sphere to talk about a world where politics should have no place: the world of scientific research."[20] The fetal research ban was originally tied to the claims of the antiabortion movement, which President Reagan supported. This is one of the most important political struggles in current U.S. politics. Now President Clinton agrees that the ban should be lifted. Is it just a coincidence that when moderate Republicans were increasingly critical of President Bush's antiabortion policy, Owen changed his position? Is this just science or is it science distorted by politics?

Is it any different from Vice President Dan Quayle's announcement that genetically altered foods would no longer be tested any differently than ordinary foods?[21] Is this the result of good science, with the only interest being to save consumers money and create jobs in biotech companies? Or is it the result of bad science influenced by a burgeoning biotechnology industry? How should we read the following statement by the Industrial Biotechnology Association?: "Biotechnology was a $4 billion industry in 1991 and is expected to grow to at least $50 billion by the year 2000."[22] Or is all science bad science because it is structured by social relations and cultural posits and is inseparable from relations of power? Is science bad whether it is used to legitimate genetically altered tomatoes for lucrative profits for biotechnology, or for the self-referentiality of science practice and the inevitable use of cultural posits to make underdetermined scientific phenomena appear as if it is determined knowledge? Is bad science "ideology" and good science "science"? These are crucial questions in a world in which science is at the center of political disputes (as in the abortion question), at the center of global economic interests (as in terms of the regulation of biotechnology), and at the center of the survival of the planet (as in global warming). Stanley Aronowitz writes:

The concept of the science/ideology antinomy is itself ideological because it fails to comprehend that all knowledge is a form of social relations and is discursively constituted. Within late capitalism and state socialism, these relations are organized according to a division of labor (principally the division between intellectual and manual labor). The rational-purposive basis of social production under both capitalism and state socialism means that science is a labor process as well as an ideology whose truth claims are entwined with the interest of domination.[23]

Here we are at the conjunture of science, ideology, and domination that we call science/power. Knowledge/power has been restructured. It is at this conjuncture that economic and political relations as well as the careers of scientists are increasingly structured by technoscience. It is in terms of science/power that professionalized technoscience posits intellectuals as technical intellectuals. This becomes the hegemonic model, that is, "the structure in dominance" for all disciplines of knowledge, sciences as well as nonsciences—even in this time of fluid disciplines, as cell physiology collapses to be reconstituted as biophysics. C. P. Snow's notion of the "two cultures" no longer holds in this postindustrial, capitalist, postmodern world of technoscience. Technoscience structures universities and public and private research institutes. It structures the professional careers of all knowledge workers, who must provide technical knowledge, that is, knowledge that is considered reliable within the historic constraints of a discipline and in terms of a knowledge community. This knowledge can be either theoretical or applied but for both it is constituted in terms of the rational-purposive requirements of producing a formal knowledge for institutional dispensing. Theoretical knowledge is not transformational, creative, or critical but instrumental, and its production maintains both technoscience and the professional technoscientists in their institutional positions. Increasingly, professional technoscientists, especially those at high-status universities and research institutes, provide their professional, technical, and intellectual knowledge for the maintenance of political, economic, and cultural institutions. These technical intellectuals, like Andre Gorz's "technical intelligentsia," produce purposively rational knowledge for corporations and the state to whom they are subordinate.[24] They do not have a transformative worldview as do public intellectuals.

Public Intellectuals and Feminist Science

The social movements of the 1960s opened up the Pandora's boxes of "the personal is the political," the issues linked to the political economics of statecraft such as the Vietnam War and black power, and the cultural issues linked to everyday life: women's rights, abortion, gay and lesbian rights, crime in the streets, the silent majority, the breakup of the traditional family, drugs, and so forth. The university became the center of the attempt to re-create a bourgeois public sphere, separate from the interests of the state and corporations. The model was the town meeting (petit bourgeois—all were equal individual proprietors). The university was the space in which the discourses of the day were open to equal participation. Theoretically, even women and blacks could participate. Participation was based on a mix of formal knowledge, experience, and common sense. Educational opportunity guaranteed access to the required formal knowledge (grounded in a community of scholars with institutional credentials). The hope was to link expertise not to domination (interests of state and capital) but to furthering the possibility of democratic participation. Thus, instead of expertise as a form of authority that legitimated hierarchy and shut off discussion (because expertise based on technoscience and academic credentials produced technical solutions to social and political problems), decisions would be based on the democratic practices of the discursive community. Against the technical expertise of scientists and scholars, many in the sixties felt that their own lived experience was sufficient.

The contradictions seemed insurmountable. First, the university demands that all of its students are at least middle class to begin with; if one does not satisfy this requirement, the second requirement is that if you are not middle class you must become middle class. Hierarchy is built into the university system at every step. The theoretical biophysicist's hierarchies are lying in wait for unsuspecting students. The new left, the counterculture, and the followers of Martin Luther King, Jr., all met the first requirement. The university is not the site of a democratic public sphere. It is not for everyone. It certainly is not the site of a working-class or lower-class public sphere. With the decrease of union influence both in terms of membership and in the transformation to a postindustrial capitalism in which nonunion spheres have come to the forefront, a working-class public sphere is unlikely.[25]

Second, the university is the bastion of technical expertise in which the rubric of science dominates and limits democratic discourse. Only certified knowledge is valued and the knowledge produced by technoscience is most privileged. As the student movement crumbled, the university as a potential new public sphere was reduced to a segmented aggregation of positions in which the conservative function of professional disciplines reasserted its power. Members of a professional disciplines decreasingly concerned themselves with questions of addressing a mass audience, the creation of a transformational discourse, of a new democratic discourse, or of the good life unless the answer was just more of the same, progress through science and capitalism. The university continues to be the purveyor of expertise, talent, meritocracy, and mastery. With funding cuts at federal and local levels, women, African-Americans, Latinos, and white working-class males decreasingly meet the requirements.

It is in this context that Russell Jacoby bemoans the passing of the political intellectual who was active in the public arena.[26] For Jacoby the academization of public intellectuals has led to their demise. We think Jacoby misses the point. He does not understand academization in terms of the hegemony of technoscience as the model for intellectual life. Without research, facts, and knowledge produced scientifically you cannot participate in the discourse. The technical intellectuals produce the scientific studies and provide the technoscience solutions for the problems of the world. Cultural intellectuals mimic this model, and their publications are taken seriously only if they are in the high-status, refereed journals and have the appropriate number of footnotes.

Many leftists are not discouraged. They believe that if the economy can be reindustrialized and if the state can rebuild the infrastruture, a new working-class base can be built. But computer-based technologies and automated factories mean that even if reindustrialization and a new federal work program occurred, a massive new working class would not be created. Further, science-based production has reorganized the economy toward a service- and information-based postindustrial capitalism. We seem closer to the "totally administered society" of the Frankfurt school than to a new working class in radical chains.

Thus the social theory of Jürgen Habermas has carried the day. Work as a sphere of production as well as the production of scientific knowledge is unquestioned as spheres of domination. The alienation of labor is normalized. The only possible space in which resistance is possible is the sphere of communicative action. It is in this context that Habermas offers the public sphere as the space where democracy is possible.[27] His problem is that the public sphere has been deformed by private interests. Capitalism's publicity has diminished the possibility for a discourse of universality. Yet possibility still exists; the public sphere is the space where discursive questions can be contested. For him democracy is possible at this location. The economy is the realm of instrumental rationality of technical expertise. For Habermas communicative action is not possible in the economic realm. He rejects it as a contested terrain. In doing this he ignores important changes. He ignores the importance of the regime of science/power for the new production processes of postindustrial capitalism. He fails to understand it as a direct threat to the bourgeois public sphere. He fails to understand that this new regime of science/power penetrates all aspects of everyday life and is incorporated in the welfare state's and corporate capitalism's penetration of everyday life as well. At its most basic, in terms of Habermas's analysis, this means that the spheres of instrumental action cannot be separated from communicative action. Though Habermas is aware of the crisis of the public sphere, he is unaware of the crisis of authority and of expertise.

At the simplest level our criticism of Habermas is that he has formulated a bourgeois public sphere outside of the experience of most human beings. Pure communicative action, like pure technoscience, excludes gender, race, and class experience. He has created a "social" that has nothing to do with the "personal is political" or the concerns of ordinary people. But, most importantly, at another level he has no understanding of the crucial space that technoscience occupies. In its penetration of all forms of discursive practices it has distorted any possibility of communicative competence without interrogating the truth claims of technoscience. Technoscience excludes Habermas's discourse from its truth claims, which are independent of social relations, cultural meanings, and thus communicative action. For technoscience Habermas is not possible because the realm of communicative action is ideology as far as technoscience is concerned.

Are technoscience and its technical solutions all that is possible? Is democracy relegated to the expertise of the technoscientist? Or are there possible cracks in the bloc? We see at least two cracks: Hegel's critique of science as claiming to be universal when it is not, and feminist science theory.

Here Hegel's critique of science in his "Preface: On Scientific Cognition," in the *Phenomenology of Spirit* advances our project. For Hegel, science exists in incompleteness; it asserts universality but is always partial knowledge. Genuine science exists as knowledge in which the active subject in the historical process of knowing constitutes the object in its becoming. Subject and object are identical. For experimental science, subjectivity is a distortion of truth. Truth exists only in its objectivity. For Hegel, this is not incorrect, but one-sided, only part of the process of knowing, only at the beginning of creating a genuine science. Science gets at only the formal categories of knowledge; this is lifeless and external. It does not get at the living contradictions that constitute the unity of the total process of knowing. This incomplete science sees only the acorns, never the whole oak:

> When we wish to see an oak with its massive trunk and spreading branches and foliage, we are not content to be shown an acorn instead. So too, Science, the crown of a world of Spirit, is not complete in its beginnings. The onset of the new spirit is the product of widespread upheaval in various forms of culture, the prize at the end of a complicated, tortuous path and of just as variegated and strenuous an effort. It is the whole which, having traversed its content in time and space, has returned into itself, and is the reluctant *simple Notion* of the whole. But the actuality of this simple whole consists in those various shapes and forms which have become moments, and

which will now develop and take shape afresh, this time in their new element, in their newly acquired meaning.[28]

For Hegel science only investigates the moments independent of the totality. Its reliance on formal mathematical representations lie outside the living process of knowing. Mathematical formalism constructs a world of true and false, good and bad, of structure independent of content and function. For Hegel these are not antinomies, not just epistemological categories, but part of the process of knowing in its becoming. Subject and object are not antinomous, as they appear to be in the natural sciences; the foundation of a true science is subject/object unity.

Science as it is now constructed is incomplete and thus must delimit variables because the whole, the subject/object identity, is beyond it. This can be seen in Hegel's criticism of Kant:

Of course, the triadic form must not be regarded as scientific when it is reduced to a lifeless schema, a mere shadow, and when scientific organization is degraded into a table of terms. Kant rediscovered this triadic form by instinct, but in his work it was still lifeless and uncomprehended; since then it has, however, been raised to its absolute significance, and with it the true form in its true content has been presented, so that the Notion of Science has emerged. This formalism of which we have already spoken generally and whose style we wish here to describe in more detail, imagines that it has comprehended and expressed the nature and life of a form when it has endowed it with some determination of the schema as a predicate. The predicate may be subjectivity or objectivity, or, say, magnetism, electricity, etc., contraction or expansion, east or west, and the like. Such predicates can be multiplied to infinity since in this way each determination or form can again be used as a form or a moment in the case of an other and each can gratefully perform the same service for an other. In this sort of circle of reciprocity one never learns what the thing itself is, nor what the one or the other is. In such a procedure, sometimes determinations of sense are picked up from everyday intuition, and they are supposed, of course to *mean* something different from what they say; sometimes what is in itself meaningful, e.g. pure determinations of thought like Subject, Object, Substance, Cause, Universal, etc.—these are used just as thoughtlessly and as uncritically as we use them in everyday life, or as we use ideas like strength and weakness, expansion and contraction; the metaphysics is in the former case as unscientific as our sensuous representations in the latter.[29]

Hegel's critique of the incompleteness of science seems to have had a rebirth, and this renewed sensibility seems to have currently three diverse sources. The first is Quine's notion of the "underdetermination" of experimental data for the theoretical inferences of scientists.[30] Scientific theories are thus based on mathematical representations that do not logically correspond to the data. Second is the feminist critique of science, which in context the process of science knowledge production seems to be approximating Hegel's position of the requirement of the subject/object unity for a genuine science. Third is the position of the early quantum physicists. As Aronowitz argues:

> It would be excessive to claim the development of quantum mechanics, especially the discovery that knowledge of the physical object entails bringing the observer into the observational field, represented a direct acknowledgement of the power of Hegel's attack. Yet although physics has recuperated this admission within realist epistemology, some of the more philosophically minded theoretical physicists still have nagging doubts that the "correction" of the principle of indeterminacy is insufficient, that physics and truth are nonidentical.[31]

Another possibility for moving beyond the hegemony of technoscience comes out of the struggles of feminist science, which like all successful social movements has inspired new theory.[32] Feminist science critique is crucial because it has interrogated the science/power regime and created the possibility of new public intellectuals in terms of the realities of technoscience. Feminists have sown the seeds for a new discursive community by which technical intellectuals can be transformed into public intellectuals in and through participation in new social movements. It is through these struggles that the crucial issues of science, of the relation of science to economic and ecological survival, can be brought into a newly created public sphere and be democratically debated, new agents can be constituted, and a world can be transformed.

We are talking with the cell biologist in her lab. She sounds like Mary Hesse in that the metaphors and models of science structure the work and the ways of theorizing and in that they are overwhelmingly masculine.[33] She says, "Science is a masculine profession. These are nice guys here, and I know that they don't mean to be oppressive, but this is a masculine profession." She talks while she's looking at the radioactively developed film from the autoradiograph of her proteins on gel. She is always rushing around—doing experiments, doing bureaucratic work, writing grants and articles, teaching, managing her lab. She continues:

> It's not easy being here. Science is very macho, internally and outside. There's always competition. . . . I don't think women and men are different. . . . It's

166

not descriptive and exploratory, it's more manipulative. And the men are better at it. . . . It's not "let's go out in the field and look for flowers."

If things don't work out women look at themselves, what's wrong with me. And there's so few of us and so little support. Even that makes a difference. Men say it didn't work out, let's do it again. For them it's more of a hunter or a soldier—let's go out and conquer—than a farmer. I see myself as a farmer but I have to be a hunter.

In a similar vein, Sharon Traweek presents the accounts of high-energy physicists who describe their socialization into the field and how they came to be successful professional physicists. For Traweek these are "male tales" that include women only as representations of their "romance of science," as the object of domination and scientific investigation. Traweek tells us about the male tales of these value-neutral scientists:

The image of real female human beings held by almost all these male scientists is that women are more passive, less aggressive than men. This socially constructed gender difference is used by many scientists to define the relation between themselves and their love object. The scientist is persistent, dominant, and aggressive, ultimately penetrating the corpus of secrets mysteriously concealed by a passive, albeit elusive nature. The female exists in these stories only as an object for a man to love, unveil, and know.[34]

The cell biologist continues:

Microbiology with the quickness of the changes with which you can do things. You're only limited by your imagination. It's great. . . . It's not the men it's the culture that is masculine. Wielding authority is quite a negotiation for me. I'm getting better. My students are the equivalent of adolescents, even though they're adults. For them to become successful scientists they have to be molded. I mold them, even though it is hard for me. I have the desire to be liked . . . it's more balanced . . . they can like me but sometimes . . . if you're a woman and get angry and scream and yell you're a bitch. The men can yell. I don't even yell when its appropriate. . . .

There's only one other woman doing parallel research. There are women faculty and the new chairman is hiring women in equal numbers. In other departments this is not the case. We get together informally; we seek each other out for solidarity.

She is laying out what Bettina Aptheker terms the "dailiness" of women's lives. In a new science, where you are limited only by your imagination, there is great potential that women can contribute new and different meanings:

The point is to suggest a way of knowing from the meanings women give to their labors. The search for dailiness is a method of work that allows us to take the patterns women create and the meanings women invent and learn from them. If we map what we learn, connecting one meaning or invention to another, we begin to lay out a different way of seeing reality. This way of seeing is what I refer to as women's standpoint.[35]

Men are the dominant group in technoscience. The cell biologist is an alien in her own environment. Even though the men are helpful and supportive, it is still their world, their metaphors and models. She is a stranger and an outsider in the world of science. The social order of the lab reinforces men's dominance and diminishes women. Women scientists feel that they are forced to play men's games and live by men's rules. This is how the lab is structured by masculinity and women feel constrained by this structure. They feel oppressed, and the male scientists just feel that they are doing science.

Sandra Harding offers an answer to this situation, which she sees as emblematic of the general oppression of all outsider groups, not just women. But in terms of women, their oppression puts them in a situation of completing what for Hegel was the partiality of science. From a feminist standpoint, a feminist science is not just about "fairness" but also about the real theoretical interventions that they can make to transform science. Harding writes:

> Knowledge emerges for the oppressed through the struggles they wage against their oppressors. It is because women have struggled against male supremacy that research starting from their lives can be made to yield up clearer and more nearly complete visions of social reality than are available only from the perspective of men's sides of these struggles.[36]

The woman scientist sees the picture that men miss. Women scientists always learn and assume the male scientist's position. This is not required of male scientists. Women must assume both insider and outsider perspectives. Men are always insiders. Thus a feminist science offers a more complete "world picture." Harding writes:

> The cumulative result is that the social order generates conflicting demands on and expectations for women in each and every class. Looking at nature and social relations from the perspective of these conflicts in the sex/gender system—in our lives and in other women's lives—has enabled feminist researchers to provide empirically and theoretically better accounts than can be generated from the perspective of the dominant ideology, which cannot

see these conflicts and contradictions as clues to the possibility of better explanations of nature and social life.[37]

The cell biologist told us that science should be overfunded. This reflects both her belief in the needs of science in a world in which it is increasingly crucial and increasingly expensive and the possibility of opening it up to diverse groups located differently in society. This is not a solution but a direction. Overfunding only potentially affords more access to oppressed groups. It does not necessarily increase the reflexivity of scientists. Nor does it necessarily transform the educational system or the media so that science becomes a normal part of public discussions. For a public that understands that it must partake in the discourse of science because technoscience is central to the workings of the world, a social movement is required. Scientists must increasingly become public intellectuals because science/power enables them to be more than servants of corporations and the state. As scientists and citizens become more active in the discourse of science, they also become agents in its construction. This creates a science in which women—who in their solidarity, within social movements, contribute to the development of science—become the model for all groups, for a science that is not separated from the historicity of its agents. A more situationalized science creates the possibility of an increasingly diverse science: a science in which the metaphors and models are not just masculine and Eurocentric, a science that contributes to democratic participation and does not shut off public discourse.

Part II

Contours of a New World

Chapter 6

Contradictions of the Knowledge Class: Power, Proletarianization, and Intellectuals

The terms *intellectual* and *intelligentsia* are foreign to the Anglo-American ear. We prefer the designation "professional" to describe those who possess credentials that entitle them to perform types of work that entail the use of legitimate—that is, academically derived—knowledge. The divergence between the Anglo-American usage and that of nearly all of the rest of the world is not merely a descriptive difference; it is theoretical and political. The problem of intellectuals is, in the first place, discovering their class position. In the history of the literature on the intellectual "question," much of the debate has centered on whether intellectuals are ineluctably linked to the core classes of any social formation. For example, Gramsci argues that every class that seeks political power must achieve ideological hegemony; it must determine the common sense of civil society. In this reprise, securing intellectual and moral leadership of society on behalf of a given class is, in part, the task of the intellectuals.

Or have they become, especially in advanced industrial societies, something more than functionaries of the leading classes of the prevailing social order, as Alvin Gouldner asserts? According to Gouldner, following a long tradition of neo-Marxist thinking, since knowledge has become the crucial force of production and administration, as the bearers of productive knowledge intellectuals have developed their own "class" identity and have separated themselves from the "old" classes of industrial society, the working class and the "moneyed" class. On the other hand, there is also a subset of people who, since knowledge equals power in some cases, have achieved power on the basis of knowledge—even some financial markets types. So one must also ask in what sense money and financial instruments are also intellectual. Of course, Gouldner does not address this issue, perhaps because, as financial markets became truly global, their association with informatics had not fully matured in 1979, when his last work on intellectuals was published. Nor does he include the policy staff intellectuals of the state and the largest corporations who, as professional servants, have been only partially assimilated into the culture of critical intellectuals. In this thesis, the intellectuals comprise many, but not all, academics, scientists, and top policy intellectuals who work for the government and the universities.

Although in their class interest intellectuals so far remain subordinated to the moneyed class, they are more capable than other subordinated classes—as a result of their unique position in the technologically advanced labor process—to mount resistance and thereby to assert their own demands. In Gouldner's view, the growing strength of the knowledge class makes their self-subordination a *strategic* consideration. However, they may seek independent political and social power *in their own name and the name of the entire society* if and when capital refuses to negotiate to meet their interests.

It is not only their control over productive forces that marks intellectuals as an emerging class, but also the fact that they have developed an independent culture of critical discourse. Gouldner writes:

> The culture of critical discourse is characterized by speech that is *relatively more situation-free*, more context or field "independent." This speech culture thus values expressly legislated meanings and devalues tact, context-limited meanings. Its ideal is: "one word, one meaning for everyone and forever." ... The culture of critical discourse is the common ideology shared by the New Class, although technical intelligentsia sometimes keep it in latency.[1]

Although Gouldner writes, in part, from within American sociological traditions, by the late 1960s he was also working in a distinctly European neo-Marxist perspective in which—in addition to economic position—language, culture, and discourse constitute class. Moreover, he had abandoned the standpoint of methodological individualism as well as stratification from which the category of "professional" derives. In sum, in contrast to the descriptive, scientistic framework of the sociology of stratification, Gouldner's investigations of class relationships are informed by political questions, specifically, What were the structural and historical presuppositions of independent social and political action by intellectuals?

The approach here toward the question of the social and political status of intellectuals connotes the emergence of critical discourse as a historically constituted *form of life* that mediates economic changes and political events. We take the notion of "form of life" from Wittgenstein, who designates *language* as the form of life. Here we will specify the discursive/social elements of the middle-class ideal, but metaphors, narratives, vernacular speech, and other speech acts are indispensable for understanding how class relationships are constituted. We use the term *discourse* to signify the ways in which we narrate and, through reflection, give meaning to our everyday relations as well as to public life. Cultural and political discourse shapes our world through the power of socially situated language to signify as well as to form experience. So it is not only that we *interpret* experience in terms of the categories of discourse, but also that discursive practices constitute what we call "experience." This perspective challenges the pristine status of "seeing and feeling" and insists that what we call "experience" is presupposed by ideas.

It may seem that introducing language into discussions of social relations runs the risk of reducing power to its discursive dimensions. Indeed, there are some who, following a certain interpretation of Foucault's conflation of power/knowledge, have jettisoned the capital/power nexus and have elided, if not renounced outright, class as a knowledge object.[2] While they proclaim the "impossibility" of the social because it is merely a construct of discourse, Laclau nevertheless retains the idea that there are agents of change.[3] We occupy positions that embody, for example, class, gender, and race discourses. These standpoints interpellate us and have consequences for what we actually do, including speech acts.

Science, for example, is a specific form of discourse that obeys the rules of any other narrative by which we describe the world, but has its specific vocabularies and procedures.[4] These have become dominant in nearly all spheres of the contemporary world. Science has become a constituent of the American dream (our imaginary life) insofar as we are collectively interpellated by narratives associated with the scientifically based technological fix and have become pluggies—that is, people who are plugged in to electronic devices such as fax machines, computers, voice mail, and so forth—at play as well as at work. Technologies mediate what we reflexively take as "lived" experience because we are surrounded by these devices and filter our knowledge through them, as does science. As many students of the social relations of science have argued, the scientist's experience of objects is mediated by the technologies of observation, the theories that she or he brings to that laboratory situation, and the opinions of colleagues whose views condition what can be seen.[5] This is no different from how everyday experience is constructed.

Since in the context of this book we cannot devote sufficient space to a fuller discussion of political and cultural intellectuals, here we will only sketch some of the issues concerning their growing social power. The social category of political and cultural intellectuals embraces not only scientists who produce narratives about nature and the producers of conceptions about human nature and the social world—physicists, chemists, and biologists—but also psychologists, social scientists, and artists of all kinds. These are the producers of knowledge that links directly as well as indirectly with power.

If power is constituted through knowledge, the conventional separation between them begins to become obsolete. Foucault disputes the claim that this fusion is a historically specific phenomenon. That is, he challenges the view that the knowledge/power axis was produced under definite economic and political conditions.[6] This view is consistent with thinking about language not as a medium but as a form of life, and it has its parallels in debates within modern physics about whether space and time can be treated separately from matter. Einstein demonstrated that space is not a void through which objects move. Nor is duration itself devoid of effects. Rather, the concept of space/time continuum as a crucial dimension of physical reality means that there are no absolutes; all measurements are relative to the spatiotemporal framework within which they occur. This analogy to analysis of social relations is surely imperfect. The social equivalent of relationality rejects base-superstructure models that relegate culture to a derivation of "underlying" structures such as the economy. Surely the knowledge/power configuration is historically specific: its forms change under different conditions. But if language is a crucial life form, if not *the* life form, it is not possible, in any historical conjunction, to separate knowledge from its vocabularies and these from its exclusions and inclusions.

Not all knowledges are accorded equal value. Practical knowledge, such as that of the craftsperson, the homemaker, and the teacher, is systematically devalued by scientific culture. And competitors to established science—astrology, holistic medicine, acupuncture, herbal medicine, chiropractic, and parapsychology, to name a few—live underground or marginal existences. Even when they gather large followings they are not considered legitimate by those who hold power over resources. Some forms of knowledge are drawn into the labor process, others are integrated within the discourses of everyday life, and still others have become part of our marginal life, the realm of the private where hopes and dreams as well as fears flourish.

The emergence of political and cultural intellectuals in the construction of new power arrangements in Eastern Europe highlights the differences between them and the technical intellectuals who are ordinarily ensconced in the prevailing technological and state systems of state socialist and Western capitalist societies. As in South Africa and the economically dependent societies of Latin America, Africa, and Asia, at the moment of the political transformation the vocabularies of political and cultural intellectuals are temporarily identical with power. They not only play a decisive part in social movements but also, after transformations in political power, frequently become key political leaders, organize and run the state bureaucracy, and, as often as not, become the surrogates for the absent entrepreneurial class as well.[7]

The rise of cultural intellectuals to power in Eastern Europe illustrates not only the moral delegitimation of professional politicos and bureaucrats of the old regimes, but also signifies the utopian dimension of popular hope. The presidents of Hungary and Czechoslovakia, immediately after the collapse of communist rule, were writers and poets. Poets and writers in power may not solve the immediate economic and social problems produced by the devastating effects of the old regime's demise, but, as in the case of Czechoslovakia, where there was some repression both of the old communist apparatus and of its left opponents who refused to embrace free-market ideology, neither can they be relied on to establish a new, more open civil society for addressing the crisis. Moreover, because a small fraction of cultural intellectuals were at the forefront of opposition to communist rule, they were among the few social categories with nearly unimpeachable moral authority and for this reason ascended to the pinnacle of political power quite rapidly.[8]

Whether or not intellectuals in power succeed in generating the conditions for a democratic polity, they are often the only source of legitimate authority capable of bringing the newly formed power coalition together. What counts is that the fusion of the cultural intellectuals—that is, the producers of all sorts of intellectual knowledge, especially scientists and artists—with political power represents, in the popular imagination, the best chance to be relieved of the oppression visited upon them for more than forty years by the political intellectuals of the communist parties. Apart from Poland, where the workers' movement assumed political power for a short time, only to be eclipsed by the Catholic Church when the uses of class rhetoric no longer retained tactical efficacy, the rest of Eastern Europe characteristically turned to formerly dissident cultural intellectuals to lead them out of the morass. But the new regimes have generally failed to resolve or even to ameliorate the economic crisis, and elements of the old apparatus have begun, slowly, to regain their former positions of power. Similarly, in the history of the postwar liberation movements in colonial and postcolonial countries, traditional intellectuals—lawyers (Mandela and Castro), ministers (King and Tutu), and poets (Cesare, Nuyere)—are selected both to inspire trust and to provide the ideological glue without which popular struggles cannot succeed.

J. P. Nettl has argued that intellectuals may not be defined solely by their *role* or function in relation to economic, political, or ideological spheres, but, perhaps more saliently, are defined by the type of ideas they convey.[9] These ideas must be made public, narrated in the social spaces within which they contend for power over the imagination as much as they contend over office and policy. Intellectuals offer theories of nature and human nature, ethical precepts upon which visions of change or stasis are based, and from these theories and precepts, policies are adduced. At the same time, other intellectuals offer specific scientific, managerial, and technical "skills" whose broad applications require the intervention of the producers of philosophic, ethical, and aesthetic conceptions.

Nettl implicitly adopts the perspective of the cultural intelligentsia for which interests are never its own but instead are always fused with the discourse of universality. Cultural intellectuals disclaim the aspiration for institutional power beyond the revolutionary moment; they almost invariably claim to want to return to their writing, painting, or music and assume political roles only to alleviate a crisis.

Intellectual or Professional?

In other countries—both in Europe and in the less industrially developed world—the category "intellectual" embraces professionals, but also virtually any occupational category that is neither peasant nor manual trades. In most countries, the intellectuals are a social category signifying those who perform conventionally defined "brain" work—whether it is technical, professional, or scientific—even if, as in the case of engineers, physicians, dentists, and experimental scientists, their routine tasks involve considerable manual labor. Contrariwise, any close observer of even the least qualified manual labor must recognize the significant role of thinking in its performance. What Sylvia Scribner has called "everyday cognition," much of which is tacit, is an integral aspect of manual labor.[10] Garbage collectors, for example, must know how to lift a full can, implicitly taking account of its weight, and economize on their movements in order to conserve body energy. This tacit knowledge is essential for both safety and efficiency. Similarly, in Scribner's own example, "manual" workers in a milk processing plant possess tacit knowledge that is essential to their work. Thus, what has been called "mental" work is not a thing of nature, the result of innate capacities that inhere either in occupations or in individuals, but a *social* category, a distinction of the hierarchically organized division of labor characteristic of advanced industrial societies.

Whether intellectuals form a class or remain a dependent social category is not a question of definition but a historical issue whose referent is the specific context. If they are able to constitute a community that shares a common discursive space that leads them to identify with each other independently of their subordination to capital, the designation "intellectual" connotes a collective that is in the process of "class" formation. They can identify independent interests and demand to share economic and political power with the conventional contending classes.

In Eastern Europe before and after the collapse of the bureaucratic communist states, intellectuals initially formed the core of a "new" political class, which, in the history of these regimes, meant they held economic as well as political power, if not ownership. For example, in the absence of extensive private ownership of the decisive means of production, the political leadership in Russia immediately following the collapse was composed of members of various segments of the *technical intelligentsia* of factory managers, economic experts, and other bureaucrats in alliance with a fraction of the old, but refurbished, Communist party apparatus, brandishing slogans such as freedom and democracy as their ideological cover. But the content of the Russian state bureaucracy has barely changed from that of the old Soviet Union. Outside Moscow, St. Petersburg, Kiev, and other relatively cosmopolitan centers, regional political and industrial leadership remains in the hands of former communist managers and politicos.

In virtually all English-speaking countries, but especially in the United States, our collective aversion to the term *intellectual* may be rooted, on the one hand, in the American revolt against the conservatism of the European (Continental) intellect, which was widely understood as the bearer of dead traditions that were, psychologically, a barrier to the development of American science and technology. The technological temper is hostile to free—rather than instrumentally guided—speculation. Above all, the traditional intellectual is a ruminator. On the other hand, anti-intellectualism in American life is intrinsic to some strains of populism, which, despite its failure to constitute a permanent force in American politics, has remained a powerful cultural influence.[11] Despite the sharply divergent visions of the founding ideologists of the republic (some, like Hamilton and Madison, were implacable modernists), Americans like to think of theirs as a frontier, even "open," society. We have constructed our myths on images of the United States as a site of tolerance and experimentation and on masculinist iconographies of conquest—of nature, and of Native Americans and other "heathens" at home and abroad who signify the untamed part of ourselves. No less than any urbanized people, we yearn for the "wilderness," which, in our cultural lexicon, stands somewhere between aspiration and nostalgia.

For example, a leading thread of the American tradition of political and ethical philosophy is the idea that economic development equals progress and the agony that accompanies industrialization and urbanization, especially its implications for democracy. In the face of the eclipse of community that many believed was a by-product of the expansion of communications technologies such as the railroad and the telegraph, John Dewey, for instance, held fast to the ideal of face-to-face interaction as the building block of a democratic polity. Despite his fairly cold-eyed recognition of the permanence of the changes that industrialization visited upon the social as well as the geographic map of America, particularly the irreversibility of urban, industrial culture, Dewey remained skeptical about the ultimate benefits to be derived from "progress." But the critique of mass society, which Dewey shared with other intellectuals of his time, did not prevent his extolling the virtues of science and the scientific method. Contrariwise, his characteristic secularism was modulated by deeply held religious beliefs, which in his friend and mentor William James bordered on mysticism.[12] Dewey's thought embodied the contradictions of American culture, which, as Michael Weinstein has shown, is still bound to the binary trope of the wilderness and the city.[13]

Indeed, Americans rarely excelled in theoretical science, but prided themselves on being masters of invention, always linked in the public mind to entrepreneurship. A pantheon of heroes including Eli Whitney, Thomas Edison, and George Westinghouse embodied a mythology of technical prowess that contrasted sharply with the European propensity for metaphysical speculation and tradition-bound craft.

The history of technology in the United States does not bear out this account. Nascent industrialists of the mid-nineteenth century not only recruited skilled British and German labor at premium wages, but also relied upon their technical and craft knowledge, which helped propel the United States to world economic dominance. And U.S. corporations have never ceased to import science and technology from Europe in the form of patents as well as people. The superiority of "American" science in the twentieth century was built on the work of Europeans, many of them refugees from anti-Semitism: Charles Steinmetz, Albert Einstein, Enrico Fermi, Leo Szilard, Salvatore Luria, and many others.

Of course, figures such as Willard Gibbs, Robert Millikan, J. Robert Oppenheimer, Richard Feynman, and a few others represent a weak but vital native tradition of theoretical physics, and Linus Pauling and James Watson were internationally known respectively in chemistry and biology. Moreover, Dewey and James, and also the great founder of American pragmatism,

Charles Sanders Peirce, were themselves world-recognized philosophers of science. Peirce's essay "The Fixation of Belief" is a remarkable precursor of Thomas Kuhn's celebrated theory of the disrupted history of scientific revolutions and influenced American sociology of scientific knowledge as well. Peirce argues for the proposition that the truth of any scientific statement is determined by whether it secures the assent of the scientific *community*.[14] In effect, he claims that the criterion of professional standing, not immutable, transcendent verisimilitude, underlies the fixation of scientific belief. And Dewey's reflection on quantum mechanics and the indeterminacy principle in *The Quest for Certainty* remains among the most acute philosophical commentaries on contemporary physics. Dewey's reading of Heisenberg's controversial discovery was that the "object is eventual" rather than a stationary entity that is "discovered" by scientific investigation.

Yet there is little doubt about the pervasiveness of American anti-intellectualism that is grounded in, among other sources, the previously mentioned periodic populism of American political culture. When farmers' and workers' movements railed against the Eastern establishment after the Civil War, they targeted multiple antagonists: the trusts whose canonical figures were J. P. Morgan, John D. Rockefeller, Andrew Carnegie, and the railroad magnates James J. Hill and Jay Gould, but for some also the anonymous "Jewish" bankers who many populists thought *really* ran the country and the intellectuals who dominated cultural life (many of whom were thought to be Jews, but until well into the century were, like Dewey, products of the small-town and agrarian heartlands). For most of the twentieth century neither the growth of the scientific establishment nor the expansion of cultural institutions signified a deepening of intellectual culture. American science was rapidly integrated into the industrial system, and even the most hallowed traditions of high culture became part of the burgeoning consumer society: the museum's audience was composed, in the main, of the growing professional and managerial strata. Far from being a transcendent experience, "art" in America has always functioned as the core of middlebrow entertainments, imbibed as much for its snob value as for its aesthetic qualities.

As we argue in chapter 8, American universities were, since the post-Civil War U.S. Congress set up a phalanx of land-grant colleges as knowledge factories, a model that spread to the Ivy Leagues by the 1930s and to the rest of the world in the past forty years. In today's universities and colleges, the humanities, the seat of traditional intellectual knowledge and non-policy-oriented social sciences survive as ornaments or, to be more precise, as legitimating discourses of an otherwise industrialized institution. At the University of California, Johns Hopkins, and Penn State, for example, the physics, chemistry, biomedical research, and computer science programs raise funds that exceed the instructional budgets of their respective campuses. And many first-tier universities have supported professional education at the expense of the arts and sciences, except those that are effective in getting grants.

Americans award little honor to their philosophers, social and cultural theorists, and critics and even less to all but a few of their artists and writers, most of whom cannot make a living at their craft. In addition to naming campus buildings, libraries, and other public facilities after wealthy benefactors, many postsecondary institutions do not hesitate to celebrate politicians (especially but not exclusively dead ones), sports figures, and entertainers (especially on urban campuses, people of color). In fact, only when an intellectual becomes a celebrity as a result of some flamboyant personal trait or, like Henry Kissinger, George Kennan, or George Schultz, becomes a player in the game of political power, do the media and other sources of popular status and honor take notice. The intellectual may be recognized as an "expert," a designation that signifies she or he is qualified to contribute to the ideology that television, print, and radio news is factual. In this instance, professional credentials are mechanisms of the media's as much as the individual's legitimacy. The United States has disdained, even when it has harbored, traditional intellectuals—those broadly familiar with the jewels of Western culture with no area of specialization. To be sure, a small coterie of traditional intellectuals was to be found among journalists as well as academics until the beginning of the twentieth century. In the wake of the industrialization of the mind, however, mass education prompted reforms that removed Latin, Greek, European literature, and philosophy—the foundations of traditional Western culture—from most college as well as secondary-school curricula. Universal specialization and its mirror, insular national culture, marked the disappearance of the traditional intellectual as a social category.

From the rationalized curriculum issued a series of anti-intellectual cultural practices: taunting academically accomplished students in elementary and secondary schools by calling them "dead brains"; branding artists "longhairs"(before the 1960s); and depicting universities as "ivory towers" and professors as perhaps lovable but impractical and "absentminded." Of course, contempt was often disguised by affection. We tolerated our scholars even as the culture reviled its intellectual critics. Veblen's contemporaries could barely hear his critique of the emerging consumer culture because his sexual practices and personal eccentricities seemed to dominate his public persona.

Similarly, Einstein was revered for his scientific accomplishments, but few knew what he had said. Everyone knew he was a genius and that he wore baggy sweaters and played the violin. But we were assured, at least until very recently, that relativity was an idea simply too obscure for the average intellect to grasp. Even the highly educated, including most scientists, were said to be incapable of understanding the full significance of relativity. Einstein might be the modern Newton, but after World War II, when Americans became aware that he was an opponent of nuclear weapons and a socialist, we condescendingly tolerated his views; in the cold war environment of the 1940s and early 1950s, the press maintained an embarrassed silence about them.

Sometimes the intellectual is something less than a figure of benign, if eccentric, character. For example, Orson Welles's portrayal of a Nazi in the guise of a young professor in a small college in his 1946 film *The Stranger* suggests what we have suspected all along: beneath the bucolic exterior of college campuses and the funky informality of the classroom and its faculties lurk evil men who are prepared to deceive their wives and betray their country. And then there are the dozens of movies about mad scientists abetted by power-hungry politicians and generals who threaten to destroy humanity. One would not wish to belittle the hilarity of the 1960s film *Dr. Strangelove* or its serious critique of nuclear policy, or H. G. Wells's stories *Things to Come* and *The Invisible Man*. Our fascination with the achievement of Dr. Frankenstein has not waned, despite its dystopian implications, especially with respect to the conventional idea that science and technology will liberate us from nature and, eventually, from ourselves.

Yet in these narratives, science and its concomitant, the "abstract" intellect, are treated harshly. We could offer more examples, but it is enough here to be reminded that film culture is suffused with an iconography that both invents and reinforces popular images—whether they are gentle, humorous, or horrific—in which intellectuals come under fire. It was only during the impressive antifascist migration of the 1930s and early 1940s that Americans were exposed to a different kind of intellect, but only in the large cities, especially on the two coasts. Artists, writers, filmmakers, and scholars and critics such as Hannah Arendt, Franz Neumann, Erich Fromm, Theodor Adorno, Herbert Marcuse, and dozens of others illuminated American academic and public life only to witness their own eclipse by the cold war.

On the other hand, the categories of political intellectual and cultural intellectual have grown in stature and importance but in a form that is sharply different from the earlier period when they were recruited from among scholars and other traditional intellectuals. "Politics" and "cultural criticism" have become specialized knowledges of statecraft and the increasingly commercialized public sphere of late capitalism. Much of this new prestige derives from the expansion of communications and information as integral aspects of economic and political power. Political intellectuals have become the technical intelligentsia of power. They are charged with broad responsibility for "public relations" in addition to their central role in policy formation and implementation. Of course, not only are policy intellectuals important in the economics and planning staffs of large corporations and government agencies at all levels, but "public policy" has itself become the core knowledge of an increasing number of relatively new academic disciplines such as business administration, and policy has become a subdiscipline in public administration, education, and social work schools. Radio and televison writers, actors, and producers and other professional and technical employees in the mass media are in a general sense political intellectuals to the degree that they produce and reproduce the ideological conditions for social rule.

The Formation of Technical Intellectuals

The vast expansion of postsecondary education in the past forty years paralleled the growth and significance of specialist intellectual knowledge in all spheres of economic, political, and cultural life. Engineers and scientists are the major specialists in production, but no leading corporation or public agency can avoid hiring teams of lawyers, economists, and political operatives to help guide policy. The line between technical and political intellectuals is blurred to the extent that specialized knowledge increasingly drives economic, social, and military policy and, perhaps equally importantly, determines major aspects of electoral strategy in most Western democracies.

The public opinion poll, along with the sociologists and psychologists who are its expert practitioners, not only determines marketing strategy for a particular product but perhaps equally powerfully has become a major weapon in political and state arsenals. Observe any U.S. election campaign. Contestants almost invariably employ polls, produced either by contract or, more and more, by a full-time staff, to determine their positions on many issues. Poll results may decide how the campaign will be conducted, which constituents and their issues will be targeted with publicity and the candidate's personal attention, and so on. Similarly, elected officials make policy decisions on the basis of polls. Officials are constantly "testing the waters" to determine whether they should support or oppose legislation; whether to pursue specific policies; and whether and how to run for public office. The public opinion poll is considered to be an accurate instrument for measuring public needs and wants.

There is a strong elective affinity between the European concept of technical intellectual and the traditional American sociological designation "professional." In contrast to the traditional intellectual who is trained as a *universal savant* and works as a journalist, an independent critic, or a professor but corresponds to Mannheim's notion of free-floating, that is, outside the class system, the technical intelligentsia possesses only specialized knowledge, usually but not exclusively accumulated through formal education and training. Further, the technical intellectual is part of the old middle class of independent entrepreneurs or is a salaried employee of the state or a large corporation.

The tendency in virtually all societies is to subordinate intellectuals to the requirements of the apparatus, to maintain their technical character. This category is not only identical to what is termed "professional"—if by this term we designate labor qualified by credentials, systems of licensure, and associations and legally sanctioned boards that establish and monitor performance and educational standards—it also comprises technically trained labor that, in American parlance, is called "subprofessional." For example, although teachers are qualified by a credential and are required to obtain a license, which in many states entails passing a test, they are still unable to claim *universally recognized* specialized knowledge. Similarly, the persistent efforts of registered nurses to elevate their occupation to full professional status have been challenged by the entrance into patient care of new occupations, such as that of assistant physician, who is given some of the physician's functions denied to nurses. And social workers have been unable to prevent technically trained counselors and those with bachelor's degrees from performing many of their tasks. As we discuss more fully in chapter 9, some writers identify the technical intelligentsia as a "new" middle class, situated in a contradictory class location between labor and capital. In this view, as salaried workers they share some of the characteristics of any group of wage labor. But, as the theorists of the new middle class argue, their position in the division between intellectual and manual labor that has become a crucial marker of class distinction places them, ineluctably, on the side of capital and thus they may be designated a "new" middle class, or, in the felicitous phrase of Sylvia Wynter, a "job bourgeoisie."[15]

Others, armed with evidence that the vaunted autonomy of professionals is eroded by managerial authority and bureaucratic organizations, have with equal force argued that—with the exception of those who as managers and specialist staffs and consultants serve as an arm of capital or the state—they resemble a new working class.[16] Still others have noted that the professions have fused with management to constitute a new class, the professional/managerial class, whose centrality to both material production and the corporate and state bureaucracies places them in a unique position: at least within the technologically and organizationally "advanced" societies, the professional and managerial class is poised to contend for political and social power, emerging as a social category that is independent of capital but never as a working class or even one of its allies.[17]

Although the ascription "new middle class" or its variant, the professional/managerial class, as the most comprehensive generalization for intellectual labor describes a considerable fraction of those who embody specialist knowledge, the claim that virtually *all* educated professionals are members of this class may describe a powerful *tendency* of contemporary societies, but is a serious overstatement. In our view, grouping the entire technical intelligentsia under these headings misses significant aspects of the historical dynamic and complexity of divisions within intellectual labor, particularly the rise of knowledge in its manifold dimensions—in production, culture, and politics. The economic/structural position of any social category underdetermines its political and cultural character. For even if a large portion of the technical intelligentsia may be properly viewed within the labor process as an arm of management's control over labor, there is at the same time considerable evidence that a majority of technical intellectuals may be experiencing relative *proletarianization* in highly bureaucratized workplaces. In these sites, management domination has extended from manual workers to both the professional and the subprofessional groups of the technical intelligentsia.[18]

Moreover, they constitute a large portion of the membership, activists, and leadership of the new social movements, especially the feminist, environmental, and consumer groups that have become politically potent in many European countries and the United States in the past three decades. These movements range from efforts to achieve greater equality of opportunity and environmental reform to demands for radical restructuring of the industrial capitalist order on ecological and health grounds. These radical demands are made by many whose function within the workplace corresponds to the task of eliminating or dequalifying labor. The significance of intellectuals—both in the labor process and those who work in cultural spheres, such as journalists, artists, and professors—in the formation of these movements may be interpreted in part as the result of an education that trains them for jobs that do not exist: occupations that offer autonomy and creative work. Having been frustrated in their quest for self-management, especially management of time, which also signifies control over their assignments, some intellectuals have turned to social and political interventions that are directed against ways in which their knowledge remains circumscribed by capital. The question raised by the tendency of important groups of intellectuals to become involved in social movements is whether they are emerging as a new, more or less independent, political and social voice in contemporary life.

However, it would be economic reductionism to ascribe intellectual categories merely to social movements toward proletarianization alone. But while the emergence of a significant number of intellectuals within some of the more vocal and politically influential movements is motivated by the erosion of their own status and of the general quality of life in the past quarter century, they are also moved by the power of alternative *ideas* of ecology, sexual freedom, and feminism. In this sense, we note the importance of "the culture of critical discourse" for the reemergence of intellectuals as independent actors, even when this independence is chiefly exercised outside the workplace.

Critical discourse, which, in the United States, has undergone only a partial development, has undoubtedly been abetted by the increasing recognition that there is little wage or salaried work that is worth doing for its own sake and that both the economic and cultural rewards of professional status have deteriorated in a period of global economic and environmental change. Yet such judgments could easily be made of earlier historical periods as well. What is new is that we have entered an epoch in which old paradigms of legitimate knowledge as well as economic and political relations are toppling.

While the constellation of new theories and new perspectives has by no means congealed into a distinct social and intellectual knowledge paradigm (indeed, the concept of "paradigm" may itself be in eclipse in the wake of the breakup of old ways of understanding the economic and social worlds), concepts such as those proposed by radical feminism and ecology are no longer regarded as completely impractical, but are taken seriously even in periods of political defeat for radicals and other innovators. For example, the prospect of global warming that might affect the survival of present and future generations of animal and plant life has produced a fairly widespread debate concerning what, and how much, production of goods is acceptable. Such discussion, while it is justified on health grounds, is profoundly at odds with the prevailing ideologies of untrameled growth and free enterprise. There is barely a major U.S. industry that has escaped scrutiny from the media, public agencies, and environmental movements with respect to the environmental impact of its products. What is remarkable about the political effects of ecological thinking, at least in the United States, Germany, and France, is that a segment of the political directorate is obliged to articulate simultaneously the utterly paradoxical discourses of ecology, free market, and growth. Hence the incoherence of contemporary political ideologies. Some, like Vice President Al Gore, claim that these discourses are not in conflict even as other members of the Clinton administration, notably the secretary of the treasury, extol the virtues of energy- driven—that is, oil-driven—economic growth policies.

Equally important, we are in the midst of a major debate concerning the effects of technologies that promise to alter the conditions of sexual reproduction: Who is the legitimate parent in an in vitro birth? More fundamentally, should government policy encourage relatively "safe" abortion procedures and especially contraceptives that make it possible for women to control their own reproductive functions without considerable health risks? These issues are being raised by intellectuals acting as feminist activists.

In this account, we see that, like workers who are also women and may identify themselves as parents or lesbians or artists, intellectuals may possess multiple identities. These identities undermine a simple determination of class by relations of production. For when technical intellectuals become ecological or lesbian or gay activists, for example, these choices bear on the relative privileges that accrue from their roles in the labor process. Lacking strong unions, they may jeopardize their jobs or other professional positions, and in more than a few instances their identity as social activists influences the kind of work they are now willing to perform.

Thus, "class" position is influenced by multiple determinations, especially the question of who really controls the knowledge-based productive apparatus. For even if the technical and scientific component of the knowledge "class" has been thrust to the center of the labor process, it remains relatively powerless, from this position, to control how the product is distributed, its effects on the physical and cultural environment, or its political uses. Within the boundaries of the profit imperative, capital may have been obliged to yield control over the management and technical composition of production, as Gouldner has argued, but its power over virtually every other aspect—distribution, marketing, and relations to state bureaucracy and public policy, which, taken together, have become an increasingly significant part of doing business—remains largely unchallenged from inside the corporation, where "democracy" has never truly flowered.

Technical intellectuals find themselves in a genuine bind. Their knowledge enlarges their *potential* power within the labor process, but, on penalty of demotion or even discharge, they are under constant pressure to renounce the relation between their knowledge and global political power. They risk being fired or forgo advancement within the corporation or public bureaucracy when they extend their reach beyond a narrowly circumscribed specialized sphere. Of course, many technical intellectuals have made the decision to become citizen-activists in the past twenty years in a wide variety of movements. Others, with whom we will be concerned in the next chapter, have responded to the limits imposed upon their autonomy by organizing unions and professional associations to advance their interests.

It is true that Mallet adopted the theoretical framework according to which the relation of a social category to ownership and control of the means of production determines, at least in the last instance, its political and ideological position. Nevertheless, the new-working-class thesis, advanced in the 1960s, had the advantage over its chief rivals of coming to grips with the undeniable proletarianization of some significant groups of the technical intelligentsia: engineers and computer programmers in industrial production, teachers, and the subordinate but well-paid salaried professionals in medicine, law, computer programming, and some branches of scientific and technological research. Mallet calls attention to the subsumption of these categories under industrial and state bureaucracies and argues correctly that bureaucratic management ceaselessly undermines the autonomy that knowledge bearers expect. Together with Andre Gorz, he argued, in the wake of the scientific and technological changes that began to transform the workplace in the early 1960s, that a fundamental contradiction developed between the emergence of qualified labor at the center of the labor process and its increasing subordination to arbitrary managerial authority.

In opposition to the conventional assumption that qualified labor is subordinated by capital or, in Veblen's judgment, bought off, Mallet documents instances in the 1960s of strikes over issues of *control* by technical intellectuals. Extrapolating from observation as well as theory, he concludes that they are likely to make new, qualitative demands that eventually extend beyond the workplace to issues such as those made today by consumer and environmental movements for corporate social responsibility and, within the workplace, for enlarged control *independent* of capital. At the same time, he assumes what thirty years ago was merely a tendency: qualified labor has become *generalized* in nearly all technologically developed industrial and service sectors of the economy.

Yet in France, much less in the rest of Europe, the countervailing social weight of the postwar emergence of consumer society scarcely detained new working-class theorists. Problems of consumption were transformed into fields of action by the new working class. Its cultural influence was persistently undervalued. The middle-class cultural ideal in its multiple aspects has, the contrary movements in economic relationships notwithstanding, maintained hegemony not only among the constituents of the technical intelligentsia but also more broadly in U.S. political culture. It has persisted throughout this prolonged period of the decline of the "old" middle class of small farmers, small manufacturing employers, and merchants. It has also survived the era of the decline of independent entrepreneurs and the growth of the phenomenon of professionals as salaried employees of small and large organizations. Some large law firms employ hundreds of attorneys who cannot expect to attain partnership, just as health maintenance organizations now hire a large corps of physicians who will never own their own practices. Deprived of a large measure of control over their own work, the professional and technical "salariat" has adopted the classical stance of many industrial workers: take the money and run (to the shopping mall).

The survival of this hardy ideology is in no small measure the result of some specific features of the economic changes that produced a *new* middle-class ideal of consumer society that has radically reduced the significance of work, especially the idea that the self may be realized through it. In a wider view we may account for the success of this development in part by referring to the fate of the workers' movements, including the trade unions in the twentieth century. In the absence of a specifically working-class political culture, which invariably entails distinctively independent concepts of the good life, American labor adopted the middle-class cultural ideal, even in the wake of the fierce struggles it was obliged to wage to gain recognition from employers and the law. Perhaps most powerfully, the cultural ideal alters the concept of property from its traditional identification with capital to one that privileges personal property, a change that corresponds to the transformation of the professional into a salaried employee.

Of course, the prevailing middle-class cultural ideal is inscribed in many practices that individuals and groups experience as beliefs and values. These values inform their implied and explicit social choices, those they call personal as well as those of a more public type such as voting and civic activity. Specifically, the displacement of work as the central moral and ethical precept for many intellectuals as well as manual workers (and for some of the same reasons) has had enormous consequences for the configuration of political and cultural relations. Consumer society has had more than one meaning: it signifies that buying and owning have partially replaced work as a crucial measure of personal identity, and the shift of emphasis from work to consumption and to the cultural, in its anthropological meaning, has widened the sphere of political conflict. For in contrast to the century after civil war when the achievement of material well-being was touted as the standard prize of hard work and personal discipline for workers and the independent middle class, the cultural ideal has undergone a crucial shift: in the past quarter century, the conditions of and assumptions about material well-being have been called into question.

Postscarcity for a considerable (now shrinking) portion of U.S. and European populations has led to movements that have challenged industrial practices such as the production of plastics, chemicals, and other toxic substances used in food and packaging. Ecologists are concerned with the debilitating effects of chemical additives on the human body, both from eating the food and from the poisoning of air and water. But beyond these specific issues, it has provided many with a much closer view of some of the underlying irrationalities of the economic and social systems. For just as environmentalists have questioned the efficacy of unbridled and largely undiscriminating production and consumption based upon the logic that equates economic growth with a better quality of life, so feminists have begun to challenge the exclusive ascription of child rearing to women, especially in light of what has been described as the "double shift" brought about by the phenomenal growth in the size of the female labor force.[19] At the same time, in a contradictory fashion, even when workers form militant trade unions, what has often been called the "American dream," the popular metaphor for the middle-class cultural ideal, has remained powerful.

Knowledge and Social Theory

The concept of knowledge and the social position of those whose principal work is to produce or disseminate it has become one of the major preoccupations of social theory in the latter half of the twentieth century. We have already referred to one of the reasons for the increasing interest in this domain: scientific and technological knowledge has become ubiquitous as the principal productive force in late industrial societies. Technology drives economic growth, that is, it has become, over the past thirty years, a leading source of capital accumulation. And theoretical science is the basis of technological research and development. As we saw in chapters 1 and 2, in the late nineteenth century, science displaced craft knowledge at the heart of the production process. The application of physical and chemical discoveries to the design of machines and industrial processes has all but dominated scientific research since the development of the oil and chemical industries and the emergence of electricity as the main source of power at the beginning of the twentieth century.

Today, the production and distribution of knowledge as an economic sector rivals in importance the production of material goods in leading capitalist countries.[20] In terms of the admittedly ambiguous category of gross domestic product (the sum of the value of goods and services in a given national economy measured in current monetary terms), scientific and technological research accounts for as much as 2 percent in some countries and its share of investment is much higher, over 20 percent. One of the principal constituents of knowledge production—information—is now acknowledged to be a major industry; the training and recruitment of scientific and technical labor is a key concern of the education system and, more generally, of government policy; and, with respect to job growth, professional and technical categories, especially those engaged in scientific work, conceived in the broad sense of the production and distribution of all kinds of intellectual knowledge have been among the fastest-growing occupations over the past three decades.

In this process, the division between intellectual and manual labor displays conflicting tendencies. On the one hand, intellectual labor of all kinds has become more ubiquitous both in the workplace and in the wider society.

In addition to state- and corporate-supported scientific research, for example, every large production and service enterprise employs substantial engineering, design, and planning staffs (or hires them as consultants). Their work consists in a large measure in increasing output through the rationalization of the labor process—especially replacing living labor with machinery. In addition, a wide array of enterprises require quality control engineers and technicians to repair the damage caused by both the rationalization and the acceleration of production. Civil, mechanical, electrical, and chemical engineers provide the technical support for the implementation of management policies for increasing labor productivity through technological innovation, much of which is linked to the prevailing organization of the labor process. And the management of systems—financial, sales, production—is a central concern of every enterprise. In the main, this describes the function of the key engineers who apply scientific research to prevailing production technologies within a regime in which labor control, recoded as "efficiency," is the imperative that shapes much, if not all, intellectual labor.

Many observers have argued that a new era of computer-mediated work reverses the historic tendency toward rationalization. According to this view, computerization facilitates the merger of design and execution as they were in the artisanal mode of production when the craftsperson both conceived and made the whole product. From the point of view of many of its pioneering inventors and developers, the promise of the computer consisted precisely in the possibilities afforded by this integrative technology to eliminate the disruption, both spatial and temporal, between design and execution. Indeed, the merger of computer-aided design and computer-aided manufacturing (CAD-CAM) in a single process has succeeded in revolutionizing the production of machine tools and aircraft engines, and some processes of car production. Depending on local conditions, shop floor technical and manual labor are able to modify and even initiate product design and feed their recommendations back to engineers and managers.

But, as Zuboff discovered, integrating design and execution is possible only when communication is a two-way street. The precondition of successful integration is the end of Taylorist barriers to shop floor control over the labor process so that the operator has a proactive voice in design, just as the engineer is intimately involved in execution.[21] The fact that this aspiration on the whole remains unfulfilled may be ascribed to the persistence of the older Taylorist model of the organization of the labor process that is superimposed on technologies that would otherwise permit a different production system. However, as we have seen, there are some exceptions to the general rule of the subsumption of computer-mediated work under Taylorist rationalization. In the perspective of *possibility* these exceptions *denaturalize* the rationality within which the detail laborer emerged.

Although computer-mediated technologies have become standard in goods production, administration, and the services in the past twenty-five years, this does not imply that the worker is becoming multivalenced or that control over the labor process has passed from management to workers. After a prolonged period during which task rationalization was the crucial marker of most kinds of industrial and service labor, the traditional hierarchies of the highly rationalized workplace characteristic of the industrializing era have, with important exceptions, been preserved. For example, computer programming is divided among systems analysts who have broad technical responsibilities, programmers whose range is bounded within the general parameters of the *system*, and operators or word processors whose labor typically is routine. In industrial production, job hierarchies are only partially integrated into the new computer-dominated era. And in the office, the "word processor" has replaced the secretary and file clerk and executes the commands written into the computer by programmers.

In fact, in the early years of computer applications to industrial production and services, the programmer tailored the program to meet the requirements of a specific labor process. Equipped with knowledge of the basic mathematical principles upon which digitally controlled information processing was based and the elementary algorithms of programming, the "hacker" or systems analyst invented the software for a *specific* set of administrative or production problems and the programmer adjusted the program to new requirements within the general parameters of the software. The movement between programmer—the line professional in the computer field—and the more autonomous systems analyst was extremely fluid before the rules of rationalized labor were increasingly applied in the late 1960s and early 1970s to the computer-mediated workplace.

In consequence, in the early years of computerization of office and shop procedures, bookkeepers wrote programs and machinists, many of whom initially opposed the introduction of numerical controls and robots atop machine tools, added programming to their craft. However, almost from the beginning of commercial and industrial computer applications, the distinction between the computer program (software) and its operation (data or word processing) reproduced the old industrial hierarchy between design and execution. By the late 1960s, academic computer science had been born: the research scientist was separated from the technical expert who, in turn, was removed from the operator. The putative distinction between science/technology and operation had been maintained.

Management "science," largely loyal to its Taylorist origins in the early twentieth century, its task to strip craft labor of its considerable control over the labor process, has subverted the radical possibilities of computerization. Put another way, the scientific and technological revolutions of the past century—at first the physical and chemical innovations whose technical applications developed commercial uses for electricity and oil and then the computerization of both administration and wide sections of production—have intensified the degree to which abstract labor dominates concrete labor. Just as the development of new materials such as synthetic fibers and plastics was crucially dependent on the discovery of the remarkable consequence of the combination of hydrogen and carbon converted to the solid state, Turing's mathematical discoveries had profound material consequences for the production and processing of information.[22] In many respects computer technology followed, in principle, the feedback mechanisms that are the basis for automatic processes such as continuous-flow operations in chemical plants and oil refineries. What unites these two different generations of production technology is that the program *is* the *manager without a personification.* Thus, as Mallet noted in the early 1960s, the generalized applications of the principle of self-activating machines transformed the idea of the self-managed workplace from a value into a practical political program. However, worker self-management, even among qualified labor, has not occurred because managers have imposed a production regime that artifically installs and reproduces their power. This goal is today more remote than during the 1960s and early 1970s because of the decomposition and recomposition of the working class, including many technical intellectuals.

The loss of operator control is evident across production and service labor processes. Managerial constructions of computer technology have replaced or otherwise altered many typical activities: the self-regulating computer-controlled lathe displaces many traditional skills of the machinist trade; and, guided by a single operator pushing preprogrammed buttons, computer-controlled robots move goods from one place to another, eliminating manual labor in handling materials.

In recent years, scientifically based technologies have begun to displace and recompose the skills of the professional, especially in medicine and engineering. For example, civil, electrical, and mechanical engineers have traditionally drawn dozens of prints in the process of making relatively minor alterations in a switch, water system, road, or building. It was not rare for them to laboriously draft hundreds of drawings for a major design. The new technology of computer-aided design and drafting (CADD) electronically reproduces the broad outlines of drawings with a program that standardizes many of their dimensions and also provides a restricted menu of standard mathematical calculations from which the operator may choose. CADD has increased drafting productivity by nearly 1,000 percent for routine design jobs and freed engineers to spend most of their efforts on "purely" design functions.

At first these engineers perceive CADD as a historic opportunity to, in the words of one engineer, "get back to what we were trained to do"—design equipment, relieved of the tedium of seemingly interminable drawing. At the same time, CADD virtually eliminates the drafter's occupation, at least for most routine design functions. Since CADD is a relatively new technological innovation, engineers have been excited and challenged to learn its manifold applications and to acquire sufficent mastery to use it efficiently.

Yet the revolutionary implications of CADD are not confined to the possibility, as yet largely unfulfilled, that designs will be more accurate, making roads, water systems, and other elements of the infrastructure safer and more durable. This potential saving is significant because drawing occupies the overwhelming majority of traditional engineering labor time. If most of this task is transferred to the machine, the economy will require many fewer engineers and drafters even if the rate of growth of construction, industrial production, and technological systems is fairly brisk. In terms of numbers of employees, among the major professions, engineering's future is increasingly limited. In response, some engineering schools, most recently the one at Pratt Institute, have closed, and others are sure to follow. We are still in the debugging stage of this technology's development, so there is likely to be as much skepticism concerning its long-term productivity as there was in the 1970s when numerical controls were first introduced on a wide scale on machine tools and robots displaced assembly-line workers. Nevertheless, just as there can be little doubt that robotization and numerical controls have radically altered the assembly line, so CADD will radically alter many subfields of engineering, especially those that involve routine designs.

Physicians who still perform direct patient care rely heavily on electronically produced test results. From the point of view of the technologically mediated diagnostic system, the patient "disappears." In many hospitals and group medical practices, computer programs such as the CAT scan are already on line and have taken over a significant portion of the physician's diagnostic and prescriptive functions. Whether there has been a rise in accuracy is still problematical. Physicians still have room for interpretation of what appears on the monitor; results do not eliminate their options. But the computerization and mechanization of diagnosis and prescription has, like CADD, tended to transfer intellectual functions to the machine. From the doctor's point of view, the patient has progressively been reduced to a series of abstracted pictures of particular parts of the body; in hospitals, patient care, always primarily the work of women, has now become almost exclusively the work of nonprofessionals—practical nurses and nurses' aides. The exceptions are patients with specific problems such as kidney and liver failure, esoteric diseases, and those who are able to afford the services of an attending physician.

From Professional to Salaried Worker

The technical intellectual is a historically evolved social category whose existence is attributable to the subsumption of science and technology under capital in the late nineteenth century. In contrast to the "traditional" intellectual, who typically was a literary/cultural figure situated either in the universities or in journalism, and the political intellectual, who practiced statecraft or led parties and social movements, the technical intellectual first emerged as a functionary of capital in spheres of instrumental activity, particularly the production and administration of things and signs. Historically, the characteristic technical intellectual was the engineer. Once one of the characteristic sectors of the independent, entrepreneurial middle-class professionals in the late nineteenth century, engineers increasingly became salaried employees. By the time of World War I, engineering, together with law and medicine, had ceased to be a characteristically independent profession.

Half a century later, physicians and attorneys found themselves unable to maintain their independence as the medical and law fields experienced, during the 1960s, their own descent from the independent middle class to the salariat. Today more than half of the graduates of law and medical schools may expect to work for salaries throughout their careers. There is almost no prospect for reversing this trend. A diminishing minority of these professionals, once among the pillars of independent entrepreneurs, can hope to become partners in, or own, their own practices. The fading of this expectation, which was accompanied by the expectation of high income, has caused significant declines in medical school enrollments and a dramatic change in the gender composition of these fields.

Today, more than half of medical and law school students are women, as are a third of all physicians and attorneys, compared to fifteen years ago when only 10 percent of physicians and about 20 percent of attorneys were women. The dramatic alteration of the gender composition of these traditional professions is, of course, not merely the result of their proletarianization; second-wave feminism has contributed to encouraging women to enter these fields, but also to lowering traditional gender barriers.

Becoming salaried employees does not at first signify the loss of traditional markers of professions: they still enjoy a high degree of autonomy in job performance and have credentials that permit both considerable lateral mobility for those who elect to remain practitioners and advancement to management or, an option for attorneys, the judicial bench. In the health field, for example, a small coterie of salaried physicians become top-level managers of the leading institutions. Even staff physicians in most hospitals still make some crucial decisions in their daily practice. Doctors decide on medication, lawyers develop strategies to resolve conflicts within the framework of both laws and the conventions of their profession, and engineers, even in CADD-mediated workplaces, must decide among (predetermined) options. In large corporations engineers retain control over a wide range of decisions in the design and processes of industrial production, even if they are subject to constraints imposed by financial and sales managers who may intervene in the design process to assure cost control and marketability.

Yet there are now two crucial constraints on their autonomy: insofar as they are employed by bureaucratic organizations—public agencies, large law firms and corporations—the range of decision making to which they have access is narrowed; and technological change has recomposed their jobs so that, for most employees, their training exceeds the requirements of job performance. For example, a physician affiliated with a large research hospital in New York City reported to one of the authors that patient care is increasingly governed by computerized information. The computer instructs the nurse when to administer prescribed medicine; the face-to-face ritual of making rounds is virtually eliminated because in many cases doctors simply read the computerized charts hanging from the edge of the patient's bed. And, more broadly, programs have been developed to synthesize information into a diagnosis from which treatment regimens are derived.

In the health industry, computerization is in part a response to chronic shortages of nurses and other trained employees. But the main impetus for the computerization of health services parallels developments in industry and financial services: introducing computers into the workplace is part of a major cost-reduction and productivity-enhancement effort. As a result of computerization, fewer professionals are required on the floor, and the training of and consequently the salaries of paraprofessionals may be reduced; computer-mediated diagnoses and prescriptions standardize patient care, even as they routinize the work of the physician. In addition, most hospitals and other medical centers have introduced management systems that also tend to remove decision making from the health providers.

The routinization of medical work is by no means a universal phenomenon. For even as many staff physicians and professional nurses have been subordinated by the centralization of patient care management and new technologies, a smaller group of administrators, computer-trained physicians, and research scientists have, through the monopolization of new knowledge, enlarged their influence and control over the system. Since the acceleration by molecular biology of medicine as a technoscience over the past decade, knowledge has become more concentrated at the top of hospital and other medical hierarchies. Many staff physicians complain that they are obliged to follow the prescribed regime dictated by the computer program on penalty of discharge or other disciplinary measures. Thus, far from providing new opportunities for staff to interact with the "smart machine" in order to provide better service to patients, the machine is often pitted against the professional as an antagonist. Like skilled craftspersons before them, physicians have suffered not only the loss of opportunities for self-employment but also the loss of autonomy signaled by the decline in their capacity to make independent judgments.

Until 1960, the very idea of public-employee unionism, much less professional unionism, was still controversial, even among hospital workers who were at the economic bottom of manual occupations and teachers who until the 1970s were among the lowest-paid professionals. "Public opinion" regarded professionals, especially those in "helping" occupations such as teaching, health, and social work, as public servants. As a result, many people, including a substantial number of the helping professionals themselves, argued that if they were not subject to the usual fluctuations of the market they should also refrain from organizing unions that could upset the economics of health care or education and disrupt public services.

The doctrine according to which public employees should not be considered ordinary workers for the purposes of unionization was codified by 1935 National Labor Relations Act, which recognized workers' rights to organize unions "of their own choosing." But there were some exclusions: the employees of very small businesses and those not engaged in interstate trade, managerial employees (a very ambiguous category), and, perhaps most important, public employees. While professionals were not explicitly exempted from coverage under the act, many fell into excluded categories. But beyond legal restrictions, although they were increasingly salaried after World War II, they were considered, and considered themselves, part of the middle class, which in the United States was almost a sacred position.

In their own eyes, professionals were an integral part of the great American middle class even though those in the education and health sectors, except doctors, suffered chronically from low salaries. Until very recently, most elementary and secondary teachers received salaries equal to or lower than those of semiskilled industrial workers in highly unionized mass production industries such as autos and steel. For example, my gross pay as a semiskilled steelworker in 1958 was about $5,000 while the pay for schoolteachers in Newark, where I lived, averaged substantially less. Since nurses and social workers were employed by public or nonprofit agencies unless they advanced to management, they shared a similar fate. Until the late 1980s most registered nurses could expect an annual salary equal to that earned by a semiskilled steelworker in the middle of the hierarchy of job classifications in that industry. Now nurses are paid about as much as a skilled mechanic in a unionized industrial plant, but not on a weekly basis as are skilled workers in the unionized construction industry.

The traditional perquisite of job security does not tell the whole tale of why many people choose teaching, nursing, and other helping professions despite relatively meager salaries. Traditionally, in addition to the privilege of relative immunity from the ups and downs of labor market conditions, teachers have enjoyed a two-month summer vacation and the prestige that accompanies the by now general requirement that they possess a postsecondary education credential. Despite low salaries and lack of autonomy in the classroom (signified by the requirement that they produce, for review by administration, a daily lesson plan), these features of their job marked them off from other wage workers, even in other sectors of public employment. A minority entered teaching for vocational reasons—that is, they felt a calling. Similarly, while nursing was, together with teaching, the best route for most women out of the clerical and manual working class, nurses were required to work all year in addition to suffering daily humiliation, not atypically sexual harassment, at the hands of male doctors and administrators. Still, nursing drew some new recruits for vocational reasons: despite low pay and demeaning working conditions, caring for the sick is regarded by our culture as holy work and has become part of our self-image as a caring society. Interestingly, teaching has no stellar, larger-than-life figures like Florence Nightingale and Clara Barton. The teacher as hero has had to be invented by fiction (not often), and in recent times teachers have just as frequently been transfigured into villains in popular culture. This is shown in the recent spate of child abuse cases in which day-care and nursery-school teachers were indicted and imprisoned before, in most instances, being vindicated by juries. Although black teachers (in *To Sir with Love*) and, in an earlier period, small-town white teachers have been celebrated in film, there is no sustained effort, at least since Eve Arden starred in the 1950s series "Our Miss Brooks," to represent teaching on television as a dignified, desirable vocation similar to medicine and law. There are 3 million teachers in the United States; teaching is by far the largest single profession. For millions of Americans, especially women, nursing and teaching are the available routes if not to middle class comfort at least to job security and a measure of respectability. Yet teaching is not drawn in popular representations as a particularly desirable job.

Until the 1950s most rural and regional school districts were composed of small towns that required of their teachers only two years of teacher training, typically provided by normal schools or teachers' colleges. Even if they were salaried rather than self-employed like their neighbors—

mostly farmers and merchants—teachers rarely suffered the uncertainties associated with hourly labor or small business. For example, even the most highly paid skilled construction workers such as electricians and plumbers could expect only seasonal employment, and laborers were often hired by the day or week. Small farming was never a stable, lucrative business. In contrast, teaching, except for blacks, was a property right: the tenure system, in addition to delivering them from market fluctuations, provided a measure of protection from arbitrary discharges by principals and school boards. Before the age of teacher unionism, however, this assurance was observed as much in the breach as in practice.

Elementary and secondary school teachers enjoyed little if any academic freedom in the classroom, or indeed in public affairs. The curriculum is prescribed, often by elected or appointed school boards. Teachers are often obliged to teach from textbooks and to use tests mandated by state education departments; they are required to generate lesson plans that implement prefigured objectives spelled out either in texts or by administrators—or both. For these reasons, teachers are typically not formally free to improvise, and when they exercise independence they may be accused of insubordination and are subject to discharge. In some school districts, teacher creativity is an underground activity. The system discourages the intellectual development of teachers but rewards them financially for completing postgraduate education courses, the disciplinary equivalent of intellectual growth.

Many school systems frown upon teachers who introduce literature rather than relying on textbooks for language development. The prescribed curriculum is frequently based upon an educational philosophy that adopts behaviorist images of child development. That is, language arts are taught in terms of "building blocks" such as phonetics and vocabulary that is graded according to presumed levels of achievement; testing is used as a system of rewards and punishments for observing the results of this perspective.

Second, especially but not only in smaller communities, teachers were considered public servants whose exercise of independent political judgment was sharply curtailed by custom, even when it was not restricted by law. Since public-employee unionism was unprotected by state and federal law (and in some states still is), membership in trade unions was often grounds for dismissal, and public "servants," especially teachers, could suffer reprimand or discrimination on the job for becoming vocal dissenters in local political struggles on issues not necessarily related to education.

Since the United States has an administratively decentralized system of public services, conditions of public employment, including teaching, nursing, and social service, differ depending on state and local statutes and administration. Additionally, many of these services are delivered by schools, agencies, and hospitals that are privately administered although they are substantially publicly funded. These organizations have their own personnel practices that are often below government standards for salaries and working conditions in the same geographic region. But it would be difficult to maintain—even today, when a national consensus exists on the need for better education—that educational policy makers should encourage independent teacher participation in school policy and creativity in the classroom as an indispensable component for achieving this aim. Instead, conservative educational policy argues that the prescribed curriculum in public schools should be strengthened to focus more sharply on certain basic intellectual skills and ethical content. Far from mandating policies that recognize differences in teachers' talents and students' cultural backgrounds, the dominant educational policy is oriented toward achieving greater cultural and linguistic homogeneity among children. Hence, after two decades during which the standard curriculum was under fire from educators, activists, and parents, there are new pressures for conformity in the classroom—in effect, for the development of national achievement norms. While opponents of testing and those who contend that teachers tend to teach to standardized tests rather than to a more profound literary and scientific content have won the battle intellectually, uniformity in textbooks, curriculum, and teaching methods and authoritarian discipline in the classroom are on the rise in America's public schools.

Within this regime, the limits of teacher autonomy are imposed from without. Teachers are made constantly aware by administrators, school boards, and the media of their responsibility to deliver a specific content that is graded according to a model of child development that divides the first twelve years of schooling into as many levels. Although *socially* accorded professional status by the community outside the big cities, in the educational labor process the teacher is a technician. As we shall see, this contradiction accounts for the erosion of the middle-class self-image of teachers. Together with other influences—notably the changing political and ideological environment of the 1960s and early 1970s and the entrance of a new generation of student dissenters into teaching—the elements of professional identity have been partly displaced by components of trade-union ideology to produce the greatest period of union growth among professionals in U.S. history.

Chapter 7
Unions and the Future of Professional Work

Engineering as a Management Tool

The New Deal consisted of a major enlargement of the government's role in economic life, although from a historical perspective it may be viewed as a continuation of practices and institutions introduced by successive national administrations since the Civil War. For contrary to popular belief, according to which the U.S. economy was based on the free market until the Great Depression, government investments have spurred economic development and growth since the founding of the republic. The government regularly used federal troops to acquire land and territories that were then exploited by agricultural, rail, and industrial capital; built roads that connected otherwise isolated regions; and, through war contracts, stimulated textile, garment, munitions, and metals manufacturing during the Civil War, helping transform the economic basis of society from agriculture to industrial production. These practices belied the deeply rooted doctrine of what C. B. McPherson has called "possessive individualism," whose cornerstone is the free market that has remained a central referent for U.S. politics and culture throughout its history.

The New Deal embodied, simultaneously and contradictorily, doctrines of the social gospel; a mainstream Protestant movement for social justice; the progressive tradition, which was largely a movement for the centralization of government powers, ostensibly in the interest of abolishing corruption; and, finally, the state-interventionist proposals of both the Keynesian wing of neoliberal economists and the social democrats. For the first time, it became *official* government policy to assume responsibility for economic growth by mobilizing the already established regulatory institutions introduced by the Wilson and Hoover governments. The Roosevelt administration enlarged Hoover's major depression-fighting tool, the Reconstruction Finance Corporation, an institutional method of increasing investment.[1] But perhaps most controversial, labor began to be regulated—its costs as much as its right to organize and bargain collectively. Although Roosevelt's major growth strategy was widely credited with providing the means of making a living to millions made destitute by the economic crisis, it was a version of what is today labeled "trickle-down" economics. The New Deal's major recovery policies were a series of administrative and legislative measures aimed at stimulating economic activity through direct government investment and a monetary policy that reduced interest rates in order to stimulate private investment. Although the electorate perceived Roosevelt's initiatives chiefly as humane relief programs, this was by no means at the center of the administration's economic approach. The primary thrust of New Deal policies was *regulation:* in the first two years of the New Deal, Roosevelt wanted to control every aspect of economic life in the interest of spurring capital accumulation. In practice, even before the Supreme Court struck down the National Industrial Recovery Act, the administration had only limited success in this effort.

Until 1970 and even today, the degree of integration of the private and public sectors through regulation called into question the dominant belief that American capitalism was a "free enterprise" economy. In addition to regulating banking, investment, and labor, the federal government broadened its economic intervention during the 1930s to include permanent rearmament, a major source of economic growth, especially after 1938; direct assistance to state and local governments to administer relief programs; and control over agricultural production and prices. In this context, Lyndon Johnson's Great Society program was merely a revival of the activist perspective of the New Deal, albeit on a more modest scale.

One of the enduring effects of the New Deal and its successors was dramatic growth of federal, state, and local bureaucracies. By 1965, public employment accounted for one-sixth of all jobs. Together with the expansion of the legal, scientific, engineering, and accounting professions that accompanied the growth of large corporations and the explosion of health institutions financed by the emergence of private prepaid health plans as well as public funds for both the poor and for capital expenditures such as hospitals, the professions of management and administration were created, virtually overnight.

To be sure, these professions were not literally born during the New Deal. Indeed, the partial separation of management from ownership was a product of the expanded family-controlled companies of the middle of the nineteenth century, and management became a typical occupation within large corporations by the turn of the century.[2] But the extension of the managerial professional to virtually every corner of economic life presupposes the abstract concept that *control* and *organization* are intellectual problems that are no longer subsumed under the older managerial rubrics. The elevation of the managerial role from production boss to planning intellectual takes place when the manager is obliged to engage in what later became known as a *science* in which elements of economic, sociological, and psychological knowledge are combined and transformed into technologies of policy and strategy.

Hence the claims of this new science to be free of particular industries or bureaucracies. Graduates of schools of management and business have learned that they are able to step into any corporation or public agency because they have mastered the basic laws of organizational systems and behavior, which vary, but only marginally, among large institutions. Particular institutions or sectors, it is believed, have special chacteristics and present unique problems, but the foundational knowledges of control and organization are universal.

Managers conventionally direct the labor forces both in production and administration. With the advent of the large corporation they have assumed operational control over nearly every aspect of production, distribution, finance, and the widening political and social functions of the corporation. These functions have been, since the beginning of the twentieth century, subject only to the constraints set by boards of directors presumably, but not always, representing only the largest stockholder(s). At first, most industrial managers were recruited among the craftworkers who, in the typical industrial plant, enjoyed greater freedom of movement and knowledge of the labor process than other workers. When science was not yet the basis of industrial production, the role of the manager, even though managers were employees rather than owners, presupposed traditional knowledge of the craft-based labor process; the knowledge required to perform the labor of management was, like any other craft, largely traditional. But as systematic, formal knowledge became the basis both of machines and of labor processes, as relations with financial institutions became crucial for procuring investment capital, and as contracts for military and other state-financed matériel came to constitute a vital part of private business—that is, when these abstractions came to dominate concrete labor—the crucial knowledge required for management changed from specialized knowledge associated with solving production problems to understanding how systems work.

Alfred Sloan, chief executive of General Motors through the 1940s, typified this development. In contrast to Henry Ford, whose roots were in the older model of the mechanic as inventor, as a trained engineer Sloan was a prototype of the new manager. But Sloan was not atypical of his generation of technical intellectuals: he rose from technical into managerial ranks, from graduate electrical engineer to salesman to chief executive officer of General Motors. Sloan's rise to the pinnacle of corporate power occurred in the context of the primacy of knowledge as the crucial productive force and the influence on the bureaucracy of those who possessed it.[3]

Until World War I, the image of the engineer as manager provided a special legitimacy to management because it seemed to confer on its authority the mantle of science. Frederick Winslow Taylor ostensibly applied scientific principles to the labor process, a claim that helped to weld managerial authority with industrial truth. Within this regime of truth, craft knowledge was viewed with condescension, as historically necessary but on the way to being surpassed by new ways of producing goods. Taylor made plain the fundamental task of scientific management: to break the crafts' monopoly over the practical knowledge required for production.[4] As the living representative of the merger of theoretical and practical knowledge—the scientist cum engineer—he would persuade workers not only that the new way of doing things was more efficient from the employer's perspective because it would increase production, but also that by giving up their knowledge to the machine and to a new form of industrial organization skilled workers would share in this expansion.

Thus, scientifically based technique became the new norm of the era of advanced industrial capitalism, and the engineer as manager became the model of a new age. Both the crafts and the mass worker were now displaced by the revolutionary force of technical knowledge because they had lost their centrality to the production system. As we have already seen, Veblen and other prophets of technocracy did forecast the emergence of engineers as a new class so long as the giant corporations were prepared to keep them loyal to the system by paying substantial salaries, providing lavish laboratory facilities, and promoting the most ambitious of their number into the ranks of management.[5] But only a minority of the scientifically based technical intellectuals were able to rise to either the top of the corporate bureaucracy or middle management. Most engineers remained well-paid members of the line salariat. The largest group of engineers—designers, quality control technicians such as equipment, materials, and product inspectors or industrial engineers—were subordinated to management, some of whom were still recruited from the bench. Engineering was subject to the new division between managerial and technical labor and the latter became subject to union organization, but because at first they were highly valued by top management, engineers did not find unionism as attractive as other groups of technical intellectuals such as health care professionals and teachers did.

With the elaboration of the corporate form of business organization on an international scale, its ties to specific industries and specific productive forces—buildings, machinery, and technology—loosened. Helped by changes in tax laws, accountants showed managers how to amortize investment in five years rather than ten, and in even shorter periods of time. Money managers—lawyers, accountants, investment specialists—partially displaced the scientific managers of the labor process in dominating corporate life in a new era of conglomerates, multinational corporations, and international cooperation and competition.

By the late 1960s the once preeminent position engineers had enjoyed in the later decades of the industrializing era had decisively eroded. Yet they were still important; their mythological significance actually grew during the 1970s in the wake of the globalization of production. Computerization of industrial production and finance simultaneously elevated technical expertise within the new corporate system and also sharply delineated a new division of labor within the technical intelligentsia. Unless engineers or, indeed, teachers or social workers were able to assume administrative/managerial responsibility, they were relegated to a position only slightly better than crafts and semiskilled manual labor employed in the key industries under the old industrial system.[6] With the maturation of consumer society the engineer lost the status of the characteristic intellectual of industrial capitalism as production took second place to distribution, marketing, and consumption. The advertising industry and other sales occupations took center stage in the corporate world along with the money and corporate managers concerned with issues of financing, corporate mergers and acquisitions, and expansion of production and investment. Lawyers, sales managers, and financial experts were elevated within the corporation over the engineers.

Even when, after World War I, engineers gradually ceased to be self-employed and became paid employees of corporations and government, many were not prone to see themselves as workers, but were afforded ample motives to adopt both the ideology and the social status of managers. To a large extent, this posture was influenced not so much by the fact that employers accorded them special treatment recognizing their specific, indispensable skills as by their actual position in the social organization of the factory or government bureau.

There are two kinds of engineers in the modern production process, including civil engineering: design and industrial. A substantial percentage of engineers design roads, machinery, tools, and production and information systems that are managed by industrial engineers on the shop floor, or, in the case of civil engineering, at the construction site. "Management" in the technical sense means laying out the work, assigning the tasks to different individuals and groups, and monitoring execution so as to ensure both quality and quantity according to technical and economic specifications. The first broad category of engineer, at least prior to computer-aided design and manufacture, set the parameters of the labor process through boundary-creating systems and machines that may be seen as the instruments of these systems. Industrial engineers are an arm of front-line management of labor and often themselves become managers.

In this instance the distinction between the technical and political function of management/ownership is ambiguous. The designer is a systems and machinery professional who, however, works within the rules established by what has euphemistically been termed the "market." This is a term for the requirement that design, whatever its purely "technical" function, obey the labor-saving—that is, productivity—imperative that rules capitalist enterprise. Therefore, technical innovation is typically constrained not only by market conditions but also by specific management objectives. We may conclude that, in this regime of production, technique is not neutral. It is pressed into the service of reducing the time required for the production of goods or services. If this is the case, technique is caught in the conflict between the "professional" norm of designing and producing quality products and the capitalist requirement of producing the maximum *quantity* of goods measured in value of product at the lowest possible cost.

The "genius" of U.S. engineering/management throughout this country's history was its ability to turn out relatively high quality goods at cheap prices even when American workers became the highest paid in the industrial world. In addition to ascribing U.S. ascendancy to the ample supplies of raw materials within its borders, historians have frequently attributed this feat to the Yankee inventors; Eli Whitney's cotton gin displaced fifty field hands as early as 1790. This engineering wonder made the United States the world leader both in slavery and in cotton production. The cotton gin was followed closely by Robert Fulton's steam engine, which provided an enormous boost to rail and water transportation; without it the continent would have remained undeveloped for decades. And, of course, there is the towering figure of Thomas Edison, a mechanic who, as the last of the great inventor/entrepreneurs, transformed Faraday's electromagnetic discoveries into commercially practical electrical energy and introduced, together with such competitors as George Westinghouse, a plethora of industrial and consumer products. Along with the application of chemistry to industrial production, these inventions revolutionized American life.

Yet, in contradistinction to the romantic image of the Yankee inventor, the magnitude of American technological achievements in this century was, of course, dependent upon fundamental advances in European and, secondarily, U.S. science.[7] As we argued in chapter 6, it was precisely the ability of American business to incorporate European crafts and some of the crucial discoveries of British and German science, particularly physics and chemistry, into industrial production that, together with the vast reserves of "unskilled" immigrant labor (unskilled relative to the regime of production), enabled the United States to emerge as a leading world industrial power.

Although the myth of the inventor/genius as an entrepreneur working in his garage has survived the passing of the people who embodied this type, after the turn of the century the salaried engineer and applied scientist became the material basis for the sustained belief that practical know-how was a unique American virtue. Even as many of these professionals were employed by giant corporations, a few others, financed by these corporations, set up small development shops. To be sure, most of them had a single customer: the home corporation. But, particularly in the burgeoning computer industry, the 1960s witnessed a revival of the small entrepreneur-inventor, a development that fueled one of the charactertistic features of the American ideology: small is the foundation of the U.S. economy.

But this type was quite different from their predecessors: the new scientists and engineers were typically not self-educated but university trained. Their laboratories and offices were not in backyard toolsheds or garages but were supplied by the corporations or public agencies for which they worked. They organized professional associations to share knowledge and to advance their economic interests and status, but they also retained the perquisites of the middle class. They conceived of themselves as an integral part of all aspects of civilized society. Despite their lack of capital (except cultural capital), they were, in their expanding domain, rule makers. They even saw themselves as the indispensable innovators of the great American industrial revolution that has shared, with its contrary image of America as a series of small towns, the position of a large cultural myth in U.S. history.

The development of mass education and social welfare marked the appearance of the helping professional as an important social type. Of course, the key ideological precept of these occupations was *service*. But insofar as the helping professions of teaching and social work were vocations, in the religious sense, their training was not purely technical but included a considerable moral responsibility for inculcating their "clients" with American values—loyalty not only to the prevailing economic system or government but also to the underlying scientific regime that counts as legitimate knowledge.

While until recently the parallels between how the private and public institutions work were largely hidden from political discourse, their affiliation with the system of administration is not born, at least in principle, out of fealty to any particular regime of political or bureaucratic power, but to a regime of truth that is inseparable from the theory and practice of science. (Here we employ the term *science* not in its popular experimental image but as any inquiry that purports to be value-neutral and is ordered according to a series of procedures recognized by those qualified to be considered professionals within a specific discipline or organized practice.)

The relationship between intellectuals, legislators, and constituents of democratic and authoritarian societies is fated to be uneasy, precisely because of the quasi-theological nature of induction into scientific work. Skilled manual workers are inducted into crafts by an apprenticeship or other training program. For more than a half century, since craftworkers attained quasi-professional status through law and union contracts, this process almost invariably has entailed acquiring knowledge by means of classroom instruction in addition to a larger quantity of practical, on-the-job training. But the formal aspect of the training is clearly subordinate to the day-to-day activity of learning on the job.

To be sure, every profession has a practicum: Recent law school graduates who wish to practice law serve their apprenticeship through clerkships in law firms or, less typically, for judges. A fairly rigorous internship is required of physicians, clinical psychologists, and social workers. Engineering and architecture, the most informal of the leading professions, still have "junior" engineer and architect categories. Student teaching is required for state teacher certification and consequently is part of the core curriculum of most education schools.

But the transmission of the knowledge connected to nearly all of these scientific professions takes place in the classroom. The acquisition of intellectual knowledge, science, is the marker of the twentieth-century intellectual. The boundaries marking the professional as cultural intellectual are embodied in the corpus of disciplinary theory, that is, the specific way that the profession constitutes its own object of knowledge, the methods by which it acquires knowledge, the canon of already legitmated knowledge and the rituals of transmission through an approved course of study. Although the curricula of various professional schools may vary in the variety and degree of innovative electives, the core required courses signify to the rest of the profession that the student has undergone an adequate intellectual preparation and, perhaps more saliently, has been properly initiated into the values and traditions of the profession.

Professional Unions: Erosion of Professionalism?

The growth of unions of professionals in the past thirty years—especially in the so-called helping occupations of teaching, social work, and qualified health care labor—is a sign that a significant portion of them have begun to question some elements of the value system undergirding the ideology of professionalism itself and their own middle-class status. But, lacking alternative visions of their own class position, many professionals cling to postgraduate education as one vital equivalent of class power. The institution of the professional school became, after the turn of the century, a crucial strategy of the helping professions, as it was for their counterparts in law and medicine. The basic route to professionalization of these occupations is to make academic credentials, as well as licenses linked to these credentials, a necessary rite of passage into the profession.

Teaching is a case in point. A hundred years ago, most elementary school teachers lacked formal credentials beyond high school. Gradually, under pressure from educational organizations, state and county postsecondary teacher training institutions called "normal" schools or colleges were established as two-year programs and awarded a certificate that qualified the candidate to teach elementary and secondary school within the state. A half-century later, many of these normal schools were transformed into four-year teacher colleges that broadened requirements to include some of the humanities and granted a bachelor of science degree in education. With the transformation of teacher training into the "discipline" of education that possesses its own subdivisions and canons has come a series of specialized knowledges that claim departmental status within education schools, now located in large universities. The most theoretical subdiscipline is the history and philosophy of education. Others include anthropology and sociology of education; administration; teaching (divided into three levels—early childhood, elementary, and secondary); and the much realigned area of "special" education, which exists as a model politically generated field.[8] These subdisciplines became nationally adopted and were widely interpreted by educational intellectuals as a signal that their profession had entered into the academic mainstream.

Americans, including industrial workers and even trade unionists, have always been uncomfortable with images of class and class conflict. The idea of class, traditionally bound up with the concept that individuals are ensconced in economic and social circumstances over which they have little or no control, seems to contravene the essential values of American culture, the most powerful motto of which is that you can be anything you want to be. A typical American narrative is that all we need is hard work and a good bit of luck to achieve success and escape the conditions of our birth. In any case, we simply refuse the European belief that collective life is prior to individual fate, that most people can only rise *with* their class, not despite it or in opposition to it. Even when we acknowledge that we are ineluctably born into a world we never made, and are reluctant to adopt ideologies of change, we are imbued with personal hope that somehow "I" will escape the limited horizons of my parents. Americans seem never to get to the "me," an object of reflection that sees itself in the world as a social being.

This mentality of, on the one hand, optimism and, on the other hand, Jobian patience, is related to the long waves of economic expansion that occurred during the course of American history. These waves, which ended in crashes, recessions, and depressions, provided a real basis for shunning class definitions of social fate, and wasn't that why this country's ideologues had pleaded exceptionalism, especially in relation to Europe? Thus, when millions of industrial workers streamed into the new unions of the CIO in the 1930s and produced historically unparalleled worker militancy exemplified by sit-down strikes during which workers occupied factories to demand union recognition from employers, the middle class was able to accept unionism by translating the struggle into a battle for *rights* waged by underprivileged groups. These dramatic events were mediated and framed by liberal concepts of "simple justice" for those who had been excluded from the benefits of prosperity or middle-class security by anomalous employer practices of inequality.

This ideological shift was accomplished in a now classic way. According to conventional wisdom, America is a middle-class society that, unfortunately, had maintained a series of unconscionable exclusions, the remedy for which was better intergroup relations, voluntary action by employers to remedy past injustices, and more responsive education systems that could provide equal, even if separately delivered, education and training opportunities to "minorities." But even before Congress enacted legislation providing labor's right to organize and bargain collectively with employers, the powerful labor insurgencies of 1933 and 1934 temporarily altered the ideological environment.

To be sure, the salaried middle class looked on as industrial workers, unskilled as well as skilled, fought for dignity and a measure of power over their own economic, political, and social destinies. Remarkably, despite the significant deterioration of their own jobs and living standards, the salariat still relied, in the main, upon its cultural capital, accumulated through schooling, and believed this to be a property equivalent that guaranteed good financial returns. Moreover, since the American Revolution, the "old" middle class of small property owners, including the independent professionals, had been considered the bedrock of the republic, its politically and socially active population, the repository of American values. These were the republicans, those who voted, maintained vigilance against government corruption, and embodied the doctrine of equality of opportunity that formed the heart of American ideology. Except for a relatively small percentage of republican-minded trade unionists who participated in politics, the middle class organized the political parties at all levels; to this day, merchants and professionals typically dominate town councils and school boards, constitute the majority of elected representatives in state and national legislatures, and, equally significantly, constitute the membership of civic organizations and professional associations that form the core constituency and set the tone of public life.

Thus, the power of the American self-conception that the United States is a middle-class society is by no means established by numerical criteria. Even if manual and nonsupervisory clerical workers constitute the overwhelming majority of the labor force they have never predominated in the *active* population—those who participate in civic and political groups. The hegemony of middle-class ideology is embodied in the usual practice, especially by the Democratic party, of selecting presidential candidates from this stratum even though, for more than a century, wealth and political power have been concentrated in the giant corporations, even at the local level. Even when either major party selects a scion of a certified rich family, none of the exceptions among recent presidents—the two Roosevelts, Kennedy, and Bush—were from families of big capital. Even so, their privileges engendered considerable popular resentment; despite the Rockefellers' success in state and local politics, for example, we are still not entirely comfortable voting for candidates with corporate backgrounds. Nelson Rockefeller could be governor of New York and was appointed vice president after the demise of the Nixon administration, but he was effectively blocked in his bid for the Republican presidential nomination. And, more recently, Bill Clinton was able to make George Bush's background as a scion of old New England money an underground issue at a time of recession. President Bush, candidate Clinton argued, was out of touch with the needs of ordinary Americans.

Within this framework, the business and professional categories view themselves as the genuine guardians of *American* civilization. For this reason, locally based civic associations are often called business and professional councils. They occupy crucial positions of institutional power, earned by virtue of either education and training or, in the case of entrepreneurs, hard work in making their small or medium-sized businesses successful. In both cases, they commonly believe they are "self-made" rather inheritors of privilege for, even when property is handed down, the survival of small business, especially in the wake of corporate encroachments, is understood to depend on hard work and acumen.

On this basis, the American middle class has aligned itself with the prevailing order. Even the historic struggles of farmers, the most ubiquitous segment of the nineteenth- and early-twentieth-century middle class, against large corporations and the government have rarely driven even them into political or ideological opposition. Rather, after the populist movement's cooptation by the Democrats in 1896 and, in many Western states, by Republicans two decades later, farmers tended to go the way of craft unionism: with the exception of the smaller National Farmers Union, which aligned itself with the New Deal and organized labor, other farm organizations, dominated by corporate and large growers, had by the late 1930s disdained all alliances with the new labor movement, the civil rights movement, and other urban constituencies seeking social justice. Most farm organizations tacitly tend to bestow their loyalty on the Republican party, whose free trade policies and other elements of economic liberalism correspond to both their perceived interests, even as they remain crucially tied to government subsidies and other state-sponsored perquisites.

Prior to 1960, in all large American cities only a minority of the technical intellectuals joined unions. Most remained tied to their cultural capital and, consequently, to the prevailing ideology that unions were appropriate only for propertyless manual labor. Those who recognized the need for collective action joined professional associations that spent most of their energy enlarging the cultural capital of their membership by trying to raise certification standards to include more elaborate credentialing procedures, a measure that usually required postgraduate work.

Despite the relative strength of teacher unions, at least compared to unions within other professions, the rise of mass unionism in the 1930s and 1940s did not help the teachers' or other incipient professional unions to win recognition from the boards of education, state and local governments, and employers in the private sector. Most unions of professionals operated as a combination of lobbying organization and professional association. Teacher unions held seminars on teaching and curriculum, lobbied on legislative questions, and engaged in political action to raise salaries.

Moreover, as in other occupations and industries in the immediate postwar period, teacher groups in New York, Chicago, Los Angeles, and other large cities were divided along ideological lines. In New York, the old AFL affiliate, the Teachers Guild, was traditionally led by social democrats, while the CIO affiliate, the Teachers Union (later expelled), was closely linked to the Communist party. Yet the groundwork for the later development of powerful national teacher unionism was laid in the 1930s and 1940s in New York by radicals of various stripes who persisted, despite adverse political conditions, in maintaining viable organizations.

Similar developments took place among social workers, organized in the main by a New York local that successfully won collective bargaining recognition from Jewish welfare agencies, including the settlement houses, but not from the Protestant and Catholic welfare agencies and not without resorting to strikes and job actions. The left also formed the backbone of a small but active union, the Federation of Architects and Engineers, which, like the social workers' union, had affiliated with the Office and Professional Workers, a CIO affiliate that was later expelled for alleged communist domination.

Not unexpectedly, the intelligentsia was a major target of government and trade union red hunters, and professional unionism stagnated during the height of the cold war period. Most of the incipient unions of teachers, architects, engineers, and social workers disintegrated in the wake of ideological split within the labor movement, political repression, and the resistance of public and private employers. Federal and state prohibition of strikes by public employees was a deterrent, but even more oppressive was the fact that public employees' unions had no legal standing at the federal, state, and local levels. So, with the exception of scattered unionized groups of engineers and drafters in large electrical and automobile corporations where unionism had sunk deep roots among production workers; a small cadre of social workers in nonprofit agencies; and two postal workers' unions that were denied recognition by the federal government, no significant breakthroughs occurred until the 1960s.

The Radicalized Middle Class

As in the 1930s, unionism in the 1960s among technical intellectuals accompanied a general radicalization within the middle class. But, in contrast to the earlier period, when mass industrial unionism among manual workers took the spotlight, popular radicalism was now a distinctly middle-class phenomenon: it was directed as much at the consumer culture that had created the hegemonic ideology of America as a middle-class society as against the proletarianization of the salariat that accompanied the growth of the welfare state. The tense political environment of the late 1940s and early 1950s was somewhat altered because many anticommunist intellectuals and professionals became convinced that cold war politics had become irrational and that the United States needed a new foreign policy that went beyond military containment. Although the communists themselves remained politically isolated, the New Left, the antiwar and civil rights movements, earned considerable respect among fairly broad segments of the new middle class. Moreover, many of the activists of these movements were recruited from among these strata.

While the peace movement was relatively small until the late 1960s, it had a substantial following among scientists, the helping professions for whom the end of the cold war might bring federal education and welfare support to underfunded programs, and the immensely important liberal

214

church, especially Protestants and reform Jewish clergy. All of these conditions, combined with the vision of New Left, civil rights, and social democratic activists in the professions, produced a veritable explosion of professional unions in the 1960s. And, of course, the entrance of large numbers newly credentialed blacks into the subprofessions of teaching and public social work, combined with the similar mass influx of college graduates who entered teaching and social work to escape the draft, fanned the flames of union organization.

The untold story of the American 1960s is that, parallel to the civil rights, antiwar, and student upsurges, public employees and technical intellectuals, especially in the so-called helping occupations, flocked into unions, which, in the post-McCarthy era, had become strongholds of progressive, social-democratic trade unionism.[9] In this period of renewed organizing, which owed much of its impetus to President Kennedy's famous 1961 executive order recognizing unions at the federal level, in the American Federation of State, County and Municipal Employees and the Federation of Teachers socialists and ex-socialists, played key leadership and organizing roles. The basic slogan of the unionization struggle was that teachers, nurses, social workers, clerical workers, and blue-collar public employees had been "left behind" during the postwar boom even as labor in heavy industry and transportation reaped the benefits of union organization despite lack of education and training. The metaphor was usually expressed by comparing the superior earnings of a truck driver to the lesser wages of a teacher. By the mid-1960s, the media played on this difference partly to legitimate their attack on the powerful Teamsters Union but also to highlight injustices suffered by teachers and other underpaid professions in public and other nonprofit employment.

Sympathetic television and newspaper accounts did not spring from the personal sympathies of journalists alone. They were stirred by the dramatic and often militant actions of public employees and workers in the voluntary nonprofit sector, especially hospitals, which by this time had been transformed by affiliation and contracting agreements into adjuncts of the public sector. But the success of the hospital workers' organizing drives was based on the perception shared by patient care, dietary, and laundry nonprofessionals that they suffered from triple exploitation: as workers, people of color, and women. Similarly, clerical workers in federal, state, and municipal workplaces partially shed the pervasive ideology of public service and mounted picket lines to protest low pay but also personnel and supervisory practices that ranged from sexual harassment and poor working conditions to the lack of opportunities for advancement.

New York was the site of the early successes of professional and technical unionism. Parallel to the stunning victories of Local 1199, the union of low-paid patient care, dietary, and housekeeping hospital workers, and the early successes of Municipal Employees District Council 37 among blue-collar Parks Department workers, in the late 1950s and early 1960s, teachers suppressed their historic differences—professional associations against unions, left and right, and high school, junior high school, and elementary school faculty—to form a united union. In 1964, union president Albert Shanker chose to go to jail rather than call off a strike of the city's 60,000 teachers. When the opposition of the mayor and the Board of Education collapsed, the long-prevailing myth that professionals, in large numbers, would never choose unionism over their conventional tactics was laid to rest, at least among public employees.

The 1960s teachers' revolt—and smaller but equally militant movements among social workers, especially in big-city welfare departments—cannot be entirely explained by economic discontent. Although this was an important spur to unionization, it also marked the first recognition among a significant portion of technical intellectuals that their middle-class status and ideology no longer worked for them. Caught in the multiple grievances of little or no genuine autonomy in their workplaces and low salaries, the teachers fairly quickly supplemented and sometimes displaced their reliance on lobbying with the strike weapon.[10] In the course of unionization, traditional professional ideologies were sometimes discarded but more often relegated to a necessary if subordinate tactic of the struggle rather than a marker of personal and group identity.

The symbol and leader of this new movement among the technical intelligentsia was a socialist and New York high school mathematics teacher, Albert Shanker. Against the prevailing view of the then leading educators' organization, the National Education Association, he became the key exponent of a brand of teacher organization that abandoned the notion of teacher as professional cadre in favor of trade union discourse in which the teacher defiantly adopted the mentality of a worker and the tactics of the workers' movement. Accordingly, Shanker proposed that the union devote itself to practicing trade unionism in the accepted sense of the phrase, that is, that it bargain chiefly over wages, hours, benefits, and working conditions. Within this framework, "professional" issues such as class size, instructional resources, and credentials were subsumed under collective bargaining. At the same time, Shanker—but not the national union of which his organization was the New York affiliate—adopted a distinct craft union persona. It was to be a union of teachers; it might organize some other school employees such as clerical workers, university employees, and even hospital workers, but not as equals.

At first, imbued with this vision, many teachers were hesitant to admit into membership in their craft union the new paraprofessionals hired with funds created by the Elementary and Secondary Education Act, part of the federal antipoverty program, in 1965. The "paras'" jobs were created to enrich classroom instruction in "inner-city" schools whose student bodies were composed of black and Latino children. They would assist in discipline, tutorials, the routine child-care aspects of classroom practice such as putting on galoshes and outerwear for small children, and maintaining instructional materials. The United Federation of Teachers (UFT) leadership insisted the paraprofessionals be organized, and in subsequent years the American Federation of Teachers (AFT)—its parent organization—has organized school secretaries, nurses, librarians, clerical workers, and others. But, despite an expanded jurisdiction, the AFT and the National Education Association (NEA), which became a union in the late 1960s, still bear the marks of their craft tradition. They are plainly *teachers'* unions.

In the early 1960s, as the leader of the AFL union of teachers, Shanker had worked tirelessly to convince the smaller but still considerable Teachers Union membership and members of the large professional associations to join the AFT. By 1964 the rejuvenated and united teachers' federation had won support among an overwhelming majority of teachers. Shanker violated a professional norm by leading the first major teachers' strike in the new era. And even more dramatically, he went to jail rather than submit to an injunction to obey New York's law prohibiting strikes by public employees. Shanker's defiance followed similar actions by hospital and municipal union leaders a few years earlier. The crucial difference was not the act of defiance, but the fact that a union of *professionals* was departing from the traditional script that wrote them into the column of defenders of law and order rather than violators of it. Shanker had boldly asserted, in action, the controversial idea that teachers were like skilled workers, not independent entrepreneurs, and were entitled to withold their labor when employers refused to recognize just demands.

The "public" could sanction strikes by blue-collar and low-level white-collar workers, especially in the private sector. After all, the traditional strike weapon has been, until recently, virtually their only real lever to counteract the superior power of the employer. Following the mass hospital workers' uprisings of 1959-61, which highlighted the substandard wages and poor working conditions suffered by dietary, housekeeping, and nonprofessional patient care workers, substantial segments of the middle-class public were impressed and shocked into support of service unionism that linked itself to the then popular black freedom movement. Moreover, in New York and most large cities the majority of bottom-level hospital workers were black Americans and Latinos. And the unions, especially Local 1199, New York's left-wing-led hospital workers' organization, adroitly made the organizing effort a civil rights struggle. Leon Davis, 1199's president, recruited the Reverend Martin Luther King, Jr., to be patron saint of what soon became a national organizing drive. This example intiated a broad-scale alliance between the black ministry and the new public-employee unions seeking to organize black workers.[11]

While in later years the union tried to attract registered nurses, social workers, and other technical employees, often in opposition to their professional associations, its fundamental appeal was on class issues rather than questions of occupation and profession. When it finally succeeded in winning over a large fraction of New York's nurses, medical technologists, and hospital and nursing home social workers, it was precisely its high-profile reputation as a militant and effective champion of the rights of the oppressed, the working poor, that attracted these newly proletarianized technical intellectuals. The success of 1199 paralleled Shanker's gamble that, at least for the time being, professional ideology had yielded to proletarian interest. The New York organizing experience established, albeit unevenly, a pattern of professional union organizing around the country. The Service Employees, which outside the northeast is the largest union of hospital workers; the State, County and Municipal Employees, which has unionized hundreds of thousands of technical intellectuals employed by state and local governments; and the Autoworkers and the Teamsters, who have had some success among engineers and technicians in the private sector and hospital employees—all downplayed, during their early organizing drives, the professional issues because these are perceived as antiunion weapons of management. Shanker's bold proclamation that teachers, as workers, had not only the right to strike but also the right to treat their situation as would any other group of workers (work fewer hours for more pay and benefits) was both shocking to the middle-class public and inspiring to the union's membership, which had suffered materially from its professionalism, not only in terms of salaries but also in terms of status. From the mid-1960s to the early 1980s, the struggle among technical intellectuals in education and health was no longer *whether* to unionize but over *professionalism versus trade unionism,* which in the context of the battle became the code name for class. The result was an ambiguous victory for both sides.

The National Education Association, which at first had obdurately resisted turning itself into a collective bargaining agent, was so pressed by its trade union rival, the American Federation of Teachers, that it became its own rival and outstripped the AFT in membership three to one. In turn, after years of opposition to professionalism, the AFT has adopted some of the salient features of a professional association. But the union is also transforming many of its characteristics: from close cooperation with teacher-training institutions on curriculum matters to becoming a teacher trainer; from an opponent of teacher participation in school administration, especially curriculum and management, to the nation's most militant advocate of what has become known as "school-based management," in which the teacher's voice can be as loud as, if not louder than, that of parents, the school board, and especially the principal. This switch was in no small measure a result of the fiscal crises that afflicted most large cities in the 1970s and 1980s, which, conjoined with taxpayer revolt, resulted in deep budget cuts for all public services and made the environment for collective bargaining bleak.

However, the new focus on teachers' professional power has other roots. In the last two decades, teachers and health professionals, including physicians, especially in the large cities, have begun to recognize that their credentials and their past experience do not really equip them to deal with a series of radical demographic and cultural transformations. In many cities and towns, teachers and administrators are on the front line of services to a massive new immigrant population from Latin America, the Caribbean, and Asia; they experience a considerable change in student expectations and values, especially with regard to respect for institutional authority, among large sections of the population. In the health professions, new epidemiological issues have replaced older problems. The spread of AIDS among both male homosexuals and people of color is highly visible, but other diseases linked to economic and social oppression—kidney and liver failure, the virtual epidemic of cancer that has become the most terrifying sign of ecological disaster, the reappearance of tuberculosis, pneumonia, cholera, and other diseases of poverty—are transforming health care and its professions.

This is not the place to discuss these changes at length, but they have contributed to the current crisis in teacher education. The old curriculum seems outdated in light of the changing face of urban America in the past two decades. And the relationships between teachers and school authorities in which teachers were content to work from prepackaged curricula, textbooks, and methods supplied by school boards and administrators is disintegrating. Many teachers now see that school administrations and school boards fall back on traditional curricula, authority styles, and so forth rather than being prepared to address the concrete issues faced by classroom teachers: cultural diversity, ethnic difference, massive economically induced health and learning problems among kids, and a new student sensibility often fueled by a widespread despair about the viability of the American dream. Further, although budget cuts have curtailed health, education, and social services, the line professionals have recognized that the budget has often been employed as a rationale for substituting managerial authority for a deep understanding of the new issues confronting the institutions. This political perception promises to change the basis of authority relationships in the schools. The teacher unions, far from maintaining the arm's-length relationship characteristic of most collective bargaining in the United States, are making a fundamental policy change in which they are demanding a share of school decision making at all levels.

Needless to say, this innovation could have complex effects: it might introduce to the education system a new corporatism in which an alliance of teachers and administrators would control, and even inhibit, changes that could threaten their respective positions. If teachers emulate the Chicago example and make their links with newly empowered parents, prospects for genuine innovation would probably improve. No doubt teachers in some places will spearhead radical curriculum and pedagogic changes in an effort to come to grips with the new requirements of cultural diversity. In other instances, teacher power could mean a return to a content-driven conservative curriculum and the introduction of a loving but strict hierarchical pedagogy, all in the name of meeting the specific needs of minorities. In short, the democratization of school authority carries with it no necessary implications for educational change.

In New York, Chicago, and Los Angeles, where strong AFT affiliates are confronting a public perception of massive school failure, the local unions have haltingly begun to support school reform efforts rather than presenting themselves, as they did in the late 1960s during a rise in educational reform movements, especially in black communities, as its implacable opponents. This shift raises the question, With whom will teachers make their alliances? The Chicago experiment with school-based management has pitted teachers against parents in some cases while in others they have close working relationships. The question is what kind of educational model is under consideration. Where school reform means strengthening the authority of the classroom teacher without altering the curriculum toward greater attention to student and community needs, the natural alliance is with administration. But if the aim of reform is to place greater power in the kids' hands, parents and teachers will have to demand a transfer of authority over the school from legislators and professional managers to those vitally concerned with the classroom.

Another sharp reversal is that the NEA has become a crucial actor in national politics. It is credited with having successfully persuaded Congress and President Jimmy Carter to separate education from health and human services and establish a new Department of Education, and it led the fight to save the department when the Reagan administration attempted to dismantle it after less than four years of existence. Today, the NEA resembles a national trade union more than a professional association, but it has refrained from affiliating with the AFL-CIO, a signal that professional identity still has symbolic meaning for an important part of its membership.

Taken together, the NEA and the AFT represent about 80 percent of America's teachers and nearly all teachers in the major cities and metropolitan areas. In the wake of the precipitous decline of production industries, they are often either the largest unions or share that distinction with other public employees in many cities and states. For example, the Nebraska and Washington Teachers' Associations are the largest unions in their respective states, and New York State's public-employee unions are, by far, much larger than industrial unions. Teacher unions are among the major labor organizations in cities such as New York, Philadelphia, Chicago, and Los Angeles. In general, the professional model has declined most severely in the big cities, where schools resemble factories more than institutes, but retains considerable appeal in the growing suburban school districts where salaries are rising and the older middle-class self-image is more powerful among teachers and health technicians.

Academics as Technical Intellectuals

The academic professoriat displays many of the same stratified features as other groups of intellectuals. Teachers at community colleges and many four-year liberal arts colleges are situated, regardless of personal traits, as privileged technical intellectuals relative to elementary and secondary school teachers. Even where their teaching loads approach those of high school teachers, the requirement of a master's degree and, increasingly, a Ph.D. has usually earned them higher pay. In elite four-year colleges, their teaching loads, although formally limited to four to six courses a year (six to nine classroom hours a week), are supplemented by a horrendous load of individual tutorials that may add up to as much as fifteen additional hours a week. At this level of time commitment, this group, regardless of credentials, training, ability, or desire, is typically turned into technical intellectuals, at least from the standpoint of the academic system. Those who manage to overcome the institutional obstacles to writing and research are often able to get jobs in "research" universities.[12]

Nevertheless, the more or less planned stratification of postsecondary education that accompanied its tremendous expansion in the 1960s has created a wave of unionization among professors. In addition to union growth among clerical workers in the 1970s and 1980s—a result partially of the spread of feminist consciousness—the other novel organizing successes were in the proletarianized branches of higher education. Thousands of community and four-year liberal arts college professors as well as clerical and blue-collar workers joined unions, engaged in strikes and bargaining-unit elections, and negotiated contracts with stunned administrations. As with elementary and secondary school teachers a decade earlier, onerous teaching loads, curriculum packages, and concomitant reliance on textbooks combined with relatively low salaries to spur trade union approaches to their professional problems. In community colleges throughout the country (many of which are financed through local rather than state government, a difference that is manifested as deprofessionalization), the degree of organization, while not as high as that among teachers in the lower grades, is impressive.

Of course, the extent of union organization declines as the school's rank in the academic hierarchy improves. Some state universities—New York's massive state and city systems, Wayne State, Rutgers, the California State University system (not the quasi-elite University of California), and a few other research universities—have recognized unions of the professoriat. And there is some union organization in private universities, notably Boston University, St. John's, Long Island University, and some others. But the organizing drive in the private sector was virtually halted when a 1980 U.S. Supreme Court decision in the Yeshiva case determined that professors in private universities were part of management, an excluded category under the Labor Relations Act. Nonetheless, unions of the lower rungs of the professoriat were part of the general unionizing upsurge among salaried intellectuals in the 1970s and 1980s. By 1992, more than 130,000 professional and technical university employees had joined the two largest college teachers' unions, the AFT and the American Association of University Professors (AAUP).

In general, professors engaged in graduate education and research have become important intellectuals who are no longer merely ornaments of the culture, retained as part of the legitimation of the social order. The older image of the scholar prowling the libraries and patiently digging through archives to resurrect a new past, or of the scientist working late at night to add to knowledge *for its own sake,* is fading as the postmodern disdain of history conjoins with shrinking support for such endeavors by both public and private sources. The scholar and the noninstrumental scientific theorist or researcher, those whose work does not lead to immediate, practical use, survive under inhospitable circumstances; at best, they are perceived by university administrators as ornaments frequently honored but rarely heard on policy matters. That is to say, they are merely tolerated, unless they deliberately employ knowledge as an intervention into the present.

The increasing concentration of intellectual skills within the universities since World War II, combined with the high cost of research, has prompted corporations to seek more formal and informal ties with major research institutions even as federal research funds in a number of fields have shrunk considerably since 1980. Together these developments have placed research universities toward the center of economic and political power. With overwhelming faculty approval, administrators have negotiated agreements with corporations wherein the school turns research programs toward product development based on basic science and in effect sells the resultant new discoveries and patents in exchange for financial support. In these cases the concept of the independent university is open to question.

Writers such as Dorothy Nelkin, H. L. Nieburg, and Martin Kenney have profusely documented the university-corporate complex, especially with regard to military-related research.[13] Some universities' geology and engineering programs have always had close relationships with oil and other older industrial corporations, but virtually *all* of molecular biology is linked to the new bioengineering industries. Computer science and the development of computer software are linked not only to the private sector, but also to the Department of Defense. This renders doubtful the claims of these sciences to be independent endeavors. Rather, there is a growing tendency for scientists and engineers in these and other fields to move freely between the universities and the corporations, perceiving little or no difference in the nature of their affiliations or their work.

Charles Derber and his associates, following the earlier work of Noam Chomsky, have concluded that intellectuals constitute a new mandarin order in U.S. society and culture.[14] Indeed, the evidence points clearly to their transformation from professional servants of power to—in the words of Konrad and Szelenyi (writing about Eastern Europe) and Alvin Gouldner (writing about Western Europe and the United States)—a new class on the road to power.[15] It is precisely their command over scientific and cultural discourse, not only their capacity to become central actors in the economic system but also their place in and the needs of cultural and political systems, that constitutes the basis for this judgment. As a knowledge class, they are relatively distinct from both material capital and labor. In this chapter, we have focused on the contrary movement—the proletarianization of a majority of intellectual labor. We have described their increasing subordination to new layers of administrative power—the managers—in both state and corporate contexts. That both tendencies have developed in the latter half of this century calls into question characterizations that try to establish either/or propositions. We conclude that social and technical divisions explain these apparently contrary developments. Scientific, cultural, and managerial intellectuals in all sectors occupy power positions different from those of workers who perform technical and ultimately subordinate functions regardless of their knowledge. So, although the idea of a knowledge class is absolutely fundamental to understanding how intellectuals tend toward class power, it is also necessary to specify where they are situated and, consequently, the discursive limits of their position.

While unions of attorneys and of staff physicians have not (yet) emerged as have unions in other sectors, the rapid transformations in these professions—more than half of all law and medical school graduates will never make partner in a firm or own practices—presage the future unionization of a substantial number of legal and medical professionals. As in other professions, union organization occurred first among attorneys employed in the public sector and nonprofit advocacy and service agencies. A substantial number of legal aid and legal services attorneys were affiliated with District 65-UAW, an old manufacturing, retail, and wholesale union that, in the wake of its shrinking base in New York City, expanded geographically and aggressively sought members during the 1970s and 1980s among clerical workers in large private universities, editors and clerical workers in publishing houses, and attorneys. The union's campaign in publishing had limited success, but it currently represents more than one thousand attorneys. Since the demise of District 65, local unions of attorneys are directly affiliated with the UAW.

Unions of interns, residents, and house staffs have formed in New York and San Francisco and a few other cities, but have not yet penetrated the burgeoning health maintenance organizations that employ thousands of salaried physicians. Doctors' unions have experienced considerable difficulty winning recognition from hospital administrations. Even when they won initial recognition and a contract, they have been forced to strike for contract renewals, and in some cases lost bargaining rights with voluntary hospitals in the New York area. But these unions have managed to survive and in some instances are growing in the wake of multiple crises of the public and nonprofit health care systems. It is not primarily the narrow economic issues that have spurred doctors to consider unions. Their loss of power over patient-care decisions as hospital managements have centralized authority, increases in workload, and deteriorating care facilities have all been interpreted as signs of proletarianization; like teachers and nurses before them, they finally shed, at least provisionally, important parts of their middle-class cultural identities.

Most doctors who have completed specialized training and passed the board examinations for their specialty can find jobs as staff physicians in hospitals and other major health care facilities. Their starting salaries are comparable to those of attorneys hired by large corporate law firms. They are still comparatively well paid, but their income, in real terms, is sharply reduced by the considerable debt they normally accumulate in completing their studies. Nevertheless, the 1980s witnessed a silent doctors' strike against the conventional hospital practice of being on call every other night. They are inundated with patients and have little incentive or time for their own intellectual development. Hospital management regards them as employees, and they retain only a fraction of their former professional respect in the workplace. Consequently, they often experience their work as "a job" to be endured while they struggle to purchase real estate or become interested in investments.

As teachers and engineers once did, and despite the increasing subordination of the traditionally independent professions of law and medicine, attorneys and physicians, except those in some public institutions, have so far preferred to pursue their collective interests through professional associations rather than unions. The possibility that a significant minority may finally buy a practice or attain middle- or high-level management positions, combined with the decline of unionism in the United States, has proven effective in preventing widespread union organization among technical intellectuals. Still, in the past fifteen years, when union organizing came to a virtual standstill, the one important site of increasing unionization was postsecondary education, both among the professoriat and among clerical and maintenance employees.

The spate of layoffs of professional and technical employees as well as manual and clerical workers by large corporations in the early 1990s signaled a decisive and visible shift in Veblen's prognostication that the knowledge "class" will remain an adjunct of capital, precisely because the privileges they enjoyed during the industrializing era and the early years of the cybernetic age have permanently eroded. Of course, the relative proletarianization of the technical intelligentsia does not signify that they have become a new working class so long as they retain the ideology and culture of professionalism, one of whose characteristic features is to foster self-blame for failure.

Chapter 8

A Taxonomy of Teacher Work

The American Denial of Class

Even before the collapse of communism and the decline of Western European social-democratic movements, the once vigorous movements for "social justice"—the American euphemism for addressing class—had already been eclipsed by the almost ritualistic litany of "competitiveness" and the urgency of "growth" and, within opposition, a resurgent discourse on race and gender. Public awareness that the position of U.S. industry in the global economy had severely deteriorated was ruthlessly exploited by corporations, conservative politicians, and economists—academic, corporate, and government. The bottom-line argument was that we could no longer "afford" social justice if America were to regain its postwar dominance. According to this view, U.S. workers had to yield many of their postwar gains, and the huge national debt meant that the social wage—that is, benefits—had to be sharply reduced.

The political environment has swung so far to the right that discussions of race and gender have on the whole refused to grant more than a passing nod to the theoretical issue of the ways in which class divisions *within* their respective domains shape the fate of large numbers of blacks and other oppressed groups, and of women. For example, the growing number of black poor have become virtually invisible within the rhetoric of black freedom. Thus a recent collection of papers on contemporary aspects of "race" theory has exactly one paper addressing the tendency of many theorists to pose "race" *against* class as defining categories of social conflict.[1]

While some radical feminists insisted in the late 1960s and early 1970s that women were a "class," with the marginalization, even the exclusion, of Marxism from feminist theory, the discourse on inequality is bereft of an explicit debate about class within feminism, even as the critique by African-American women writers of the exclusion of race from feminism has achieved a fairly high level of visibility.[2] At the same time, feminist issues within the discussion of racial "formation" are, for the most part, muted.

Race and gender are in effect being treated as class questions, if by that term we imply the definite relation of a group to the panoply of exclusions and inclusions within any social formation. They have displaced the category of economic class and obscured the fact that the overwhelming majority of African-Americans have suffered huge losses in living standards over the past thirteen years. Since the demise of the Reagan-Bush administrations in the early 1990s, we observe some resurgence in the discourse of civil rights for both blacks and women. These rights have been asserted by the respective movements almost exclusively in terms of social and cultural issues such as abortion and, for blacks, police brutality. At best, the "urban" question raises the question of the degree to which the cities have been bypassed in education reform, economic development, and rebuilding infrastructure such as roads and housing. Yet commentators ascribe the rotting of cities to neglect and hesitate to link the fact that most African-Americans live in cities and are working class with government policies of "neglect." Tacitly, "whites," and especially white men, have been judged as a privileged class, barely differentiated by whether they are owners, managers, professional/ intellectuals, or manual workers. It is the phrase "panoply of exclusions and inclusions" that we want to explore briefly as a clue to the class question.

Class-ification is among the chief conventions of human understanding. It is a differencing machine whose implied method is to select among a plethora of characteristics a *standard* upon which to ground boundaries in the natural world and the social world. These boundaries may be vertical or horizontal. Horizontal boundaries connote difference but imply no necessary hierarchy. For example, Marx speaks, without the necessary connotation of inequality, of a *social division of labor*—different industries produce different goods. Vertical boundaries, the usual activity of classification, however, are not imposed merely to differentiate but to signify superiority and inferiority, better and worse, high and low. Plato's classifications of human activity were keyed to precious and base metals. Aristotle's schemata of classification in nature and human affairs were invariably directed to establishing differential *value*. Classes are based upon a *technical* division of labor that spills into social divisions: male and female, town and country, intellectual and manual. For example, peasants and farmers produce food while workers make finished products, but agriculture has rarely been accorded the same place in the pantheon of social structure as the crafts or manufacturing. Contrary to common belief, following "classical" political economy (a normative term derived from classification systems) Marx held that classes are formed by the technical division of labor as much as by relationships of ownership. That is, occupational differentiation implies class boundaries; manual labor is typically *excluded* from access to occupations classified under intellectual labor, which, in turn, *entails* differential relations to production. Intellectual labor, of which the function of management is a very early form, is not merely accorded higher status than manual labor, it is a form of *domination*.

Class theory has suffered from mistreatment by not only critics of Marxism but also its most passionate proponents. Until now, the debate has been framed around a dubious proposition: class as a *political* category is equivalent to its economic dimension. Class formation is reduced to what happens in the process of capital accumulation between those who produce surplus value and those who expropriate it, not merely owners of capital but also their retainers, the preponderance of categories of salaried intellectual labor. The presupposition of this formulation is that the labor process and the concomitant division of labor is the site of class formation. From this *monostructural* view derive efforts to describe occupations such as teacher, scientist, and engineer as situated in a "contradictory" class location between capitalists and workers or to define them as a "new" working class, a "new" middle class, or a professional-managerial class. All of these presuppose that there is a single class system that subsumes all *politically significant* social relations; within this framework, even if they are important, race and gender are, in the last instance, subordinate to those relations formed by the value-producing labor process. Of course, within feminist theory and theories of racial formation, the Marxist emphasis on relation of material production to gender and race has been displaced without fundamentally altering the structure of the Marxist argument. In some feminist and race-formation theory, women, blacks, or the "third world" become the oppressed class and politics is defined as a gender, race, or north-south struggle where almost the entire southern half of the globe is designated as subaltern.

Without surrendering some kind of classification schemata of difference, having shown that many of the political predictions emanating from economic reduction have proven historically false, many have accepted the warrant that class relations, even if they "exist," are irrelevant. For example, the historian Joan Scott argues to the extreme that to speak of class involves a discredited Marxist orthodoxy.[3] In the main, however, the articulation of the relationships between class, race, and gender are simply evaded by most writers. Perhaps one may ascribe this blindness to the bitter legacy of Marxist dogmatism, which, at least until the 1930s, stubbornly refused to grant the category of racial oppression any degree of autonomy; and even as women's oppression was acknowledged, the leading theorists of the old left never went beyond Engels's flawed but pathbreaking *Origin of the Family, Private Property and the State* and, with the exception of August Bebel, the German Social Democrat of the Second International, marginalized gender within class theory. For the past sixty years, the left has, with mixed success, addressed the specific character of race and gender discrimination but with few exceptions has failed to adequately theorize it. Contrariwise, opponents of class reductionism have ignored or otherwise demeaned the issue altogether, but have not asked what the consequences of eliding classification are.

In this chapter and in our fuller discussion of class in chapter 9, we take classification seriously, not in its normative aspect but as a necessary tool of description, since this category remains the crucial sorting device in contemporary society. We argue that class relations are as multiple as the institutional complexity of late or advanced industrial societies. Perhaps the fundamental characteristic of class is that all institutions, most of which are closely integrated but have strong elements of autonomy, are structured as hierarchies of inclusion and exclusion. Consequently, class identity is not fixed. For example, it can hardly be denied that in the post-World War II era, industrial workers in the large-scale industries of the U.S. economy enjoyed considerable job security and income in comparison to white and black workers in highly competitive industries such as garments and textiles and retail and wholesale services. Moreover, despite many strikes and other industrial conflicts, millions of white workers identified their interests with their own employers and the state. Exploited in the labor process, they were nonetheless prone to support U.S. foreign policy and the competitive advantage of their own employers as the best hope to achieve job security. If class connotes more than a determined position in relation to ownership of the means of production, contrariwise, it implies a definite political and cultural identity. The mutiple identities of American workers rendered problematic this "always already" definition. Did African-American workers in these key industries, some of whom managed, during the war, to become craftspersons, share the fate of their fellow (white) workers? Or were they identified with the mass of blacks who remained excluded from the highly paid basic industries and the crafts? Clearly, to be a black industrial worker in a major unionized plant posed some significant conflicts with respect to class identity. To the extent that "worker" was posed against "black" economically, politically, and culturally, black workers occupied a liminal position. Black steel and auto workers were, with difficulty, able to buy homes, new or late-model cars, and appliances and could send their kids to college— privileges that were simply unavailable to the majority of African-Americans. They proved to be among the most militant trade unionists and in many instances were in the vanguard of industrial struggles. At the same time, they suffered discrimination in the plant; they were on the whole unable to advance to the skilled trades or to supervision and, despite considerable seniority, many were consigned to the dirtiest and heaviest jobs, even in the old "progressive" union plants.

Likewise, even if in the labor process groups such as engineers are situated on the "side" of capital, since their work consists in developing technologies to eliminate or control labor, in other power relations they are considered, and consider themselves, workers. In all advanced capitalist countries, even in the United States, where the extent of union organization among professional and technical employees is relatively weak, they form unions to address their grievances against high-level management. The engineer who develops a robotic system that displaces human material handlers does not necessarily enjoy control over her or his work, except in its technical aspect. Orders are handed down to work on this labor-displacing machine, and the rewards of successful accomplishment may or may not be more pay.

Again, many professors in major research institutions have direct authority over the knowledge they produce and disseminate, most particularly to students, but have lost power over university policies, which are increasingly directed by government, private corporations, and the university administration. This triumverate arrogates to itself determination over finance, research, and even curriculum. In recent years, owing to the centralization of educational power in the hands of legislatures and administration, the professoriat, especially in state technical and liberal arts colleges, has flocked into unions. In these unions professors promote their economic interests, and they increasingly use the unions as vehicles for academic citizenship as well. These are cases less of contradictory class location than of differential location within multiple structures.

Thus, organizational complexity—in our definition, individuals and groups affiliated with overlapping power systems and situated differentially within them—becomes the crucial marker of subordinate classes in late capitalist societies. We propose a shift from conceptions of class that focused exclusively on relations of material production in the narrow sense to one that comprehends that *situations* alter class identities in an era marked by globalization in which production has been radically deterritorialized but ownership has become more highly centralized in a fairly compact transnational corporate system. If we retain the clarity of the two-class model on a global scale, proletarianization extends to virtually all strata of the productive system, except a tiny group at the top. Except for periods of fundamental social changes, of which class simplification is a cardinal feature, this monostructural framework renders class relations meaningless because it conceals more than it reveals. However, we do not accept Weber's view that class relationships are confined to what determines the access of a specific group to goods and jobs. Rather, we maintain that class determinations produce significant political as well as social effects. We argue that class is operative in the multiple relations of economic, political, and cultural power that together constitute the ruling systems of production and *reproduction* of goods, services, and knowledge.

Universities as Knowledge Systems

The complex of institutions that constitute the knowledge system—the universities and research institutes, "think tanks," government agencies, and corporations, especially their research and planning and engineering departments (including computer programming and systems management)—are related to global and national capital in specific ways, but they also constitute themselves as hierarchies of power; groups have differential access to goods and job opportunities and, depending on both rank and location within knowledge production, also to the power/knowledge that is produced by the institution. The fact that class power is multilayered complicates the conditions for action, both discursive/communicative and political.

We cannot adequately explore questions of race and gender with respect to the organization of institutional power here, but we know that in the crucial areas of knowledge production, women and African-Americans remain underrepresented, especially in the sciences and engineering. Even as they have made considerable strides in the broad domain of culture over the past twenty years, they have not, with the partial exception of women in molecular biology, succeeded in gaining access to those key areas of knowledge production that bear on capital accumulation: scientific discovery, invention, and their practical applications. And compared to the position of women in knowledge fields, the situation among African-Americans is substantially worse except in chemistry, where they have traditionally concentrated. While African-American students are around 9 percent of the total enrollment in colleges and universities—about three percentage points under their representation in the population as a whole—their participation in the sciences and engineering is about 4 percent of total employment, including categories such as quality control and line supervision, which are the least well paid and are of relatively low status. In research categories the proportion is about 2 percent and is declining.[4] In contrast, although they are underrepresented, women have made significant strides in medicine, including medical research, and in law, both in the professoriate and as practitioners.

Of course, the entrance of women into these fields corresponds to the relative proletarianization that has accompanied the decline of private practice and the expansion of the salariat as a result of the growth of large corporate law and medical firms and quasi-public health maintenance organizations. Women are still lower paid than their male counterparts and are less likely to win partnerships or become top managers. Yet compared to the situation of most women and blacks, which has eroded since the last half of the 1980s, the *relative* position of African-American and women professionals is quite favorable. Their substantial political and cultural capital, derived from university credentials and membership, however subordinate to the political and corporate hierarchies, makes them leaders within their respective communities. In addition to the traditional entertainment and sports routes to membership in various institutional elites, academic prestige and rank have become an important vehicle for achieving influence and even power at the middle levels, if not the pinnacle, of corporate and state bureaucracies.

Certainly, women and African-American intellectuals have become "opinion leaders" because of their ready access to television, newspapers, and other media. As emerging public intellectuals, they tend to influence, if not define, the acceptable *discourse* about race and gender, but only so long as they do not violate too often the unwritten consensus that these "problems" can be resolved within legitimate channels such as legislation and persuasion. And in addition to their function, at a fairly high level, as public intellectuals, some are influential in the political and public policy arenas, even when their power within the university is limited by the fact that they have little, if any, institutional base. The development of this class of public "subaltern" intellectuals is a measure of the significance of these spheres as public issues, but also is a sign that the universities produce many types of legitimate knowledge: in addition to scientific and technical knowledge, they produce ideological and political knowledge—and, perhaps equally important, they produce a *class* of intellectuals for both the pinnacles of economic and political power and for other layers of institutional hierarchies.

Professors under Siege

Despite dramatic changes in the conditions of professorial work during the past quarter century, college and university teaching remains a high-status occupation in the professional panoply. Professors typically are required to earn a doctorate, a credential that requires, on the average, eight years beyond the baccalaurate to complete. The large difference in income notwithstanding, only physicians spend a comparable amount of time achieving their credentials. Additionally, in contrast to most other professionals, who increasingly function as purely technical intellectuals for commercial or public bureaucracies, the professoriat, mainly in elite and major research universities, may adopt the role of public intellectuals, either in politics or in administration. The news media frequently seek their expert views on a wide variety of public issues; since they possess and otherwise represent the authority of legitimate intellectual knowledge, their books and articles are often taken more seriously than those of journalists and other nonfiction writers. Moreover, at least in elite universities, professors have much more control over their time than other professionals.[5] Their relatively light teaching duties and modest departmental administrative tasks leave them free to attend to research and scholarly work in which the subject matter, methods, and outcomes are largely under their own control. (Of course, this is a phenomenological observation that excludes the fact that their standing in the profession depends on the degree to which their work is governed by the rules of their respective disciplines.)

And most (but not all) college and university teachers who achieve tenure after six years of publishing, constant peer reviews, and often backbreaking committee work are rewarded with a lifetime job that can be rescinded only under circumstances of extreme malfeasance. The tenure system, originally designed to protect the intellectual freedom of faculty from incursions by the state, also once protected teachers from economic vicissitudes. The professoriat, though only in some quarters, was relatively immune to labor market conditions. Although this protection still obtains for most faculty, significant shifts in the tenure system are already under way. For example, the San Diego State University administration, facing massive budget cuts, in 1992 proposed layoffs of tenured faculty in several academic departments, and other institutions are considering similar measures.[6] In response to its budget crisis, New York's City University administration proposed cuts that included eliminating several foreign languages and arts departments.[7] Moreover, as we shall show, the relative security of tenured professors is under siege in nearly all universities and colleges. Not only job security, but perhaps equally saliently, working conditions—the very conditions that justify the enterprise of college and university life in terms of knowledge production and scholarship—have substantially eroded in the last two decades, and there are few signs of reversal.

As tenured professors in large universities, the authors of this book for the most part share the conditions of academic teacher work that we describe in this chapter. Our use of the third-person voice should not be construed as naive distancing from our own circumstances. For the last ten years, DiFazio, an associate professor, has worked at a third-tier private university. (We will explain the notion of "tier" later.) Most semesters he teaches four classes and almost invariably teaches two summer sessions in order to meet his expenses, many of which are not covered by his nine-month salary. Occasionally he does adjunct teaching at other schools, including an elite university, in order to earn extra income, and he occasionally consults with labor unions.

DiFazio participates in regional, national, and section professional conferences, where he delivers at least one paper, and frequently more, every year. In addition, he is active in departmental affairs and was nominated by his colleagues to become department chair. He has also served on search committees for new faculty. Since the appearance in 1986 of his first book, a study of longshoremen, DiFazio has earned a reputation as an expert on work and has appeared on television and been interviewed by the press. His current research is about poverty; for six years he has served in a church-run soup kitchen, where he spends two days a week.

Aronowitz is a full professor at a second-tier graduate school where his most important teaching takes the form of dissertation supervision for about twelve students during any single year (although his formal load of dissertation research and advisement and students taking independent studies is about twice that). He teaches two courses a semester, one of which is from time to time at an undergraduate institution. He has been relieved of many departmental committee assignments because he directs a center and coordinates a curriculum in cultural studies—for which he receives no additional compensation and, for most of the five years he has been in this post, no course or supervision relief.

As a center director he is obliged to perform as an academic entrepreneur—that is, to raise funds—as well as an administrator, organizing conferences and colloquiums and participating in research projects associated with the center's activities. Aronowitz gives about two dozen talks a year to professional, labor, and educational groups. He is active as a public intellectual, appearing on radio and television and granting extended interviews to reporters and other researchers interested in labor, education, science and technology, social movements, and general political issues. In addition, he has participated in the development of new educational programs, the most recent of which is a public high school project.

DiFazio attended graduate school at City University of New York (CUNY) and like most of his fellow students was for years an adjunct instructor in one of the university's undergraduate colleges and served as research assistant to several professors. In the early 1970s, while DiFazio was a graduate student, Aronowitz taught full time at a community college, where he also directed a youth and community studies program at four New

York City sites, and taught as an adjunct instructor in several New York institutions including the New School, New York University (NYU), and the College of New Rochelle in order to earn additional income to supplement his relatively low salary as an assistant professor. After publishing two books, in 1977 he became a full professor at the University of California at Irvine, where he taught for four years. He taught for two years as a visiting professor at Columbia University's political science department before returning to CUNY in 1983. Since his first year at Columbia was a part-time appointment, Aronowitz taught three courses a semester as a visiting professor at Rutgers-Newark and another course at NYU's School of Continuing Education. The five-course load, which is also not unusual in DiFazio's academic career, is typical for many professors and adjunct lecturers in community and four-year colleges.

The fact is that only a small fraction of the professoriat enjoys working conditions that correspond to those that have marked the putative professional culture of elite and other major research universities. Today, most professors work at the third and fourth tiers of the academic system: the state and private four-year colleges and universities that are designated primarily as teaching institutions, and the community colleges that claim a growing proportion of students enrolled in postsecondary schools. At all levels of their occupational hierarchy they are obliged to teach between classes and tutorials the equivalent of three or more courses a semester; they frequently perform extensive administrative duties such as serving on committees, engaging in academic advisement, and managing fairly large classes; and despite these tasks they are often constrained, on penalty of being denied promotion and salary increases, to perform research and scholarship, give professional lectures, publish books and articles, and remain moderately to substantially active in their professional association(s).

A 1992 study conducted for the State Higher Education Executive Officers found that faculty workloads had increased from forty-four hours a week to fifty-two between 1977 and 1988. During this period, and especially in the past four years, the rate of full-time faculty hiring has declined in U.S. universities while the number and proportion of part-time instructors has dramatically increased. For example, at Rutgers University's main campus at New Brunswick, there were just 10 part-time instructors in 1988, while in 1992 the number had jumped to 113. In many teaching institutions, the proportion of classes taught by part-timers has reached 50 percent, and the ratio is even higher in community colleges.[8] At the same time, while the fear of unemployment is not (yet) the rule, many professors in relatively marginal disciplines such as linguistics and the fine arts, and even some in major academic disciplines such as sociology and anthropology, have reason to be anxious about their collective and individual future as the fate of their disciplines hangs in the balance in an age of fiscal austerity. Increasingly, university administrations in third-rank institutions are choosing to shift from the arts and sciences to vocational areas.

The deterioration of the conditions of academic work may be illustrated by the decline of support services since the late 1970s in many research universities. For example, more than fifteen years ago, when Aronowitz went to Irvine, the School of Social Sciences maintained a "Wang" room where a group of women at word processors typed professors' papers and full-length book or report manuscripts. The faculty's office telephones were linked to a receptionist who answered them; professors answered only when the office worker rang them. Now, the caller to a central university number is likely to hear recorded instructions for reaching the desired party; if the person is not available, the caller electronically records a message. Human agency has been eliminated from the transaction. At most colleges and universities, departmental secretaries are unable to provide substantial assistance to faculty; they serve the departmental chair in the performance of expanded administrative tasks.

Except in universities with graduate programs, the faculty assistant is a series of electronic instruments. Enter the vast majority of professors' offices, especially in research universities, and you are likely to encounter an answering machine, a personal computer, and perhaps a fax machine astride the telephone. There may be a photocopier, but it is more likely that the professor is at the department's photocopier duplicating class assignments. Today, most professors are their own typists, secretaries, and receptionists. The new computer-mediated technologies have transformed the academic as well as the commercial office; the quantity of clerical labor has not been reduced, but the job is rarely done by a separate individual performing these specialized functions. A significant amount of clerical work has been transferred to the professor.

In 1987 Aronowitz completed the last book he drafted on a typewriter and paid to have it transcribed on a word processor. In fall 1988 he purchased a computer, a 286 IBM clone. Including the present work, he has published four books drafted, corrected, and edited on this computer. DiFazio bought his computer in 1990 and does all of his writing on it. While the new computer-based word-processing technology may have enhanced the writing and communications process, as well as the dissemination of information through such devices as fax machines, modems, and E-mail, the overall effect has been to make possible the integration of clerical and intellectual labor, expanding the tasks of professors and at the same time eliminating jobs for clerical workers.

The effect of technological change—not only machines but also organization—on the professoriat has been to increase workload. The restructuring of postsecondary education currently under way is aimed at dramatically downsizing the academic and clerical workforces, if not the student populations, of the institutions. The professoriat is increasingly subject to productivity norms comparable to those that are part of the taken-for-granted conditions of manual and clerical labor. Despite steady tuition and other fee increases that have routinely exceeded the inflation rate, the student population may stabilize at current levels rather than decline as legislatures and the trustees of private colleges and universities reduce funds available to the faculty and administration. Although the size of the full-time teaching force in higher education is clearly dropping, a *New York Times* article points out that the same currents, especially economic stagnation and decline, that have forced educational austerity make it likely that student enrollments will remain high: education is a better alternative than complete idleness, and many hope that better credentials will yield better jobs and higher income.[9] If this prediction proves correct, professors are bound to suffer further increased workloads; in many cases, even the remaining amenities will disappear for all except a relatively few stars and successful academic entrepreneurs.

The American Academic Occupational Structure

Many, perhaps most, university and college teachers suffer from overwork and underpay, and their job security is more tenuous than at any time since the 1930s. At the same time, working conditions and salaries vary with the type of institution (private or public, research university or teaching college); with whether they enjoy protections and benefits of union organization; with whether individuals have considerable administrative duties such as committee work or whether the school is run from the top. In general, the institutional context determines how much time the faculty member devotes to teaching, research and writing, administration, and other tasks.

Of course there are always individual factors that influence the kind of work a professor does. Even when teaching duties are relatively light, some faculty at major universities do almost no research and even less published writing. And on the other end of the spectrum, a few indefatigable souls regularly overcome formidable barriers of heavy teaching loads and administrative duties in four-year liberal arts colleges and community colleges to publish books and articles. Yet these people are exceptions. The rule is that the institutional setting is crucial in deciding the kind of work the professor tends to do and becomes a fairly accurate predictor of career trajectory. There is considerable evidence that, as in other sectors of society, the cream is more likely to sink to the bottom than overcome institutional obstacles.

Our fundamental contention in this chapter is that, under conditions of austerity (which for some disciplines, especially the humanities and social sciences, have obtained since the early 1970s), academic credentials are the necessary but not the sufficient condition for a successful career. In the past twenty years, receiving a degree from an elite or major research university is certainly the prerequisite for an appointment to a similar university. Yet job scarcity obliges many Ph.D. holders from these institutions to accept initial appointments to third-rank universities or four-year colleges. Often these appointments shape their careers; having landed in these niches, many primarily perform teacher work and only occasionally have time for research. As we have already noted, they may be forced to accept teaching chores in summer rather than having time to write. Since sabbaticals are granted only once in seven years, their ability to do their own research and writing depends on whether they receive grants that relieve them of all or some of their teaching duties. Since in the wake of reduced funds competition for grants has greatly increased, few are successful, and teaching tends to absorb nearly all of the working time of most professors in these third- and fourth-rank institutions.

Five types of academic professionals have emerged since World War II. All have typically earned a Ph.D. in a field of the arts and sciences where the presumption is that they are familiar with the core knowledge of the discipline that is embodied in its canon of theoretical works, specific studies of at least one of its conventional subfields, and its methods of inquiry. These are pure or ideal types—that is, the attributes of each type may not correspond to the work of any particular individual, which may cross over into other types. From the individual's standpoint, however, the work generally corresponds to one of the following types. They tend to spend most of their working time performing as: (a) scholars and/or intellectuals; (b) administrators and managers; (c)entrepreneurs; (d) teachers; or (e) the academic proletariat. Increasing numbers of adjunct lecturers have earned Ph.D.'s in the social sciences and the humanities and are neither specialists in the private or public sector nor graduate students teaching their own classes rather than acting as teaching assistants for a full-time faculty member. These people are not included among the "professionals" because their working conditions are entirely described as one would any category of casual labor.

Higher education is like the mainstream occupational structure in which part-time labor accounts for only a fraction of the labor force: the academic proletariat is not the majority of working teachers, except in an increasing number of community colleges. Most working academics are full-time teachers in the sense in which we use this designation: they teach three or more classes each semester, perform extensive counseling and committee work within their departments, and do very little research or writing.

A Taxonomy of Teacher Work

As we have indicated, the "teacher" in a state college or a four-year private college spends most of her or his time in the classroom, advising students and correcting papers in introductory or lower-division courses where frequent writing has become the norm and, in private colleges, directing senior theses. In addition, the teacher is required to perform departmental committee work.

The following example is a professor at a state college who is required to teach a four-course load, four days a week. This requirement makes it very difficult for her to engage in the research and writing that would enable her to get an academic position at a more desirable university. We interviewed her just after she started working at a private university in New York City. Despite a long commute and a heavy teaching load, she produced two books and a few articles. She discussed her academic experience:

State college was not the kind of environment that I wanted long term. I would have left even if I stayed there. I would have left for someplace else. If I didn't get a job someplace else, I would have killed myself [she starts to laugh]. . . . I think I might have looked for a nonacademic job at that point. I don't know how long I could do that commute. Their requirement was four days a week on campus. It was really teaching, teaching, teaching, semester after semester. It didn't have the variation. I was also concerned that the longer I stayed there, the longer I would eventually be stuck there. I wouldn't have the time to do the writing that I would need to do to get out. A catch-22. So I'm not sure that I would have stayed there long term.

How did you do your writing and research? we asked, and she replied:

I used my fifth day all the time. That one day off I certainly used. I used vacation periods all the time. For three of the four years I didn't teach summer school there. There was one summer that I taught there and I swore that I'd never do it again. Another summer I taught a grad course in the university because that was close. And it was another opportunity to do something else. But then the other two summers I used for professional work. You're right— when I think about it, I did a lot given the kind of commitment I had there. I think that was in my head. I was driven because that was the only way I could get out of there. That drove me, that sort of gave me motivation to write and write and produce. . . .

My first course would be at eleven. I would leave between nine-thirty and ten. The other thing at state college was meetings, you were required to do lots of committee work. The more committees you were on the better light you were in in terms of the administration for tenure. So there was lots of committee work going on as well. In addition, the requirement was five hours of office hours a week, and nobody came. Well, I find at private university [I] get a lot of paperwork done because nobody comes here either. But five hours was just a lot . . . But I was on a million committees at state college. I was on [a] collegewide promotions committee, I was on school-level committees, and on every goddamn department committee. You know everything the department did you had your nose in, or you would have to do some of the work.

Now at a private university she has a three-day-a-week, three-course load. Though she has fewer courses and less committee work, she still must publish to obtain promotions and salary increases, and time must be managed very carefully. She knows this all too well; one of her new colleagues was fired last year and she is in the last year of her contract at the private university.

Scholarly and intellectual work is done, for the most part, by professors at first- and second-tier research universities where teaching loads are confined to one or two courses a semester or quarter. In these institutions, support services such as photocopying and computers may be relatively more available to professors, teaching loads may be reduced so they can complete research and writing, paid leaves are more frequent, and internal and external grants are more plentiful than in other colleges and universities.

We distinguish scholars from intellectuals by the kind of work they do. Scholars may also be intellectuals, but their major work is adding to the incremental knowledge within the "normal" science framed by the dominant paradigm in their discipline. They neither challenge the dominant paradigm nor work outside it. Their work corresponds to Thomas Kuhn's "puzzle-solvers" doing normal science—that is, they are engaged in getting the bugs out of the paradigm and drawing its implications for empirical, concrete problems. "Intellectuals" are defined here as theorists, critics, or scholars whose methodological or ideological orientation may or may not depart substantially from the dominant paradigm. Theorists and critics can be entirely ensconced in the paradigm and they almost invariably use scholarly tools, but they tend to be engaged in a wider scope of reflexive knowledge production. As we shall see, this work requires that they have large quantities of time unbounded by institutional constraints such as excessive teaching, administration, and entrepreneurship. Since the projects associated with producing new knowledge and other types of intellectual inquiry constitute the legitimating sources for the development of the modern university, this activity retains a high degree of prestige and yields substantial honorific, and sometimes financial, rewards compared to more insecure or encumbered types of academic work, especially teaching.

Having engaged in significant scholarship or critical intellectual activity remains a crucial qualification for those who elect to become entrepreneurs or administrators at major research universities. They are recruited from academic ranks, and—especially in elite or second-category universities whose research mission, as opposed to teaching, remains primary—it is important that administrators and those who direct knowledge factories, the entrepreneurs, be or have been legitimate scholars and/or intellectuals. This qualification is necessary to maintain the image of the university as a community of scholars. Front-rank scholars, intellectuals, and administrators may perform different tasks in the academic division of labor, but the presumption is that of being part of the same enterprise of adding to the fund of human knowledge. In recent years, university administration has become an extremely unstable position for several reasons, among them the substantial realignment of the U.S. academic system from its expansion phase to a period of contraction. But it is still an important destination for a fairly large number of former scholars who, having achieved a reputation in their respective fields, want to advance to positions of power or greater financial remuneration rather than remain in the professorial ranks.

Administrators who lose their political power for reasons such as changes at the top, personal or policy differences with presidents or provosts, or, in the case of presidents or chancellors, disputes with boards of trustees or with governors and legislatures rarely return to the professoriat; more typically, they seek positions at or above their administrative level at other universities, usually of the same category in which they have served, or with private corporations, where they often run education and training programs. The exceptions are scientists and humanists who continue their writing and research and manage to keep up with developments in their fields while they are performing as administrators. In sum, administration has not only become one of the major functions within universities and colleges; it has merged with management as a separate career. Many professors actively seek posts as deans after relatively short careers as scholars and often see themselves as full-time decision makers or administrators rather than as scholars or intellectuals.

In this long-term period of contraction, many middle-rank scholars and intellectuals have achieved economic improvement and mobility by securing departmental chairs because their own work is not sufficiently recognized to gain either increased financial advantages or academic recognition without administrative rank. They leave major research universities to accept administrative or teaching jobs in lower rank institutions that offer more money and power. To some extent this new development represents a change not only in their personal career trajectories but also in the opportunity structure of academic institutions.

The entrepreneur as an academic type is among the premier products of the "contract state." This role dates back to the nineteenth century, when land-grant colleges were established, in part to assist the development of American agriculture through research and scientific training of farmers and agronomists, but entrepreneurship has grown exponentially since World War II. Leading universities have become major contractors of federal and state agencies, providing such services as research for the military and for civilian agencies such as the U.S. Departments of Labor, Agriculture, Interior, Education, Health and Human Services, and Commerce, among others. For example, professors have assisted the Department of the Interior, which regularly contracts for research ranging from geological surveys to studies of national parks as sites of consumer leisure, and the Department of Health and Human Services, which administers the still huge social welfare programs of the federal government. More recently, MIT, several University of California campuses, and other schools have conducted research in biomedical fields, especially the bioengineering of DNA molecules. Although the federal government remains an important source of funds for biomedical research through the National Institutes of Health, to the degree that this "science" has been transformed into technoscience whose "products" are commercially viable, large corporations have become important funding sources as well. This development is part of a major trend toward the privatization of scientific research within state universities as well as private institutions of higher education and research. These relationships are spreading from MIT to other private universities such as Columbia, and also to public schools such as the University of California campuses at San Diego and Irvine.

The academic entrepreneur, typically although not invariably a tenured professor, often sets up a "shop" within the university that functions virtually exclusively on the basis of contracts obtained from public agencies and private corporations. In the public sector, the agency sends to likely contractors a request for proposals specifying the kind of research expected, providing guidelines, and setting funding limits. The entrepreneur and staff submit a proposal in which she or he is to be named principal investigator on the basis of presumed expertise, although the entrepreneur's actual relation to the research may vary from intense personal involvement to a largely managerial position. Most of the work is usually done by junior and senior faculty. But a considerable and perhaps crucial role in academic research is performed by professional researchers who (usually, but not always) have Ph.D.'s but either cannot get teaching jobs or prefer, for reasons of higher pay and greater flexibility, to remain researchers. They usually have titles such as research associate and may or may not enjoy health, pension, and disability benefits.

Academic entrepreneurs retain a closer relationship to their fields than administrators but cannot, except under unusual conditions, remain scholars or intellectuals in their disciplines because most of their work is devoted to fulfilling the projects of others. Since they may employ a large number of research associates and graduate students, the survival of the shop often takes precedence over original work that may not be fundable. Moreover, there are considerable financial rewards for academic entrepreneurs, in addition to which they may do little or no undergraduate teaching, confining their pedagogy to directing dissertations linked to shop projects. As the shop becomes larger, academic entrepreneurship itself becomes an occupation whose connection to theoretical developments in the field is increasingly tenuous, although a good entrepreneur must stay abreast of methodological innovations. An academic entrepreneur who is a sociologist describes his problem: he is torn between being a grantsman in order to bring in money for the department, graduate students, and advocacy groups, and for himself and his intellectual work. He says:

I got up this morning at six-thirty so that I could get to my computer by seven-thirty, because I had promised somebody who is writing a proposal that I'm working on that I would finish a piece of the proposal about the needs of this population based on other research that we've done. I spent the first two hours of my writing time doing that, and I sent it to her on electronic mail over the network. And then I started writing a second chapter of a three-chapter monograph that I'm working on certain aspects of for an advocacy group. They gave me a small grant of $15,000 to put together the three chapters of what will eventually become a monograph, written by quite a number of others as well, and cover the economics and management and so on. My chapters are drafts towards a little bit more of the intellectual and graphic reasoning behind what this group is doing. I had a chapter of my own to write and then two other technical chapters to write. By four I had worked enough so that I could meet with you. I've also been working with a colleague on a planning grant from [a] famous foundation.

I have a lot of problems about the whole business of grant money and writing grants and proposals. I really don't spend a lot of my time doing that. I do and I don't. I mean I don't do it as much as I might or some people think I should . . . In the sciences there's pots of money. A scientist has four or five agencies to write to. If he's done his work every year he's got a really good chance of getting money. A scientist's odds are very good. He knows what grants he's going to get. He's spoken to the grants officers. These are all government grants mostly. They're government grants with a lot of overhead money and money for helpers. Anyway, I don't want to make it too rosy because I know they have problems. In sociology, if you apply to a government agency such as NIMH [the National Institute for Mental Health] or NSF [the National Science Foundation] you have to do what they want you to do, the way they want you to do it. And after getting it, it's still very small. Not so much with NSF. You don't go to NSF except for the kind of thing they want you to do, and the way they want you to do it, and so on and so forth.

If you apply for an NIMH grant, which is one of the only agencies nowadays that still pays the overhead and does it like they used to, in a top-shelf way, your odds are very small that you are going to get a grant. And you use a lot of time putting together a grant like that, a lot of hours with very little prospect of getting . . . that money. . . . I try not to do anything if I don't think I'm going to get the money before I start doing the work. My time is getting scarce; I fuck a lot of time away. I mean, I work a lot. I'm still a workaholic. I squander my work hours doing things, writing things that someday I'm going to regret I did. Because if I use my academic position and skills better—differently—I wouldn't do a lot of the grant writing I even do. I wouldn't do a lot of the report writing I even do. I wouldn't have taken that $15,000 grant and spread it out with these graduate students and faculty members and ended up working for the last week and a half and having the responsibility of putting it all together. Because I would have used that week and a half on my magnum opus or whatever the hell it is that's going to be around after I'm dead. Whereas what I'm working on now is sort of going to be a scene or two but it's not any main event. It's the work I do and I'm proud in many ways, but it has a lot of problems connected with it.

Just last spring the president [of the university] called a meeting of all the center directors, and I direct a center. Her message to us was do a lot of work, raise more money, and not only that, what we'd like you to do is make sure you have a really strong advisory board of eminent and rich people from the community—powerful and rich people who then will become benefactors of the university. We definitely want you to have an advisory board. The last thing I want is to have an advisory board of anybody. Because I don't want their advice . . .

I'm very ambivalent about doing this. I can always have some money. Somebody is always going to call me up and say, "We'd like you to do some-thing." And I'm always going to get money. . . . I've been on NIMH panels. I've gotten Ford Foundation, U.S. Department of Labor grants. I've written all kinds of proposals. I've written every kind of proposal. And I'm still writ-ing them now. I have a proposal for [a national institute] . . . I've got a $3 million contract with [a large corporation]. I've done this in cooperation with other people from my center. But they wouldn't have gotten the grant with-out me because they wouldn't have any credibility, they wouldn't have had anybody who knew how to do it. I'm not blowing my own horn. I'm just saying that this is how it is. These are things that happen because of pump priming, because of going to lunch.

Now I'm tired of it. I've been doing it for nineteen years and I'm in my fifties and every time you get a grant, you have to write a report. And that's always one of the things that I've been good at, writing some goddamn report. So instead of polishing up my own more serious writing or writing a casual piece about this or that, or a book review, or whatever, I never do that. I write by the pound. I write proposals. I write memos about proposals. I look at questionnaires. I talk to other people about their research and their proposals. And I've been doing that for a long time. And I'm going to do it because I don't have much choice. But my heart is not in it. I try to be more selective but I don't know if I'll be able to. It's hard to ignore a good idea when it comes along, or a plea to civic virtue.

As a rule, theorists and critics seek neither administrative posts nor grants, except individual stipends such as Guggenheim or National Endowment for the Humanities fellowships, because they conceive their work to be generally satisfying, and the prestige gained by the major figures exceeds that of most university presidents. Many who choose administration are in fact middle-ranked scientists and scholars who have already exhausted their intellectual potential with a modest contribution to the dominant paradigm. One would be hard pressed to name a significant number of intellectual leaders who have chosen academic administration above department chair level. For this reason, administrators are increasingly political actors within the institution rather than intellectual innovators. There are exceptions to this pattern: Derek Bok, past president of Harvard University, is a major scholar in the study of industrial and labor relations, although of a previous era; the late Columbia University Teachers' College president Lawrence Cremin was perhaps America's preeminent education historian; and several prominent scientists such as John Toll of the University of Maryland and Harold Brown of the California Institute of Technology have become presidents of leading research universities.

Institutional Contexts

We have proposed distinguishing among four different institutional contexts within which the various types of academic work goes on: elite universities, second-rank universities, purely or predominantly undergraduate teaching institutions, and the community colleges and two-year technical colleges. The elite universities include the Ivy Leagues and major private research universities (Johns Hopkins, the University of Chicago, Emory, Stanford, Duke), certain campuses of the Big Ten (Michigan, Illinois), and two campuses of the University of California (Berkeley, UCLA).

Most of the second-rank universities aspire to the first rank, and some departments may lead in their respective disciplines. This group includes campuses of state universities such as the Big Ten and the University of California, North Carolina, Rutgers, some campuses of the State University of New York (Stonybrook, Albany, Binghamton, Buffalo), the City University of New York (CUNY) Graduate School, Virginia, New York University (NYU, which is private), Texas, Penn State (a leading defense contractor), Georgia, Maryland, and some Catholic universities (particularly Notre Dame, Georgetown, Catholic University, and Loyola of Chicago).

In these first two categories are found most of the scholars, intellectuals, and academic entrepreneurs. With some exceptions, undergraduate teaching until recently was not conceived by legislatures that funded them, or indeed by faculty and administration, as the core "mission" of the institution. Rather, the faculty is expected to perform research and publish articles and books, and the institution traditionally releases certain faculty from (most) teaching duties when research projects are internally funded. In addition, many faculty are expected to seek and procure outside funding for their research or grants used to obtain release from teaching and administrative duties within their departments.

Typically, the humanities faculty receives little or no outside funds for research and writing. The two-course teaching load is meant to facilitate time to write and perform research. Still, in recent years humanities faculty members have been encouraged by administrators and by funds made available by federal and state governments and private foundations to establish centers, seek grants for some types of research, and otherwise emulate the practices of the social and natural sciences. These centers have given humanities professors an opportunity to accumulate some power and prestige and to establish networks within and beyond their home universities.

Sometimes the attraction to establish and maintain centers and institutes is linked to curricular and other intellectual innovations such as the recent explosion of ethnic and cultural studies among the language faculties. For example, centers and Ph.D. programs in cultural studies and related fields have been established at Pittsburgh, Santa Cruz, CUNY Graduate Center, Minnesota, Irvine, SUNY Stonybrook, and other universities, and there are relatively new ethnic studies departments at the University of Washington and the University of California at San Diego, among others. Similarly, scholars and intellectuals have felt obliged to accept administrative positions to run women's studies, African-American studies, and, more recently, gay and lesbian studies programs. These developments are linked to a need many feel to secure legitimacy for new forms of intellectual knowledge. In many instances, these faculty are reluctant administrators and entrepreneurs and, like rotating department chairs, these posts are accepted out of duty rather than ambition. Natural and social sciences faculty within research universities are routinely expected to seek outside funds, a process that most often entails combining scholarly and entrepreneurial functions and, equally importantly, tailoring projects to the priorities established by government and corporate agencies. In some fields, especially the natural sciences, the ability of a candidate for tenure to secure outside funding may be taken as the moral equivalent of publications or used as a criterion, even if it is an unstated one, for determining tenure.

The ethical questions posed by the increasing collaboration of scholars and intellectuals with government and corporate policy makers has called into question the status of the university as a site of critical and "disinterested" inquiry. More to the point, the transformation of universities into producers of useful knowledge—for the development of military matériel, for "products" that can be sold in the market, and for social and other aspects of public policy—has also raised the issue of institutional independence. Needless to say, the process of educating and training scientists and humanists rarely includes a discussion of these ethical issues even as graduate students are regularly exposed to the practical consequences of the articulation of university life with the needs of the larger society. Their income is frequently derived from grants that fund research projects that employ them as assistants, a practice that has become more common since financial aid and educational grants have been cut back by all levels of government. On the other side, the scientific and humanistic disciplines assume that scholars pursue knowledge "for its own sake," a pursuit that has become virtually impossible in most of the natural sciences because as collective enterprises requiring complex and expensive equipment they are entirely dependent on outside funds.

Naturally, a fairly high proportion of scientists share the values of their benefactors, a happy coincidence that ameliorates some of the ethical agonies suffered by dissenters. And, on the other hand, many active scholars and intellectuals are able to work within universities without external support. This is particularly true of faculty doing theory and criticism—social scientists, and humanists such as philosophers and literary critics. The position of historians and of those working in empirical subdisciplines in sociology, anthropology, economics, psychology, and political science resembles that of the natural scientist. For although, with the exception of experimental psychologists, their research requires no laboratories or expensive equipment, they need such amenities as access to computer databases, travel funds in order to work in libraries and other sites where documents are available, and released time for ethnographic or survey research. Increasingly, the universities do not provide this support; at best, faculty rewards may take the form of reduced course and administrative loads. Travel funds, except for attending professional meetings, are increasingly in short supply.

These problems are exacerbated for research-oriented faculty at third- and fourth-tier institutions.

The third rank are the purely or predominantly undergraduate teaching institutions. These are usually called colleges; if they have adopted the university designation it does not routinely entail postgraduate Ph.D. work, although they may grant these degrees in certain fields for historical or institutional reasons. The administration, the faculty, and, in public schools, legislatures perceive that the key function of the institution is *teaching*. At the top of the third rank are the important small four-year private schools, the so-called seven sisters: Barnard, Smith, Mount Holyoke, Bryn Mawr, Vassar, Radcliffe, and Wellesley. Bennington, Sarah Lawrence, Bard, Mills, Trinity (Hartford), Occidental, Haverford, Hampshire, Wesleyan (which occupies a somewhat more complex and ambiguous position), and Earlham are of similar stature. These are elite schools from the perspective of the social composition of the student body, which, except for a few scholarship recipients, is recruited from among the upper-middle-class and ruling-class socioeconomic categories. From the standpoint of faculty, however, the teaching responsibilities in these schools are often onerous. The heavy workload is not chiefly signified by the number of courses faculty are required to teach; on the surface, the teaching load resembles that of major research universities. In addition to classroom teaching, however, they are required to perform tutorials with seniors who are completing their theses, and some schools have a tutorials system throughout the four-year curriculum. Faculty may be required to carry from ten to fifteen tutees a year. As often as not, they meet with students at least once a week. Moreover, their office hours for students enrolled in their classes may be five hours a week rather than the one or two hours typical of faculty in research schools. In addition, they are required to participate in campus administrative and social functions; at major universities, faculty participation in these functions is far more restricted.

Here are two testimonies from this tier, one an art professor in the New York area, the other an English literature professor from the Midwest.

The art professor says:

The week before the term begins students come to interview me for two days. They make selections based on these interviews as I do. These are fed into a computer and that's how the students in my class are selected. Fifteen students in a class. I have two classes. I've got excellent students in my classes this term. I spend a lot of time with them both in class and out of class. Every two weeks I have to meet with all my students for a half an hour. It's a lot of work and I have my own work to do. I have two shows of my own coming up. It's not only the students but my work is demanding too.

The English literature professor says:

Listen, I love my students. They're not just spoiled brats. Very few of them are spoiled brats. They're hardworking, they care about their work. I teach them English literature and they care about it. They take it seriously. There's a real community here and I get to know them. That's very nice. I don't mind having to spend so much time with them. What I mind is the overzealous, self-sacrificing martyr complexes of my colleagues. I mind that they see my publishing, my own ambitions, as a lack of commitment to the students. They boast about what shining lights they were, what great careers they could have had. They boast about how they sacrificed their own success for the good of the school and their students. They're sincere about this, I don't doubt them. But because I publish, get grants, and take my own work seriously, they ignore my own teaching. They ignore that I'm a good teacher. They see me as betraying their sacred school traditions. This is what I hate.

On the second tier of the third category are the majority of state colleges and universities. Of these, most City and State University of New York campuses (their combined enrollment is nearly 500,000), the California State University system (as distinct from the University of California system), and most New England state universities possess some of the characteristics of second-category schools because they have somewhat lighter teaching loads, corresponding to the fact that some departments, especially in the "hard" sciences, are designated as knowledge-producing faculties. Yet because many faculty, regardless of rank, are increasingly called upon to carry heavier teaching loads in a time of budget austerity, any pretense that these schools promote an active intellectual life has rapidly eroded.

However, except for the Rutgers University campuses, all of the campuses in the New Jersey college system and the preponderance of the campuses of the State and City University of New York campuses and hundreds of others are more or less explicitly designated as knowledge transmitters, technical training programs, and, at worst, holding pens for a surplus labor force that would otherwise be counted as unemployed. Most Catholic colleges and universities fall into this tier of the third category. In these schools professors are required to teach between three and five courses a semester; they have large enrollments and few, if any, seminars for advanced undergraduates or graduate students. They are frequently obliged to share offices, have limited amenities such as photocopying, computer, telephone, and fax services, and may be required to perform extensive student advisement. There is little intellectual life on these campuses. In contrast to the first two categories and the elite private colleges, they are off the circuit of prominent academic lecturers because they generally do not have funds to bring them (although student funds are often sufficient to underwrite the cost of rock concerts). Characteristic of the attitudes of administrators toward the faculty is the statement made by a dean at one of these campuses: "If you're here you must not be good." Despite the fact that teachers employed in this tier of the academic system typically hold Ph.D.'s and began their careers hoping to join the top ranks of their profession (their referents are Yale, Harvard, Berkeley, and other major research universities), in relation to the broad division between the professions and subprofessions, their work, status, salaries, and lack of autonomy lead us to conclude that, from the perspective of income as well as status, they are more subprofessionals than professionals. How do these professionally trained academics survive under subprofessional conditions?

As we have already mentioned, some become administrators or aspire to administrative positions.

Some, like Robert Merton's ritualist category, become "teacher-teachers," that is, the teaching function becomes all-important and scholarly work becomes almost totally unimportant. As one of the academic professors told us, "I love to teach. The only reason I think about publishing is that I'm coming up for tenure. . . . As you know, my normal load is five courses. I'm tenure track. I'm an assistant professor. And I'm up for tenure, I guess, in a couple of years."

We asked him what he would do if he didn't get tenure, which seems possible because it is very difficult to find time for writing when you are teaching a five-course load. He replied:

> I would certainly try to find another job in academia. Yeah, I love teaching, it's my first love, but I couldn't go back to adjuncting. I couldn't afford to. If I didn't get tenure and I couldn't find another full-time teaching job I would have to go to another field. My first choice would be some kind of research job, like drug research. There's a lot of federal money going into drug research. I'd probably look in that area. And if that didn't work, there's a lot of other possibilities like teaching in the public school system, the private schools, something like that. . . . I'm a teacher. I'm a researcher. That's what I want to do. I don't want to do anything else. What would it take for me to go out of my field?

Some become involved in civic life and make their teaching just a job, while their real passion is local politics, church work, union or community activism, or, as one person told us, "caring for the halt, the lame, and the blind. We'll never do great intellectual things, but we're the people that make society function." As Leon Trotsky once said of the socialist leader Norman Thomas, "he is a socialist out of a misunderstanding." We might parody this remark by observing that some academics wandered into sociology when they were really looking for social work, political science when they really intended politics or policy, economics when they really wanted policy consulting. Of course, we might apply this type to English as well. Many professors take teaching writing and communication skills seriously not so much for pedagogic interest (although there are those who have become intensely involved in composition for this reason), but because they are dedicated to helping people function better as literate workers and citizens. We do not mean to deny the intrinsic intellectual fascination of language concerns. Surely, learning a second language is an increasingly important area of criticism and scholarship. In the present context, we address those for whom teaching becomes a "helping" profession rather than an object of formal intellectual inquiry.

Some connect themselves to subsections of their fields. They become active in affiliated organizations of the major professional association and publish in less prestigious journals and follow specialist literature, even when their teaching assignments do not afford them the chance to teach in their own specialty. They are engaged at this level with little prospect of professional advancement within their own institution even though they may enjoy a modest reputation among fellow specialists. Their choice is to seek specialist rather than institutional or larger-scale professional gratification and recognition. Through specialist networks they may gain consultancies with corporations, hospitals, and local government agencies. This path is especially relevant for social scientists, although composition and educational curriculum specialists often get similar opportunities.

Institutional dissatisfaction leads some to attempt to switch professions. Two examples are law school and psychotherapy, each of which entails between three and five years of training. Success ratios are fairly modest. For most professors, even after completing an alternative professional program, may remain full-time teachers because it is not easy to attract clients, to get jobs at equal or comparable pay and security in law firms or agencies. Often they are confronted with the prejudices of managers and clients about their advanced age; whatever their life experiences, they are obliged to enter the new profession at the entry level. Some end up practicing their second profession on a part-time basis. For many others, retraining does not result in actually leaving teaching or practicing the new profession.

An unfavorable institutional situation leads many to a vigorous pursuit of the private sphere of family, friendships, and avocations. For example, a political sociologist whose early work received some recognition in his field in middle age became a fanatic maker and flyer of model airplanes; this intense interest far exceeded any lingering interest in his academic field. Similarly, a promising organization theorist became a devoted tennis player and gambler in his later years. For others, parenting becomes the defining activity of their lives; sadly, they rarely write about this abiding passion, about which they obviously have great expertise.

Needless to say, all of these accommodations also take place, but in a more restricted way, in leading academic institutions. It is not uncommon to find a full professor at the top of the profession, having "run out of gas," will, or other incentives to scholarly and critical pursuits, attending to otherwise "private" pursuits or operating as a veritable full-time consultant in the private sector. There is no need for faculty of elite institutions, unlike those who are propelled to leave academia, to change venues. They can retain their professorships and make a lot of money in business at the same time. For example, one of the most famous studies of patterns of consumption among working-class housewives was carried out in the context of a market research survey for a soap company; the principal investigator operated a private market research company while teaching at the University of Chicago. This is not an unusual case. Similarly, many economists at leading universities own consulting firms or work for corporations on a more than casual basis.

On the second rung of the third-tier campuses, one may observe the most virulent form of the primacy of the administrator over the faculty. Although these tendencies are present in virtually all levels of academic life, administration tends to dominate here in an almost pure form. The traditional conception of the university as a community of scholars is far from the self-definition of these institutions. They are, and conceive of themselves as, vocational schools that train subprofessionals and technical knowledge employees as we described them in chapter 4. Where the third-tier university has a law school, graduates find jobs in law enforcement agencies or open neighborhood practices and rarely rise to major corporate or criminal law firms.

The convergence of some areas of law with the subprofessions may at first appear anomalous, but on closer examination one may discern that even the traditional high professions such as medicine and law are hierarchically organized. At the lower levels of legal practice, the functions, financial rewards, and status are scarcely above those of subprofessions such as pharmacology, social work, and teaching. A graphic representation of this convergence may be observed in Sidney Lumet's film *Q and A*, in which a new assistant attorney assigned to the homicide division of the police department becomes embroiled in a case involving a police lieutenant who is accused of using excessive force and killing a suspect. The young man, a former cop and the son of a policeman, is advised by the head of the division to bury the case. When he refuses, he is, like Frank Serpico, an example of a previous generation of whistle-blower cop, treated as a pariah. Lumet invokes the banality of the ordinary work of low-level public attorneys who are expected to do little but certify the policies of the departments to which they are assigned. The young man's refusal to acknowledge this role propels the narrative and provides the dramatic impetus for the film. Relevant for our analysis is the fact that he is depicted as a night-school graduate of a third-tier New York law school. Here Lumet is drawing on tacit knowledge about the close relationship between third-tier law schools and law enforcement agencies. Many of their students are recruited from the uniformed services of local urban communities. In addition to cops, probation officers, parole officers, and prison administrators may return to law school to either change (sub)professions or to enhance their positions.

Many state colleges were originally founded as teacher training schools; their elevation to liberal arts institutions accompanied the higher education explosion of the mid-1960s when the demand for teachers was supplemented by a new demand for technical and administrative labor, veterans of the Vietnam War were seeking mobility, and policy makers and politicians were beginning to face the surplus labor force issue. We would not like to be interpreted as saying that these schools are, in the emerging postwork world, primarily aging vats, but there can be no doubt that in the context of a fairly unpopular war, war-induced economic growth, and technological transformation the consequences of which are only now being fully felt, the expansion of state-supported higher education was a felicitous way of retaining a large degree of legitimacy for the state and its policies. Seen as a welfare agency, these schools performed brilliantly to provide income, take up students' time, and provide training for jobs in the burgeoning service sectors.

With the exception of many faculty in technical or subprofessional fields such as nursing, management, medicine, and business—where the professoriat is routinely recruited from the professions or occupations and the job brings considerable prestige and security—the faculty in these schools is fairly dispirited, if not actively angry. It is from these ranks that academic unionism has grown in the past two decades. As we argued in chapter 7, the typical response of a faculty that experiences its work as a degradation from the expectations many had in graduate school is to focus its energy on making more money, devising ways to reduce the actual teaching requirements by duplicating sections of the same course, a maneuver that reduces preparation time if not time in class and, for a minority, to leave teaching and join the ranks of administration.

The fourth category is the community colleges and two-year technical colleges, many of which are private, for-profit institutions that offer associate degrees and certificates. These schools have two separate and sometimes conflicting missions: historically, in their original "junior" college form, they were either finishing schools for young upper-class students or prep schools for senior colleges. When the community college movement began to gain steam in the 1920s and 1930s, and especially after World War II, they gradually added vocational courses and programs and relegated the arts and sciences to "service" departments. Students and faculty who recognize the limitations of vocational degrees have, however, resisted this trend. As Brint and Karabel have argued, vocationalization was largely an invention of administrators that was approved by legislatures rather than a direction dictated either by economic trends or by student preference.[10] Consequently, the concept of the associate degree as a transfer rather than a terminal credential has gained strength. Today, the community colleges are by far the fastest-growing segment of postsecondary education, and enrollments are approaching those of four-year colleges.

Arts and sciences faculty in community colleges work in "divisions" such as science, social science, and humanities. The courses offered in these divisions are frequently justified as breadth requirements for otherwise uninterested students destined for low-level medical, business, technical, and administrative careers. Consequently, the proportion of introductory courses—or in the case of the humanities, of writing courses—to electives is extremely high. Many community college faculty *never* teach electives or courses corresponding to their particular scholarly interests. They almost invariably teach four or five courses a semester. In effect, their course load is equivalent to that of a secondary school teacher, even if the hours spent in the classroom are somewhat shorter. On the other hand, they are more extensively responsible for keeping the institution running and, as counseling staffs have been reduced, spend a considerable amount of time on academic and career advising.

A professor at a community college told us:

> I can't continue to teach here. I can't continue to teach five courses a term. I can't stay here. I want to publish. I want students who really care. Very few of my students care about what I'm teaching them. They complain about reading books. Most of my colleagues assign only one book in a course. Because I assign more books, because I care and I make them read, they treat me like I'm the enemy. It's very discouraging. This isn't why I became a professor. I thought of myself as a scholar. I want to publish scholarly work. I'm trying to get my dissertation in shape so it can be published. But it's hard with five courses a term and students that don't care. I tell my students—my good students—that they have to transfer out of here.

At the bottom of the teacher category is the adjunct lecturer. In most state colleges and third-category universities, at least half of the basic teaching force is now adjunct faculty. Adjuncts teach as many as 70 percent of the sections in community colleges, and between 50 percent and 60 percent in many third-rank institutions. (In a large third-tier university in the New York metropolitan area, adjuncts teach 71 percent of the sections.) Before the 1970s, the adjunct lecturer was usually a specialist who supplemented regular faculty expertise. Thus, in some areas of the law—maritime law, for example—accounting, public administration, journalism, and medicine, the adjunct was conceived as a necessary *supplement* to the faculty. For the past two decades, however, the adjunct has become the key actor in many teaching universities and colleges.

As the academic job market dried up in many liberal arts fields by the late 1960s, adjunct lecturers with Ph.D.'s became more common, especially in the large metropolitan centers and major university towns. At first this phenomenon was ascribed to a refusal to truly "graduate" and travel to other parts of the country. In the past fifteen years, however, the number of adjuncts who are fully credentialed for university teaching has grown dramatically. The plain fact is that there is an overproduction of Ph.D.'s in many fields in relation to the number of full-time teaching positions available. This problem was aggravated during this period by the unusually large number of *young* tenured faculty whose positions were created in the academic expansion of the late 1950s and 1960s, but it also reflected the general cutbacks in funds for public higher education.

Many in this new academic proletariat work as adjuncts full time and teach between five and seven courses a semester, frequently in different schools. And most teach one or two sections in each of two summer sessions, bringing their load to twelve to eighteen courses a year. For this work they earn between $18,000 and $30,000 a year, depending on the per-hour or per-section rate. The range is between $1,000 and $2,000 for a 40- to 45-hour course in a semester. In a few instances, especially the highly unionized community colleges and some third-category schools, they are qualified to receive health benefits if they teach a minimum number of courses and have seniority. For most adjuncts, however, there is absolutely no tenure or other form of security. Regardless of their long experience in teaching within a specific institution, their income is at risk from semester to semester. Often they are rated by students as the best teachers.

Needless to say, the use of adjuncts for many introductory courses and a substantial number of lower division electives has further reduced full-time positions. Equally important, it has effectively reduced services such as academic advisement, since adjuncts who because of economic imperative teach five, six, or even seven courses a semester in widely disparate institutions are hard pressed to work with individual students. Their classes are usually large, and they are not typically provided graders or graduate assistants, especially in third- and fourth-category schools. Despite serious reservations about using multiple-choice questions for grade determination, circumstances often compel them to use this pedagogically suspect tool of evaluation. In turn, students are deprived of the chance to write papers, perform research, and make class presentations in the lower division courses. And in many instances even upper division electives have the same conditions: large classes and adjunct teachers.

In her excellent study of the new academic proletariat, Emily Abel found, in dozens of interviews, that many adjuncts are persuaded by the American ideology that they are personally responsible for their own failure to obtain a full-time academic teaching job.[11] The pervasiveness of self-blame reveals the degree to which the self-perpetuating features of the academic system are introjected by one group of its victims. Abel details the large number of explanations adjuncts give themselves for their situation. Rarely do they perceive the political, economic, or cultural features of the academic system as contributing to their circumstances.

Why do many Ph.D.'s hang on to adjunct status given the uncertain employment, low pay, powerlessness, and meager—if any—benefits? Some leave the profession after three to five years of unsuccessful job searches and drift into word processing, computer programming, and other occupations they learned during their academic careers. Or they work in an entirely unrelated job such as sales, business, or public administration. Yet a large number are tied, emotionally and intellectually, to the academy and its ideal of teaching and writing as a vocation. They suffer extraordinary deprivations: they have virtually no prestige among their colleagues, who are grateful to be relieved of some of the burdens of composition courses, introductory courses in the social sciences, and the like, but because teaching is not valued, except in the elite liberal arts colleges, their work is not valued, other than in crass economic terms. Professional adjuncts who have Ph.D.'s become progressively isolated from the latest developments in their disciplines unless they make strenuous efforts to keep up, and they have no power over what they teach, departmental curricula, or, indeed, their own schedules. Ironically, in contrast, adjuncts recruited among students doing dissertation research and writing are probably among the most up to date of all teachers.

One adjunct with his Ph.D. told us:

> I've taught seven courses in a term as an adjunct. I'd really prefer not to have to run around like that. But to finish my dissertation I taught only one or two courses for a year. I spent more time doing my dissertation and I was able to

accomplish more than I ever did before. In the end it was very helpful in getting the thing done. But it's the typical dilemma for adjuncts, on the one
hand you need time to finish your dissertation, on the other hand you need
money if you have a family. And because I'm in debt I am adjuncting eight
courses in one semester at three different places. That's my record.

Another adjunct described the difficulty she is having trying to finish
her dissertation and be a mother after she spends her days teaching at two
different universities. It is a difficult task, and the monetary and status
rewards are few:

Whenever I can scrape up some time, I'll do it. And sometimes I'm in the
state of mind where I can push other things out of the way and sometimes I
can't. It's a lot like the dilemma that Virginia Woolf talks about in *A Room
of One's Own,* because when I'm finally not too sleepy and everything's done
at night and I'm wide awake still and I go downstairs and I sit at the computer and start working, in twenty minutes my daughter walks in with something, or my son comes in with a question, or my baby wakes up. Something
is always interrupting me, and I've learned to work in little spurts. It's not a
good way to work but it's the only way I've got. I've got a lot of anger. I've
got to deal with the anger of not being able to work, of not having any free
time.

Many profess to love teaching, especially those who are at schools
with large numbers of adult students who bring their rich work and life
experiences into the classroom and are often more eager to learn than traditional students. Others are enamored of the *idea* of the life of the mind,
reflecting a strong belief in education and knowledge for their intrinsic pleasures. A third group entered teaching as a political vocation, not in the sense
that they wish to "convert" students to radicalism of the left or the right,
but in order to stimulate critical thinking against what they believe to be the
deleterious effects of mass culture and official knowledge. Finally, many
become uneasily "comfortable" in the adjunct niche. They may earn a meager and insecure living, but it is enough and it beats the work offered to
them elsewhere. For many, especially those in college towns, the work is the
best they can get without migrating. As remarkable as it may seem, and all
complaints aside, the adjunct instructor is today a regular type of academic
in the third- and fourth-category institutions. And although the official position of the leading academic unions—the American Federation of Teachers
and the American Association of University Professors—is that the category
should be abolished in favor of full-time positions, little has been done to
arrest the substantial growth and perpetuation of this teaching category.

The Fate of the Professoriat

Like engineering and the sciences, college and university teaching cannot be regarded as a unified profession in which equal credentials enable individuals to share equal opportunities. The job destination of successful new Ph.D.'s is determined by factors such as the status of the institution that granted their degrees; whether they work in a mainstream subdiscipline of a field that is currently in fashion; whether their work is considered within the bounds of "normal" scholarship or is on the margins; whether their dissertation advisers have good networks to propel their careers; and, of course, the general economic climate for academic work. Department faculties still generally do the searches for new faculty and are influential in deciding who gets hired, even though deans sometimes intervene on behalf of preferred candidates. Although it is not invariably true that faculties tend to duplicate themselves, this is the general rule. The Ivy Leagues and other elite institutions rarely hire candidates from the second category and almost never hire from the third category. In turn, second-tier research university faculties prefer candidates with first-tier credentials since most of them are graduates of Ivy League or major research universities. In the economic climate that has prevailed for the past two decades, graduates of Ph.D. programs in third-tier institutions generally do not compete successfully for third-tier openings. When they are successful in procuring academic jobs, it is likely to be in community colleges or technical institutes. And recently even these jobs are difficult to get because second-tier graduates are competing for them.

This pattern effectively reproduces the stratification of universities, not only in terms of the continuity of faculty from certain institutions across academic generations, but also in terms of intellectual issues. Our survey of some leading research faculties in the social sciences, culled from catalogs and the magazine *Lingua Franca*'s periodic listing of recent academic appointments, confirms the conservative patterns of faculty hiring. Scholars and intellectuals working in new theoretical modes may occasionally be hired by leading universities if their credentials are appropriate, but unless they are in literary studies, where a veritable intellectual revolution occurred in the 1980s, their chances of getting jobs in leading research universities are slight. Consequently, dissonant intellectuals are likely to land in four-year liberal arts colleges, third-rank state universities, or Catholic universities. They may be able to get a job in a major university if their books and articles receive attention within their academic discipline; if they work in emergent discourses such as feminism, ethnicity, and race, which are periodically in favor; or if they enter the university as well-known professional practitioners in fields such as journalism, accounting, administration, and education. Often public attention is interpreted by faculty as evidence that the scholar is merely a "popularizer" and probably not up to the "standard" of the elite institution. The scientific or scholarly community rather than the public is the significant referent for the professoriat, marking the partial eclipse of the older public intellectual. The notable exceptions today are blacks, Latinos, feminists, gays, lesbians, ecologists, and the dwindling number of labor and other organic intellectuals of social movements. Consistent with the fragmentation of the alternative and adversarial culture as well as what we have described earlier, following Lyotard's characterization, as the postmodern rejection of totalizing discourses, the public intellectual exists only as an organic intellectual of power (Henry Kissinger and Samuel Huntington; in a somewhat more "liberal" vein, Arthur Schlesinger; and, more recently, Robert Reich) or at the margins within or related to social movements.

In some fields, the job market is such that only degrees from elite or second-category institutions afford the candidate a teaching job of any kind. History, philosophy, English, and smaller fields such as geography, linguistics, German, and French—among a declining group of once-prominent languages—have found themselves in this position for nearly twenty-five years. Thus, many graduates of elite schools got jobs in the 1970s and early 1980s in third-category institutions and were unable to move even when they succeeded in publishing their dissertations and a second book: a prominent historian of immigration teaches at a community college; a leading scholar of seventeenth- and eighteenth-century English literature taught at Staten Island Community College for many years; a promising, well-published anthropologist lands in a third-tier college.

Attributing career chances to individual merit is a uniquely American trait. We argue, on the contrary, that the chance occurrences that put many new Ph.D.'s in teaching institutions are overpowering for many not only because of oppressive teaching loads but also because they are isolated from both broad cultural developments and intellectual currents in their discipline, especially if they work in schools located far from metropolitan centers or major universities, where some intellectual life goes on. Put simply, the point is this: the size of this country as compared, for example, to France or Great Britain precludes a single city like Paris or London remaining a center for intellectual endeavor. Nor has the U.S. academy developed major intellectual centers beyond the two coasts and Chicago. Rather, the major universities—which are, relatively speaking, geographically isolated, while they often are on the "circuit" and have major resources—are simply unable to provide the diversity needed to constitute an alternative intellectual community. Some medium and large cities such as Pittsburgh, Minneapolis, Atlanta, and Cincinnati have become regional but mainstream arts centers, and Pittsburgh is one of the major science and technology research centers.

But these places are not conducive to nonacademic artists, writers, and independent intellectuals gathering and making a living there.

Thus, the degree to which faculty function as scholars and intellectuals, administrators, entrepreneurs, or teachers correlates with the category of institution where they find a job and is conditioned by a host of determinations that are independent of the quality of the person's scholarship or ability. It is highly unlikely that a person situated in a third- or fourth-category school will be able to work consistently as a scholar, intellectual, or academic entrepreneur, although there are infrequent exceptions to this general rule.

This trend describes a process that we call the "Europeanization" of the U.S. university system. In Great Britain, for example, the distinctions among academic institutions are widely acknowledged. Oxford and Cambridge and the London School of Economics, the "red brick" universities, the arts colleges and polytechnics are plainly arrayed along a vertical grid. In France, the Sorbonne, the Ecole des Hautes Etudes, the Ecole Normale Supérieure, and the Grandes Ecoles Administratives et Techniques occupy the pinnacle of the system, followed by the rest of the University of Paris (each of whose campuses is internally ranked) and the regional campuses. Although a parallel system has emerged in the United States, it is only recently that faculty, administrators, and the public have broadly acknowledged it.

Austerity and Decline

By any quantitative measure, most academics earn an upper-stratum working-class wage, higher than that of many public school teachers but equivalent to the wages of unionized craft and technical categories in the private sector such as production engineers, computer programmers, and nurses and high-category production workers in the most unionized mass production and service industries such as autos, communications, long-distance trucking, electrical manufacturing, and steel. The $61,000 average full professor's salary—calculated across the entire academic system—does not reflect the differentials among the various categories. In 1992-93, the average salary of the ordinary full professor in third-rank institutions was about $50,000 a year. In large cities, salaries are between 10 percent and 20 percent higher. In the leading first- and second-rank universities the annual salary for a full professor was about $80,000 for a professor at the top of the scale (except the few who are in special overscale categories). In the community colleges and many state four-year schools, especially in the South and West, the rate was considerably lower. In four-year colleges, for all ranks, the average yearly salary was about $43,000.[12]

Many teachers with tenure remain at the associate or even assistant professor level for much of their careers. Their salaries range from $30,000 to about $45,000, adjusted for regional living standards, union contracts, and economic circumstances in a given state or community. In fact, the starting rates for assistant professors have risen faster than salaries for upper-level assistant, associate, and full professors, resulting in a narrowing of salary ranges between entry-level and the overwhelming majority of tenured faculty. The reason for this inequity is that, as the job market has shrunk, many third-rank institutions can choose among a large number of elite school graduates and Ph.D. candidates who may be more qualified than tenured associate professors and lower-level full professors. Consequently, university administrations in second- and third-rank institutions have been willing to pay more for published, high-status younger faculty. This disparity has produced considerable resentment among older, tenured faculty. They complain that the administration has, to their disadvantage, shifted the rules in recent years. In many third-rank institutions, excellent teaching and longevity no longer merit salary increases. Despite reports, such as that of the Carnegie Foundation, that stress the importance of teaching and urge universities to change their reward structure so that excellence in teaching may be elevated to the level of research and publication, few major schools have followed this advice.[13] Rather, excellence in teaching is viewed as a necessary but by no means sufficient basis for tenure, promotion, and salary increases. Articles, books, and, increasingly, grants have remained the crucial criteria for determining rewards.

In sum, contrary to conservative charges that professors have succeeded in making academic work a "scam," we find that, like much of the rest of the workforce, they are working harder than ever and are being relatively less well paid. Course and advisement loads have increased, support resources have dried up, layoffs are routinely threatened as departments are either abolished or faculty positions disappear, and the dramatic slowing of salary increases in the normal school year have made teaching an eleven-month job for many. As we argued in chapter 4, these conditions in large measure account for the dramatic growth of faculty unionism in the past two decades, when organizing in other sectors has slowed to a crawl.

During this period, once-prosperous academic entrepreneurs have found that the grant world has become almost intolerably competitive. Where once one-third of all research proposals in natural science were funded, recent data show that the proportion was only about one-tenth by the late 1980s. As we have seen, faculty in the sciences whose fields or methods are no longer fashionable are almost invariably relegated to undergraduate or medical school teaching. Lacking grants for marginal research projects, they lose access to laboratory facilities; they have few if any research results to report and, consequently, they are unable to publish in scientific journals.

Without publications, their hopes for promotion or salary increases shrink nearly to zero. Under these conditions, many attempt to find jobs in industry or government. Science administrators do not mourn or even notice the loss because they are committed to the proposition that the university is a site for "cutting-edge"—that is, eminently fundable—research. Others need not apply.

Austerity has tested the diversity and liberality of the U.S. system of higher education. The bottom line for most first- and second-rank institutions is research in activities that produce knowledge useful for the economy, particularly for large, technologically cutting-edge corporations, and for the state. The social sciences are most likely to be supported when the knowledge they generate is articulated with public policy or, as ethnography and survey research can be, the requirements of industrial or commercial development. The humanities may prosper only if they are ornaments of elegant scholarship, in which case they are legitimating disciplines that maintain the status of the university as a repository of Western high culture—except for historians and social scientists engaged in area studies of Latin America, Asia, Africa, and Eastern Europe.

The tendency in both state and private universities has been to respond to the long-term downturn of the U.S. economy by cutting the number of staff and students. In the face of draconian budget cuts in many state systems and the threat of similar, although by no means equally steep, reductions in many private institutions, not only in the third tier but in the top two rungs as well, administrators, state legislators, and boards of trustees increasingly invoke the criteria of "relevance" and "productivity" to play their triage game. In this environment, some academic programs are destined to decline or even disappear; those that can pass muster according to these norms may actually expand.

In December 1992 a committee of college presidents and professors at the City University of New York issued a report calling for the "consolidation" of some academic programs, especially foreign languages and the arts and humanities, among the system's nineteen campuses. Instead of retaining the traditional autonomy of these colleges, faculty in some disciplines would be concentrated in a few colleges, and students who wished to study these subject areas would be required to travel to the central programs, even if they were enrolled at a campus nearer to their homes. This consolidation continues a process that has been going on since 1988. Between 1988 and 1991 the system lost more than 10 percent of its faculty, primarily to attrition, while student enrollments grew by 8.7 percent. Despite the nearly 20 percent productivity increase, CUNY Chancellor W. Anne Reynolds argued that further cuts were necessary because of "changes in the marketplace and the reality of CUNY's fiscal condition."

A year earlier the president of San Diego State University proposed cuts of tenured faculty in sociology, comparative literature, and other programs. Yale University's president unsuccessfully tried to implement reductions in linguistics and sociology, and Columbia University's president floated a proposal to drastically reduce language programs. At the same time, a new, fiscally conservative Massachusetts governor virtually gutted the state's major university at Amherst and made even deeper cuts at the Boston campus.

Higher education is under siege everywhere, but not as a result of drastic declines in student enrollment; indeed, the widespread perception that more credentials are a prerequisite for larger numbers of occupations combined with the recession helped swell the student populations at universities and colleges throughout the country. What is at stake in the budget cuts is linked to the fact that in a technologically advanced economy, the "marketplace" dictates the need for fewer professional and technical categories. Whereas in the prolonged period of U.S. world economic hegemony and the subsequent era of cybernetic technological development there were significant labor shortages in computer-mediated occupations, a combination of factors, especially the nature of the new labor-saving technologies that affect intellectual as well as manual labor, has sharply reduced demand for programmers and other categories of technical work. While there is no one to one correspondence between university financing and economic trends, to the extent that legislatures once approved higher education expansion because they believed that it served the vital function of supplying technically trained labor, many legislators are now reluctant to maintain the current level of funding because they perceive that the marketplace is glutted with trained workers.

Beyond market vicissitudes is the widespread trend of justifying expenditures on higher education on the basis of the degree to which a school passes the criterion of relevance, which means that its products—both knowledge and trained labor—are useful to the state and to industry. Put another way, the point is that the main drift of higher education is toward a closer and more overt articulation with business than has hitherto been deemed necessary or discreet. In these terms, we can see that the vaunted academic autonomy of higher education institutions is rapidly eroding. While in earlier periods it was necessary to "penetrate the veil" of the ideology of knowledge for knowledge's sake to reveal the underlying economic and political function of universities with respect to the larger social system, the overwhelming power of business as a *moral* pursuit increasingly makes this deconstructive work superfluous. Only those who refuse to see what is in front of them can maintain the fiction that the higher education system of the United States retains more than a trace of the community of scholars.

Part III

Beyond the Catastrophe

Chapter 9

The Cultural Construction of Class:
Knowledge and the Labor Process

Working Class: No Exit or Only Exit

The working class has been decaying in postindustrial America. It has no place to go. For Karl Marx the proletariat was the "class in but not of civil society,"[1] which could resolve the antinomies of capitalist society.[2] The victory of the working class was inevitable if it became a class "for itself."[3] Workers were the agency of change, the movers who would transform the world and overcome alienation and exploitation in industrial capitalism. Though in the past the working class has demonstrated a formidable ability to organize itself into labor parties (though not in the United States) and powerful trade unions, it has not proved to be the class that would make the revolution that would resolve the contradictions of capitalism. Increasingly, the working class finds itself a casualty of capitalism. It has not been the agent of change. Now even "actually existing" workers' states are opting for market capitalism. Agency itself is suspect, and only a limited equitable redistribution seems possible within the parameters of global capitalism. In fact, there are those who have long been either denying that class itself exists or diluting the concept until it virtually fades before our eyes.

We begin with a few concrete examples: analyses of three movies that represent possible strategies for working-class people to deal with a restructured industrial workplace. Our examples assume strategies of class stratified along gender, racial, and ethnic lines. These movies represent crises in the lives of working-class people as skilled and manual work opportunities decline and high-wage jobs become dependent on going to the right university and becoming a professional knowledge worker. The problem for most working-class people is that they lack both the economic and the cultural resources necessary to become professional knowledge workers—that is, they have no exit. The minority of working-class people who can become professional workers must leave their class culture behind—that is, they can *only* exit. There appear to be no alternatives. The successful motion pictures *Do the Right Thing* and *Working Girl* exemplify the "no exit" and "only exit" scenarios of the lives of working-class men and women.

In *Do the Right Thing* we see the conflict between two class cultures, one African-American and the other Italian, in Bedford-Stuyvesant, Brooklyn. The movie is set in a black ghetto that is degraded economically, politically, and culturally. In a working-class/poor neighborhood in crisis, there appears to be no way out. This is not unlike the actual tensions between working-class Italians and poor blacks in Brooklyn that resulted in the murder of Yusef Hawkins in Bensonhurst, Brooklyn, by a group of angry young Italians. Working-class whites and blacks in general are having a tough time in the postindustrial 1980s and 1990s. This is an information society, and neither group is doing well in acquiring the educational credentials that are required for full participation in it. African-Americans have a high school dropout rate. In New York City, Italians have the highest school dropout rate of any white group. Their similarity to blacks does not end with school failures. Traditional working-class jobs are increasingly hard to come by. Factories are closing, construction trades are no longer booming, and semiskilled jobs with decent wages like longshoreman have been almost completely automated out of existence. Clerical jobs either pay low or are being replaced by high tech. Both Italians and blacks are having trouble breaking into the relatively high-paying professional-managerial occupations. These are tough times for Italians and blacks.

Sal in *Do the Right Thing* is a small businessman who owns a pizzeria, which he works with his two sons and Mookie (Spike Lee) the delivery boy. Economically Sal is "petty bourgeois," but culturally he is working class. For our purposes, the crucial scene in the movie takes place in the pizzeria between Sal and Buggin' Out, who argue first about the price of the pizza and then about the amount of cheese on the pizza. Buggin' Out finally pays and sits angrily in a booth. As he starts to eat he stares at Sal's Wall of Fame.

All around Buggin' Out peering down from the WALL OF FAME are signed, framed eight by ten glossies of famous Italian Americans. We see Joe DiMaggio, Rocky Marciano, Perry Como, Frank Sinatra, Luciano Pavarotti, Liza Minnelli, Governor Mario Cuomo, Al Pacino and of course, how can we forget Sylvester Stallone as Rocky Balboa: The ITALIAN STALLION, also Rambo.[4]

Buggin' Out asks, "Sal, how come you ain't got no brothers up on the wall here?"

Sal answers, "You want brothers up on the Wall of Fame, you open up your own business, then you can do what you wanna do. My pizzeria, Italian Americans up on the wall."

Buggin' Out responds that blacks spend a lot of money in Sal's and they should have some say in who's on the wall: "Put some brothers up on this Wall of Fame. We want Malcolm X, Angela Davis, Michael Jordan tomorrow."[5]

This conflict portends the concluding disaster of a riot, the burning of the pizzeria, and the murder of Radio Raheem. (And yet it is only with the burning of Sal's Famous Pizzeria, when no one is looking, that Smiley finally hangs the picture of Martin Luther King, Jr., and Malcolm X, which he's been selling, on the wall of fame.)

Though blacks are more disadvantaged in the United States, working-class Italians and blacks both feel exploited, oppressed, and ultimately left out of the "new information society"; both groups have a sense of stagnation and failure. The evidence is that for the Italians the Wall of Fame is necessary. The blacks don't even have one. In popular culture blacks and Italians are represented in similar ways: as either celebrities or gangsters—as Frank Sinatra or John Gotti, Bill Cosby or Nicky Barnes. An entertainment elite and a criminal elite are making it. Everyone else is failing. Everyone else is hurting. Two class cultures with no exit. When a culture becomes "practico-inert"[6] all that is left is stagnation, decline, and, too often, crime and violence. It was not always this way; as knowledge work displaced both skill and manual work, the working class was displaced as well. In the past it was possible to get a good union job in which educational credentials were not required—only hard work was required. Sebastian DiFazio, the father of one of the authors of this book, left high school in 1937. He became a long-shoreman and shaped up on north river pier forty in Manhattan. He worked hard and made at least sixty dollars a week and often more. He was a worker without a high school education and he made ten dollars a week more than his college-educated high school teacher. Since the late 1970s unionized jobs that require little more than the willingness to work hard and reward the worker with a decent wage no longer exist. In the 1990s there are almost no opportunities for workers with just high school degrees to make a decent wage and even fewer for those like Sebastian DiFazio without a high school degree but willing to work hard. Hard work does not guarantee a good job at a good wage. Non-college-educated workers are doomed to the low-wage sector. The African-American and Italian class cultures of *Do the Right Thing* both have high dropout rates and thus poor occupational outlooks. Lawrence Mishel and David M. Frankel state that the annual wage of a full-time year-round high school graduate fell 8.6% from 1979 to 1987. In contrast the wage of a full-time year-round college graduate rose by 9.2% in the same period. As a result, the wage gap between high school and college graduates doubled from 16% in 1979 to 33% in 1987.[7]

In *Do the Right Thing,* working-class whites and blacks are victims, and too often they victimize each other. Neither the blacks nor the Italians portrayed in the movie are members of a class in "radical chains."

In *Working Girl* we have a different scenario. A working-class woman can "make it" only by leaving her working-class culture behind ("only exit"). *Working Girl* is a paean to free-market capitalism. Feminine success and liberation correspond to climbing the corporate ladder. Tess McGill is an ambitious secretary who has completed a bachelor's degree in business as a night school student. Her hope for "making it" is to get into her firm's "entrée program." Mobility is going to be offered only to the graduates of elite colleges. Certainly not to a secretary. The men in the firm offer her admission to the "entrée program" as a sexual carrot, but Tess refuses to bite.

Tess's frustrations temporarily come to an end when she is reassigned and her new boss is a woman. Tess becomes Katherine's apprentice. As Tess helps Katherine put on her ski boots, Katherine gives Tess some personal advice: "Tess, you know you don't get anywhere in this world by waiting for what you want to come to you. You make it happen. Watch me, Tess. Learn from me."

Though Katherine has just stolen one of Tess's ideas and hopes to steal more (without giving Tess any credit), she continues to act as if she is helping Tess. "You just keep plugging and bring me your ideas, and we'll see what we can do. Look at me. Tess, who makes it happen?"

Tess: "I do, I make it happen."

Katherine: "That's right. Only then do we get what we deserve."

Katherine has given Tess a pep talk in how to be successful in the dog-eat-dog world of free-market competition. In the 1980s this is the hegemonic worldview in the modern world of fragmented worldviews. The movie never embodies a critique of this dominant ideology. Instead, Tess is going to take Katherine's advice seriously and beat her at her own game.

Katherine breaks her leg skiing in Europe and asks Tess to take care of both her office and her apartment. After seeing her idea on Katherine's home computer, Tess begins what Harold Garfinkel calls a secret apprenticeship.[8] She begins to learn the cultural codes of the upper class and leaves her working-class background behind. She studies Katherine's audiotapes and learns to mimic the speech patterns of the upper class. She begins to wear Katherine's clothes and to use makeup in a non-working-class way. She has her best friend, Cyn, cut her long frizzy hair: "You want to be taken seriously, you have to have serious hair." Tess chooses one of Katherine's gowns, worth six thousand dollars. She explains her choice: "It's simple, elegant. It makes a statement. It says to people, confident, risk taker, not afraid to be noticed. Then you hit them with the smarts."

She manipulates herself into the business world and the life of ex-hotshot business dealer Jack Trainer. She accomplishes this by taking the part of both the secretary and the successful businesswoman in phone conversations. She speaks like a working-class secretary and then like a successful businesswoman, or she has Cyn play the part of the secretary.

She continues her secret apprenticeship with Jack, who "trains" her in the specifics of making a deal. Tess's transformation is successful, and she increasingly becomes an outsider in her own working-class world. She gives up her man (who has been cheating on her), her apartment, and her job as a secretary. As Tess says, "I'm not going to spend the rest of my life working my ass off and getting nowhere just because I followed rules that I had nothing to do with setting up."

Tess is leaving the world of the working class where you are always doomed to failure and playing by new (to her) class rules. She has successfully learned the cultural codes of professional career women through her secret apprenticeships. Tess has learned that the cultural codes of class are as important as the material bases for class structure. There is an implicit "practical" critique of the orthodox Marxist base/superstructure distinction in which cultural forms of life are seen as a reflection of the economic base. Instead, class consists of reciprocally determining instances. Culture, economics, and politics are all social forms, and they are mutually determining. Class is overdetermined, but there is no necessity of economics in the last instance as in Althusser's formulation. Tess's economic capital is meager, her "human capital" insufficient for her to compete with the high rollers on Wall Street; her only power lies in her ability to manage what Pierre Bourdieu calls symbolic power.[9] Tess has learned to act as if her artificially acquired middle-class cultural practices are natural; they become, in Bourdieu's words, "second nature, a habitus, a possession turned into being."[10]

Working Girl concludes with Tess making the big deal and getting the deceitful Katherine fired. Jack and Tess fall in love. Tess, because of her competence, her successful risk taking, and her pledge of loyalty to her new boss, is hired for an entry-level management position. The ending of the movie is ironic. Tess begins her new job in her own office, "all the way at the end of the hall." She has her own secretary and sets up rules in which there is more equality: "I expect you to call me Tess. I don't expect you to fetch me coffee unless you're getting some for yourself."

Sitting in her new office, Tess calls up Cyn at the secretarial pool in her old office and tells her of her success. Her former co-workers cheer because Tess' success is an indicator of their own possibilities. Tess has made it out of the working class into the middle class, and she is a model for all of them. In an ironic ending, the camera moves slowly back. We see Tess in her office talking on the phone; we see that Tess's office is one of a beehive of other offices, one office building among a multitude of office buildings. Tess disappears, and we realize that her success is minimal when we consider what moving up the corporate ladder from the bottom up means in a city full of office buildings, in a world full of office buildings. Tess is better off, but she has sacrificed a whole way of life. She is still in a working world dominated by men, where women are forced to fight with each other for the few slots available to them. With the continuing recessionary trends and the 1987 Wall Street crash and the rule of "last hired, first fired," her new position is at best tenuous. Tess has moved up, but just a little, and her situation is still precarious. She is still dependent on men. What might make her more independent is the concept of comparable worth.

In June 1984, Clarence Pendleton, Jr., then chairman of the U.S. Commission on Civil Rights, called comparable worth "the looniest idea since Looney Tunes came on the screen."[11] This was not simply a remark made by a conservative to defeat a liberal reform measure. It was made with the understanding that comparable worth is a radical proposal. The liberals seemed unaware of how radical the proposal is. Comparable worth calls for nothing less than the complete redefinition of the occupational structure. The criteria for this redefinition is not simply economic. Comparable worth assumes that inequities in pay are based on a cultural bias that devalues "women's work." "Women are systematically underpaid," and the basis for this is not simply the economists' rational calculus that "women's jobs are less valuable than men's jobs."[12] For economists, wages are the result of the functionality of jobs in the economic system in relation to market value. Thus wage rates are an economic fact, a scientific object. The women's movement has deconstructed this objective, economic fact and demonstrated how culture and gender bias have been involved in the construction of the wage system. Women's jobs were not just less valuable, they were structured that way. In our chapter on architects and engineers we demonstrated how the category of skill is gendered and not naturally masculine. Comparable worth along with affirmative action were developed as solutions to occupational and wage inequities.[13] These policies both attempt to create occupational justice, one by opening up occupations that have been traditionally closed to women and certain racial groups, the other based on a seemingly just assumption of equal pay for work of comparable worth without regard to gender or race. Comparable worth assumes that cultural as well as economic criteria are a constitutive feature in the value placed on a specific position within the occupational structure. Thus, from his conservative position, Pendleton was correct in calling comparable worth "Looney Tunes." In demanding comparable worth, the women's movement is demanding a total redefinition of the class structure. The reconstitution of the occupational structure is conditioned by other developments as well, including the high-tech revolution and the increasing centrality of knowledge and knowledge workers in the production process, the restructuring of world capitalism, and the struggles for liberation in Eastern Europe and the "third world."

Working Class: The Last Exit

Here it is important for us to define class clearly. When we talk about class, we are always talking about it in relation to the labor process. Class is never just a stagnant category based on occupational position, income, or wealth as it has been typically defined by sociologists. Class assumes an active relationship to a labor process. Nor is class only a political category, as Marxists have often seen it. Thus the bourgeoisie, the proletariat, and landowners are not automatically conscious of their class position. Marx's great achievement in relation to class is that it is a project constituted by the labor process but never independent of social and cultural relations. Class certainly has to do with inequality, but inequality grounded in one's location in a labor process, attached to community and political only if (Marx's "if" clauses have almost always been missed by orthodox Marxists) the class becomes conscious of itself as an agent of change. This crucial component of class, becoming conscious of itself, moves beyond the spatial location of a labor process, located in a workplace to the communities and cultures who share a common place in relation to a labor process. Marx, writing in *The 18th Brumaire of Louis Bonaparte* on whether "small holding peasants" are or are not a class, makes his clearest statement on the relation of community and culture to class consciousness:

> In so far as millions of families live under economic conditions of existence that separate their mode of life, their interests and their culture from those of other classes and put them in hostile opposition to the latter they form a class. In so far as there is merely a local interconnection among these small-holding peasants and the identity of their interests begets no community, no national bond and no political organization among them, they do not form a class.[14]

This is not an aberration in Marx's thought; he continues to make these points in relation to the importance of science, the general intellectual base, and the arts in all four volumes of *Capital* and the *Grundrisse*. Discussing the relation of scientific knowledge to production and machinery (fixed capital), he writes:

> The development of fixed capital indicates to what degree social knowledge has become a *direct force of production* and to what degree, hence, the conditions of the process of social life itself have come under the control of the general intellect and been transformed in accordance with it. To what degree the powers of social production have been produced not only in the form of knowledge, but also as immediate organs of social practice, of the real life process.[15]

The labor process is not independent of the knowledge process, which is constituted culturally, nor is it independent of the social "life process." And beyond Marx's intuitive understanding of the relations of everyday life to the labor process, our current multicultural concerns have made us understand that the labor process is structured in terms of gender, race, and ethnicity. Class is multidimensional, not one-dimensional. Class, race, gender, and ethnicity mutually determine one other. Which is more determining is relative to historically specific situations. It always has been so, from shopgirls to African slaves to Max Weber's Protestant capitalists. Historically this is continually in transformation.

Here we contend that there are at least four spatial class locations in relation to the labor process. We use Marx's language because the analysis of class and labor process is central to his analysis as it is important to ours, though for us class analysis is not the only form of distinction that has an impact on social life. It is crucial but not determining. Thus, first, class members can own and control the labor process. This includes not only capitalists and executive managers, but also their servants: lawyers, certain intellectuals, and political figures.

Second, class members can participate in the labor process and be either directly or indirectly involved in the valorization process, in the production of value: the classic proletariat and workers involved in industries related to distribution and consumption (from truck drivers to store clerks). It must be remembered that it is the nature of capitalism—and here even very few capitalists disagree with Marx—that it must continuously revolutionize the means of production. Thus, as the labor process changes (that is to say, as labor goes into the machines and fuels and runs them), there are changes in class structures and class positions. Scientific and technological innovation is a constitutive component of the capitalist labor process. At this historical moment scientific knowledge is a crucial commodity, and many technicians, engineers, and scientists are involved in the valorization process.

Third, class members can be involved in the administration and maintenance of the labor process. This includes many of the new service industries, the new skilled workers who repair and maintain the high-tech machinery, accountants, middle managers, teachers, and the majority of academics.

Fourth, one can be excluded from the capitalist labor process because of gender, race, or ethnicity. The homeless, those on welfare, and the unemployed are excluded, as are those who work in alternative labor processes such as housework, criminal work, and volunteer work. Here we, unlike Marx, argue that the bourgeoisie and the proletariat are not the only agents of social and political change. For us agency of change is possible in all of the four class locations. In the rest of this chapter we will elaborate on the relations of class, culture, and knowledge in a gendered, racial, and ethnic work world.

All of this makes class as determined solely by economic relations problematic, yet many social theorists continue to analyze capitalism around the old orthodoxy. In another Hollywood film we see the contradictions of the orthodox representation of working-class life. *Last Exit to Brooklyn* exposes the limits and possibilities of class culture and points in the direction of the required cultural transformation of class for a new class-based social movement. *Last Exit to Brooklyn* is about a working-class community in Brooklyn during a strike. Vinnie and his boys, neighborhood hoodlums, hang out at a bar. Tralala, a whore, also hangs out at the bar. She seduces soldiers and sailors from the navy yard and takes them to a vacant lot to give them a blow job. Just as she begins, Vinnie and the boys mug Tralala and her customer. They split the money four ways.

Harry Black is the shop steward who runs the strike office. He has an expense account that he spends liberally:

> Whatever I want, I order and the union pays for it. We're a big union, you know. Look at us, we're in the paper every day. I'm making more than when I was working and it's tax free. And the wife don't know nothing about it, heh. I run the whole show around here. I'm on the payroll and I got an expense sheet too.

The strike has made Harry a big shot in the neighborhood. Like Harry, Joe is a member of Metal Workers Local 3392. He is on strike as well. We first see Joe in his shorts and underwear. He has to go to the bathroom, but his daughter is there, trying to hide her pregnancy from her father. Upset, she stays in the bathroom. He is complaining but she still does not come out. Joe, obviously suffering, urinates out the window of the tenement building.

Georgette is a homosexual who is in love with Vinnie. Vinnie uses Georgette for drugs and continuously taunts and teases him. He never has sex with Georgette.

Working-class life is presented as uncivilized. The workers are dirty, crude, violent, criminal, and perverse. They seem to be dogmatic in Habermas's sense: "He leads an unfree existence because he does not become conscious of his self-reflecting, self-activity. Dogmatism is equally a moral lack and a theoretical incapacity."[16]

Somehow, in 1952, against all odds and in seemingly degraded and oppressed conditions, they wage a successful strike. Somehow these uneducated and dogmatic workers take on capital and win. These workers actively participate in a working-class community. In the context of this strike they actively construct a class culture in which they struggle to redefine the conditions of their lives. There are limits to this transformation, but in the strike they have transcended the "in itself," class as a reified category. In actively constructing a class community they become members of a class "for itself." Here class is a living movement, a continuous project. They have become conscious of themselves as men who are powerful, men who will win the strike, who are taking control of their lives. These are men doing class struggle. But again there are limits.

The limitations of this class movement can be seen in the parallel lives of Tralala, Harry, and Georgette, all outcasts in this working-class community. Harry, unable to express his sexual desire with his wife, has sex with her in a brutal and violent way. He cringes at her touch and is alienated from his child. Georgette has invited Vinnie to a party. Vinnie and the boys bring Harry along to the party because they need someone to pay for the cab. Georgette introduces the boys to the queens at the party. They start to get high and violent. Harry is immediately attracted to Regina. As the party gets violent, Regina and Harry leave. Harry has his first homosexual experience and it is love at first sight. He is finally able to truly express his sexual desire. His disgust for his wife is explained.

At the same time, Tralala is having her first experience whoring out of the neighborhood. She cons a soldier into taking her to a bar frequented by officers. She gets him drunk and rolls him. She leaves with Steve, a first lieutenant. She goes to his hotel room. Her intention is to roll him as well, but instead she stays the weekend. Steve is romantic, treats her well, and tells her she has "the best tits in the Western world." She hopes to be paid by Steve when she sees him off at the ship. Instead he thanks her for the weekend and expresses his fear that he will never be with a woman again. He is worried about dying in the Korean War. This was her chance to leave the neighborhood, but now she has "no exit." She is a neighborhood whore with no place to go.

At the party, Georgette is rejected by Vinnie. Overcome with despair and love, Georgette shoots up. Full of dope, he nods. When he awakes, Vinnie has left. In his grief-stricken, semi-drug-induced state, he runs into the rain-soaked street yelling, "Vinnie, Vinnie where are you?" He is hit by a car and thrown into the air, his fragile body broken. He is killed instantly. Georgette's death foreshadows Harry's and Tralala's inevitable end. Georgette's and Harry's personal tragedies are representations of how only male heterosexual desire can be legitimately expressed within working-class culture; there is no space for homosexuality. For Tralala as for all working-class women, the free expression of sexual desire can only be a form of degradation. You can only be a whore. Donna, Joe's virginal daughter, is pregnant but she is not a whore: she is legitimized as a good woman when she marries Tommy, through her formal dependence on a hard-working man.

The morning after the party, scab-driven trucks break through the picket line and get into Brickman's Metal Company. Harry is late to the confrontation—the beginning of the end for the homosexual and good union man. He climbs the fence outside the plant and screams, "You scabs, you punks, you came in here but you sure as fuck ain't gonna get out." He shouts, "No trucks out. No trucks out." The men join in: "No trucks out."

A cop telephones the precinct: "This thing is going to get out of hand here. We're going to need backup, and I don't mean just a few men. Send everything you got. Send men, send gas, send horses."

The war occurs at night. Brickman's scabs and company goons attempt to move the loaded trucks with the protection of the police. Before the war breaks out, Tommy tells Joe that Donna had the baby. The proud father is beaming, but Joe is angry at Tommy for defiling his daughter.

The plant is part of the community. The workers live near the plant. They walk to work. As the violent confrontation develops, their families watch. The company goons armed with nightsticks and water hoses, the cops with tear gas on horses try to get the trucks through. The strikers, with two by fours, rocks, and superior numbers, first try to prevent the gates from being opened, and when they fail they try to prevent the trucks from leaving. Harry jumps on one of the trucks and smashes the head of the scab trucker through the cab window, but he is not able to stop the truck. The confrontation is short and brutal. The trucks get through.

Harry suggests to Boyce, the head of the local, that Vinnie and his boys can help the union cause. Vinnie tells Boyce that he knows who supplied the scab trucks; for two hundred dollars he will stop Bulaggi Brothers from supplying trucks in the future. Boyce pays Vinnie, and Vinnie and his boys blow up the Bulaggi Brothers trucks. This act of sabotage is the beginning of the end for the company.

In the morning Boyce rebukes Harry for failing to tell the union that the trucks were coming through and accuses Harry of robbing the strike fund. Boyce tells Harry that he is stepping down as shop steward and that he will pay back every cent that he took from the strike fund. Harry protests, but Boyce scolds him, "Don't beg, Harry, and don't slam the door." A broken man, Harry seeks solace in Regina's arms, but Regina rejects him because he no longer has access to the strike fund: "If you have money you can see me later . . ." He is alienated from his wife and child, and the man he loves has rejected him. Drunk, Harry tries to seduce a teenage boy, Bobby. Vinnie and the boys come to rescue Bobby. They stomp Harry and crucify him and leave him bleeding on a wooden fence.

Meanwhile, Tommy and Donna's wedding is going on. In the midst of this happy occasion Boyce announces that management has collapsed. The strike is won.

Like Georgette and Harry, Tralala is down on her luck. Drunk and realizing that her only status in the neighborhood is as a whore, she exposes her breasts at the bar: "Best tits in the Western world." The men crowd around her and grab at her breasts. One of them asks, "What is this, all tits and no cunt?" Tralala replies, "I'll show you, all of you." A brutal gang bang begins. Only Joe's son tries to protect Tralala. She is badly battered, yet she says to him, "Don't cry, don't cry."

On Monday the men return to work empowered, victorious, and with a better contract. Joe and Tommy are standing together, happy with the strike victory and ready to go back to work. One of the men asks, "Tommy, how's the baby doing?" Tommy, with a big smile on his face, replies, "Ahh, he cries all day. Shits all over the place. It's real good."

The whistle blows and the men go back to work. The movie ends with a shot of the masses of men returning to work. Instead of "no exit" or "only exit," this movie as a representation of working-class life offers an alternative, social movement as the "last exit." In the space created by their class community and their union organization grounded in their class culture they successfully struggle for higher wages and control over the conditions of their work, and thus their lives. The possibility of control exists, but they must struggle for it not only at the economic level but at the political and cultural levels as well. But *Last Exit to Brooklyn* is also a powerful representation of the limits of working-class culture in Brooklyn in 1952. In that last long scene, as the men triumphantly go back to work, the music seems mournful. Is it mourning the loss of an empowered working-class culture in the 1990s? Or is it mournful about what is absent in the final scene? Where are Harry, Tralala, and Georgette? Donna is at home with the baby, as are the wives of all the men from the plant. Where are the black workers? The Hispanic workers? They are completely absent from the movie. These simple facts illustrate the limitations of the labor movement. With few exceptions, the U.S. working-class movement is white, heterosexual, and male. The solidarity of the working-class is empty if it excludes workers on the basis of race and gender. The class-based union movement has to be recoded in terms of race and gender. Class struggle must be recoded in terms of the new social movements of identity politics. The cultural aspects of class become increasingly crucial.

The End of Class?

Though we believe that we have made a sufficient argument for the cultural construction of class and that class is still an important category for social analysis and political organization, it is still a term that is under attack. This attack takes at least two forms; the first privileges the political, the second privileges the scientific. Class is now seen as an obsolete concept. Class, grounded in a capitalist, industrial modern world, seems to have outlived its usefulness in the postindustrial, postmodern world. Class becomes a secondary concept, either decentered by race and gender or subordinate to political ideological relations. Ellen Meiksins Woods summarizes this position, with which she is in disagreement: "The formation of a socialist movement is in principle independent of class, and a socialist politics can be constructed that is more or less autonomous from economic (class) conditions."[17]

Ernesto Laclau and Chantal Mouffe write in *Hegemony and Socialist Strategy: Towards a Radical Democratic Politics* that the Marxist conception of class has "exhausted its productivity." The class struggle model is untenable and the consciousness of the proletariat is historically unsubstantiated. What is necessary is the creation of a political imaginary that is not founded on an a priori class-based agent but on the pluralistic principles of a radically libertarian democracy. Not that a class movement is not important, but it is only a component of a movement for radical democracy. Class as the basis for a movement was too narrow. Radical democracy opens the possibility for a broad spectrum of social movements. Laclau and Mouffe's thesis is that it is only from the moment when the democratic discourse becomes available to articulate the different forms of resistance to subordination that the conditions will exist to make possible the struggle against different types of inequality.[18]

These struggles are subjectless and occur outside the social body. Agency that requires a subject like Marx's universalistic agent, the proletariat, is "essentialistic" for Laclau and Mouffe. Marxists reduced all oppression to inequities in the capitalist production process. It is only by constituting subordination in new discursive formations of radical democracy that they can begin to do a nonessentialistic politics against all forms of oppression. Thus, "from the point of view of the determining of the fundamental antagonism, the basic obstacle, as we have seen, has been classism: that is to say, the idea that the working-class represents the privileged agent in which the fundamental impulse of social change resides."[19] They come to this conclusion via a survey of soviet and nonsoviet scientific Marxism. Their reading of the history of the "crisis in Marxism debate" demonstrates that for scientific Marxists class is an unalterable fact; it is determined by the economic base, and it discloses the underlying foundation of all superstructural phenomena (cultural and political). They reject this reading of class as deterministic and insufficient for explaining twentieth-century political and cultural movements. This is particularly true in a world in which differences of nationality and gender have become crucial and in which it seems that economic reductionist explanations based on a rigid base/superstructure analysis are simplistic. Class is a term steeped in orthodox determinism; it is incapable of sustaining solidarity. In this postmodern world the political is increasingly autonomous and pluralist. Laclau and Mouffe privilege the political and cultural over the economic. Since they view class as an economic and not a cultural category, their analysis is a continuous attack on class.

We agree with Laclau and Mouffe's claim that class mapping is insufficient. Our analysis of *Last Exit to Brooklyn* has already made this point. But, though they take on the history of Marxism, they never take on Marx. As a result, they reduce and thus distort the notion of class, flattening it to an economic category, missing Marx's cultural and social components. Thus they reinforce the economistic analysis of class of which they are critical. In continuing with this line of thought they substitute a Marxist essentialism of the working class as the a priori agent for an essentialism of a particular discursive formation, the principle of radical democracy. Whether Marx, Marxist, or Laclau and Mouffe, all are trying to reduce social theory to its essential space and action, the determining instances of all other spaces and actions. They all resemble physicists looking for a unified field theory. Laclau and Mouffe's political imaginary views the discourse of socialism as a limited voice because of its claim to be the single, universal voice. Instead they offer radical democracy, which they claim is prior to the limited socialist vision. They fail to see the hierarchy in this. They write:

> The discourse of radical democracy is no longer the discourse of the universal; the epistemological niche from which 'universal' classes and subjects spoke has been eradicated, and it has been replaced by a polyphony of voices, each of which constructs its own irreducible discursive identity. This point is decisive: there is no radical and plural democracy without renouncing the discourse of the universal and its implicit assumptions of a privileged point of access to the 'truth' which can be reached only by a limited number of subjects.[20]

In resisting the Marxist privileging of the agency of a single subject, their analysis denies any agency to the proletariat. Laclau and Mouffe allow the working-class struggle as a discursive formation, but this struggle takes the form of a transformation that will occur without any human agents. As a result, their argument denies the autonomy of the plurality of the oppressed groups they are defending. They deny the subjectivity of groups that have only recently constituted their own subjectivity, whether they are women struggling for equal rights, gays in ACT UP, or the Brazilian working class. What they have unintentionally posited is a new vanguardism of "intellectual and moral leadership." With their claim that domination and oppression do not exist until a discursive formation labels them as such, the actual experience of ordinary people is negated. Nor can these people struggle to formulate their own discursive formations. Radical discourse seems to exist independent of their lives. But ordinary people, as subjects of their own lives, do reflect on, create, and produce their own cultural worlds, their own communities of resistance, both materially and discursively.

Nicos Poulantzas, in a large body of work, also calls into question the nature of class and its relation to socialist politics. Class is not simply a category, a position within the economic structure and an expression of material interests, but is manifested by political class struggle: "Political class struggle is the nodal point of the process of transformation, a process that has nothing to do with a diachronic historicist process 'acted' by an author, the class-subject."[21] Classes are not subjects but formations that contend for power within the state. For Poulantzas the state is the dominant institutional sphere. Class as political class struggle is a component of the state:

> Contrary to conceptions that treat it as a Thing or a Subject, the State is itself divided. It is not enough to simply say that contradictions and struggles traverse the State—as if it were a matter of penetrating an already constituted

substance or of passing through an empty site that is already there. Class contradictions are the very stuff of the State: they are present in its material framework and pattern its organization; while the State's policy is a result of their functioning within the state.[22]

For Poulantzas the space of political class struggle is a formal institutional space. Oppositional movements or the dominated classes can only contest for power at the level of the state. The structural grounds are set. The state even with an "explosion of democratic demands"[23] can only create a new state in which political class struggle would function. Class is always a function of the state. This is problematic in the way that Laclau and Mouffe point to, in that there are a plurality of struggles and a plurality of sites. The state is not the only institutional sphere of struggle. The economy, though decentered, is also a sphere of struggle, as is the cultural world as well.

From our standpoint, Laclau and Mouffe, and Poulantzas, are guilty of (1) analyzing class as simply an economic category and rejecting it as a result (Poulantzas's solution to this is to talk about class as political class struggle in which the economic is negated; but, more importantly, none of them views class as a cultural construct) and (2) reifying class as an abstract category without any relation to living human subjects. The rejection of class as reification is central to the analysis of class as an economic, political, and cultural construct—the production of a form of life by active human participants, that is, class as a continuous human project. Thus the opposite of this, class as reification, is most interesting to us, especially in an era of major transformations.

Class as a reification takes two forms. The first is exemplified in the work of Eric Olin Wright: class as a sociological reification. Second, in the work of Guglielmo Carchedi, class is a dialectical reification.

Wright and Carchedi and the Reification of Class

Wright in his important book *Classes* has made an attempt to save Marxist class analysis from two of its most important criticisms: (1) that the concept of class no longer has explanatory power because it has failed to solve the problem of "the middle class" and (2) that Marxism is not a science because its propositions cannot be verified: its key concepts such as class, class formation, and class struggle are vague and random and have little predictability. Wright counters these arguments by demonstrating that a middle-range Marxism that has theoretical propositions that can be empirically proven with scientific rigor is possible.[24] His task is to build a Marxian scientific base for what he calls radical democracy.

Marxist theorists assume a polarized class model. One is either a member of the class of producers of surplus value or one is a member of the class of the appropriators of surplus value. One is a worker or a capitalist. The middle class is neither. Most Marxist solutions to this problem have been unsuccessful. Poulantzas's "new petty bourgeoisie"[25] and Mallet's and Gorz's "new working class"[26] are class fractions of already existing classes and reconstitute Marx's antagonistic class model. Gouldner's "new class" constitutes a new entry into the class structure, but for Wright this is an unnecessary complication of the class structure. For Wright these new categorizations have the same weakness that Marx's own class analysis suffers from, that is, they do "not provide a set of precise concepts for decoding rigorously the structural basis for most of those categories."[27]

Wright's solution to the lack of a rigorous and coherent analysis of class structure is to develop the concept of "the middle class" in terms of what he defines as "contradictory class locations":

On the one hand they are like workers in being excluded from ownership of the means of production; on the other they have interests opposed to workers because of their effective control of organization and skill assets. Within the struggles of capitalism, therefore, these "new" middle classes do constitute contradictory locations or more precisely, contradictory locations within exploitation relations.[28]

Middle-class, "contradictory class" location is defined in terms of relations of exploitation. Wright argues that an exploitation-centered definition of class is superior to a domination-centered definition even though his earlier work on the middle class stressed a domination-centered definition. His move toward John Roemer's exploitation-centered analysis of class was facilitated by two weaknesses in the domination-centered definition. First, the domination-centered concept of class relations proved less coherent because "it weakens the link between the analysis of class locations and the analysis of objective interests."[29] Second, it is too relativistic and falls into the trap of "multiple repressions," which are relations of domination centered on gender, race, class, and so forth. This weakens the explanatory power of the theory. For Wright as for Roemer, even Marx's theory of exploitation based on the labor theory of value is insufficient. In Marx there is no real analysis of the middle class, which seems to be an "empty space" in his theory. At the same time, Wright wants to retain the centrality of class struggle in his own analysis. His solution is to turn to Roemer's theory of exploitation, which provides the foundation for his empirical mapping of class structure in modern capitalist society. It also provides the solution to the problem of contradictory class locations: the middle class is now defined in terms of relations of exploitation.

Roemer's Analysis of Exploitation

John Roemer's analysis of exploitation is centered on relations of property. Exploitation-centered class locations are defined in terms of productive assets. For Roemer, asset inequality structures class relations. This assumption is rooted in G. A. Cohen's defense of Marxist base/superstructure analysis in which the base, the economic foundation of capitalism, determines the superstructure, the political, legal structure.[30] To relate this to class analysis, a domination-centered definition is too superstructural. For Wright the emphasis on relations of exploitation is true to Marx's historical materialism, in which actors are rational and knowledgeable about the importance of their class locations, which are determined by relations of exploitation that affect their interests:

> In Marxian exploitation one class appropriates the surplus labor performed by another class through various mechanisms. The income of the exploiting class comes from the labor performed by the exploited class. There is thus a straightforward causal link between the poverty of the exploited and the affluence of the exploiter. The former benefits at the expense of the latter.[31]

Wright judges Roemer's analysis of exploitation to be more pervasive than Marx's. Roemer's twofold analysis consists of the labor transfer approach and the game theory approach. In the labor transfer approach, the labor theory of value is minimalized, and Roemer attempts to demonstrate that the institution of wage labor is not crucial to his analysis of exploitation. His analysis of the exploitation (labor transfers) from the asset poor by the asset rich is illustrated by his construction of the

> class structures and patterns of exploitation on two imaginary islands, "labor market island" and "credit market island." On both islands some people own no means of production and other people own varying amounts of means of production. The distribution of these assets is identical on the two islands. And on both islands people have the same motivations: they all seek to minimize the amount of labor time they must expend to achieve the common level of subsistence. The two islands differ in only one respect: on the labor market island people are allowed to sell their labor power, whereas on the credit market island people are prohibited from selling their labor power but are allowed to borrow at some interest rate the means of production.[32]

Roemer draws two theses from his imaginary islands. First, the class exploitation correspondence principle: "There is a strict correspondence between class location (derived from ownership of differing amounts of the means of production, including no means of production) and exploitation status (having one's surplus appropriated by someone else)."[33] Second, he argues that the class structures on the imaginary islands are isomorphic: "Every individual on one island would have exactly the same exploitation status and class location on the other island."[34] Roemer argues that he has demonstrated that property relations and class locations, which are a function of the means of production, determine market-based exploitation.

The game theory approach compares different systems of exploitation: feudalism, capitalism, and socialism. Here exploitation is a game defined by withdrawal rules. Rational actors who are exploited assess whether they would be better off if they left the game and played an alternative game. In capitalism, workers are exploited and would be better off if they played an alternative game:

> What Roemer demonstrates is that if the coalition of all wage-earners were to leave the game of capitalism with their per capita share of society's assets, then they would be better off than if they stayed in capitalism, and capitalists would be worse off. The "withdrawal rule" in this case—leaving the game with per capita shares of physical assets—then becomes the formal "test" of whether or not a particular social system involves capitalistic exploitation.[35]

Roemer's discussion of feudal, capitalist, and socialist exploitation generates two analytic forms of exploitation. The first is skill assets exploitation, which is based on credentials. Those with credentials get paid above their marginal product and thus have a vested interest in maintaining skill differentials:

> "Credentials . . . are not the only way the price of skilled labor power may exceed its costs in production; natural talents are a second mechanism. Talents can be viewed as affecting the efficiency with which skills can be acquired. A talented person is someone who can acquire a given skill at less cost (in time, effort and other resources) than an untalented person.[36]

For Roemer, skill is an inalienable asset; physical assets are alienable. A third and overlapping asset is labor power, which is a productive asset. In capitalism everyone owns at least their own labor power; in feudalism the lord owns more than his own labor power while the serf owns less:

Reformulating feudal exploitation in this manner makes the game-theoretic specification of different exploitations in Roemer's analysis symmetrical: Feudal exploitation is based on inequalities generated by labor power assets; capitalist exploitation on inequalities generated by ownership of alienable assets; socialist exploitation on inequalities generated by ownership of inalienable assets. Corresponding to each of these exploitation generating inequalities of assets there is a specific relation: lords and serfs in feudalism, bourgeoisie and proletariat in capitalism, experts and workers in socialism.[37]

Class structure is determined in these different forms of societies by inequalities in assets. Class formation and class alliances follow in terms of their location within the structure of exploitation relations. The middle class is still a problem, but a problem that can be resolved once it can be located within this classification system.

The second form of exploitation is "organization asset exploitation." This form can be found in both capitalist and socialist systems, that is, corporate managers and state bureaucrats, "but organization—the conditions of coordinated cooperation among producers in a complex division of labor—is a productive resource in its own right."[38]

In contemporary capitalism the organization is owned by capitalists, but managers have control over the organizational assets. In state socialist societies there is no separation of ownership from control, and centralized bureaucratic powers control organizational assets.

The Middle Class as Contradictory Class Locations

With his conceptual framework articulated, it is now possible for Wright to deal with the problem of the middle class in detail. Contradictory locations in class relations pose specific problems for a Marxist analysis. The middle class does not fit neatly into the category of a "class in itself" because structurally it has characteristics of both workers and capitalists. Managers do not own the means of production, and thus they can be categorized as workers. Yet they do have transfer rights over both organizational and skill assets, as capitalists do. Nor is the middle class clearly a "class for itself." What are its interests and does it consciously construct itself as a powerful agent for achieving these interests? What optimizing strategies can the middle class generate to take advantage of its organizational assets and skill credentials? Does it try to overthrow the exploiters based on the power of the scarcity of its skills assets? Or does it try to use its strategic expertise to demand a larger share of the distribution of assets in society? Or does it allow itself to be co-opted by large salaries for its scarce skills? On the one hand, capitalists are dependent on middle-class skills; on the other, ever-increasing labor costs can cause economic decline. Does the middle class align itself with the exploited because it is being proletarianized? For Wright it is crucial that the proletariat has been moved from center stage and that contradictory locations in class relations have complicated the whole picture of class relations. There is now a profound indeterminacy, and that class struggle may have taken a new turn. Agency may have to be redefined. As Wright states:

> Within capitalism the central contradictory location within exploitation relations is constituted by managers and state bureaucrats. They embody a principle of class organization which is quite distinct from capitalism and which potentially poses an alternative to capitalist relations.[39]

It is in terms of an exploitation-centered analysis in which class structures determine class formations, which determines class consciousness and the trajectory of contradictory class locations, that Wright's class theory attains a coherence and clear theoretical status that it never had before.[40]

Wright's concern with "coherence and clarity" is central to developing an empirical, scientific concept of class that can be applied to a broad-based analysis of class structure in different societies. He wants to demonstrate the power of his exploitation-centered definition of the middle class in terms of forming alliances and developing class consciousness determined by their location in the class structure, and this is the result of the methodological power of his analysis. The coherence of this analysis depends on the coherence of his analysis, which depends on the coherence of his methodology. For Wright this relation is reciprocal: "To the extent that a particular concept is more coherent than its rivals, meshes better with the overall theory of which it is a part and provides greater explanatory leverage in empirical investigations, it is to be preferred."[41]

Wright has created a middle-range theory of the middle classes that explains a range of phenomena and can be operationalized and empirically tested. He demonstrates the power of this theoretical and empirical trajectory in an analysis in the United States and Sweden. Marxists have often been concerned only with global analysis and have failed to link their theoretical analysis to micro-level outcomes. Wright shows a causal link between an asset-based exploitation model of class structure and the actions of individuals in society. He does this by demonstrating with scientific rigor that class structure is in the last instance determined by property relations:

> Class structure is of pervasive importance in contemporary social life. The control over society's productive assets determines the fundamental material interests of actors and heavily shapes the capacities of both individuals and

collectivities to pursue their interests. The fact that a substantial portion of the population may be relatively comfortable materially does not negate the fact that their capacities and interests remain bound up with property relations and the associated processes of exploitation.[42]

The implication of this analysis of class structure for political change specifically through class struggle is that it provides a space for the middle classes in this struggle—not only in terms of the formation of class alliance but also in the formation of an object of struggle, that is, radical democracy. This is a positive struggle in terms of the assets linked to the production process: physical, organizational, and skill. This is a society that is for democratic control over consumption, the quality of life, reducing the massive waste of capitalism and creating the possibility of class struggles for an expansive socialist future.[43]

Guglielmo Carchedi's Critique of Wright

For Guglielmo Carchedi, Wright's analysis in *Classes* is part of a long socialist tradition dating from the Second International in which the relations of production and class struggle are determined by the forces of production. The forces of production are treated as a neutral, objective economic fact. A scientific socialism that can be studied as a positivist science was created. In this vein, Carchedi quotes Engels: "The productive forces become an exogenous variable in the process of human history."[44] He continues in terms of Kautsky's positivism: "Rather, they are understood in terms of cause and effect, in terms of dependent and independent variables, i.e. in static terms."[45]

Carchedi argues that what is true for Engels and Kautsky is true for Wright. Scientific determination is substituted for dialectics and class analysis becomes static. Wright's discussion of class assumes an individualistic epistemology in which individuals are "systematically affected in various ways by virtue of being in one class rather than another."[46] For Carchedi, Wright is always the sociologist who assumes that the social is the aggregation of individual units, a key assumption for his "careful statistical research."[47] But class as a social phenomenon is not a simple aggregation of individual units, and thus this is insufficient for Wright's claim that he is doing class analysis in the Marxist tradition.

Consistent with Wright's individualistic epistemology is his use of John Roemer's game theory approach to exploitation. Wright's notion of exploitation is a matter of distribution in which one class is "asset-rich" because the other class is "asset-poor." Wright, following Roemer, generalizes this in an ahistorical manner to all classes in all forms of societies as relations of exploitation. In their view, the problem of exploitation can be solved by the equitable redistribution of assets. Carchedi contends that exploitation is not a problem of inequitable distribution. The problem is not selfish, asset-rich capitalists. Carchedi says that exploitation is structured by relations of production. Thus, equitable redistribution is not the solution; "the fact that workers would still have no say as to what to produce, for whom to produce it, and how to produce it would not have changed."[48] Equitable redistribution as a solution to class exploitation is for Carchedi just another example of Wright's social democratic reformism.

Wright's concern is not only reformism; it is also to carry out a scientific analysis of how "class locations structure the objective interests of actors."[49] Wright accomplishes this by reducing class consciousness to attitudinal analysis.

It is no longer important to be part of a workers' movement. What is now important is the response of individuals in occupational categories to the key question, Is your net orientation procapitalist or proworker?[50] Carchedi says, "This is revealed by the respondents' answer to eight questions. On the basis of these answers a measurement of consciousness is constructed, which goes from maximally procapitalist to maximally proworkers."[51]

Carchedi argues that consciousness is not a simple aggregation of individual responses to the questions on a sociologist's questionnaire. Wright's use of an attitudinal analysis suits the objective requirements of statistical tests of significance and not the requirements of an analysis of the relations of class structure to class consciousness. Wright's analysis negates the dialectically determined relation between the supersession of capitalist production and capitalist class relations. Carchedi concludes:

> My thesis is also that Wright has chosen a frame of reference in which the Marxist unit of analysis (classes) has been replaced by individuals . . . the Marxist Method of historically determined abstractions by the fantasies of game theory; the Marxist concept of classes rooted in production relations by a concept rooted in distributional and occupational categories; Marxist dialectics by determinism; and the Marxist concern for social change by the concern for individual consciousness.[52]

Carchedi's Class Analysis

Is Carchedi's class analysis less static and less deterministic than Wright's? Or is it the substitution of one form of determination for another?: A determination centered on a game theory analysis of exploitation for a determination centered on a dialectical analysis of class struggle? A determination of a distribution-centered class analysis for a determination of production-centered analysis? Is Carchedi as guilty as Wright in separating theory from the lives of people who live in history? Is his answer to Wright's statistically significant science an abstract dialectical science? One explains class only in terms of the distribution of assets (economics) or as a production relation (economics as a determinant instance), both in the mode of Althusser's positivist Marxism. Both ignore the cultural component of class. Class is a cultural category as well as an economic, political, and ideological category. But class is more than a reified category. Class is a conscious community producing its own life in struggle. Carchedi seems guilty of his own claim against Wright; his analysis is equally deterministic and static.

Carchedi's class analysis is an attempt to remain true to Marx's theoretical and concrete analysis and to develop the dialectic as a tool for social research. But his dialectical class analysis suffers from the same problem that Hegel saw in Kantian science—that is, its starting point is the "in itself." It is only a theory of knowledge and not a theory of becoming. Carchedi is not wrong but his dialectic, like Kant's, is one-sided. The determinant instance is always class-rooted in production relations (read economic relations). Thus, for Carchedi the determinant instance is always economic. Though Carchedi posits a unity of determinant and determined instance, this is a unity in which the economic is ultimately determining. He thus reconstitutes a base/superstructure relation in which the superstructure, though not simply determined by the base, is subordinate to the base. Here the difference between Carchedi and Wright becomes minimal: both seem interested in developing a science of class analysis in which cultural and collective relations are never autonomous but at best are dependent variables and in the last instance are determined by the independent variables of the capitalist marketplace and the material basis of life.

Carchedi's dialectic is reduced to a system of typologies. Thus the principles of class analysis have epistemological and ontological postulates. The ontological postulates are the material basis of life, the social basis of material life, and the basis class of social life.[53] Here social life is defined not in all its richness but in only one sense. Ontological here means material and for Carchedi this means economic relations of production. All other social relations are superstructural: "production relations, i.e. who produces what, for whom, and how. It is for this reason that production relations are the most important of all social relations. To understand society we must understand production relations."[54]

The epistemological postulates are the material determination of knowledge, the social determination of knowledge, and the class determination of knowledge.[55] For Carchedi, each class sees reality through the prism of its location in the class structure. This is his determinant instance in his dialectics of class structure. In his two-class model there is the collective laborer and the nonlaborer. The collective laborer produces surplus value while the nonlaborer functions in control and surveillance of the labor process and the extraction of surplus value. The collective laborer has the privileged epistemological position and the possibility of correct knowledge: "The laborer's view is also *the only one* which has the objectively determined possibility of being correct, of finding a 'match' with the reality it depicts, of knowing the essence (nature) of things, not in spite of, but because of, its being class determined."[56]

His use of the collective laborer includes both material and mental work; knowledge is given a crucial position in the production process, but it is never autonomous and always determined. He never understands the centrality of knowledge in the production process or that knowledge is a cultural construct. Still, he understands the necessity of creating a broader category—that is, collective labor—because it includes primarily mental laborers as part of the production process:

> Thus in this work the concept of collective laborer will indicate all those engaged in the transformation of material use-values, and thus in the production of knowledge determined by it, under capitalist conditions of production. It therefore encompasses both the material and mental laborers employed by capital.[57]

Corresponding to the division of collective labor into material and mental labor is a material and mental labor process. Whether a labor process is mental or material is dependent on the "determinant aspect" of that labor process. It is a material labor process if the commodity's use values have been produced and exchanged because of its material qualities. It is a mental labor process if the commodity's (mental commodity) use values (mental use values) have been produced and exchanged mainly because of the "knowledge it carries":[58] "Consequently knowledge becomes a commodity, something which has both a use value and an exchange value but which is produced in order to be sold, i.e. basically because of its exchange value."[59]

Ultimately, knowledge as a commodity is used to produce surplus value in a material labor process. Knowledge is an "economic activity" that is a determined instance in a capitalist labor process. Knowledge, though seen as possibly being crucial to working class resistance, as a determined instance can never become a phenomenon that is autonomous from the capitalist labor process. Even more important, in Carchedi's analysis there is no space for the working class to become conscious of itself; thus it cannot achieve the autonomy necessary for it to successfully resist the structure of domination in the capitalist production process. Thus, the supersession of capitalist history is unlikely. This is the case for both Carchedi and Wright: class as a reified category negates social change. Class consciousness and class struggle are determined by class structure. There are no subjects. There is no agent. The systemic nature of class structure within an economically determined society has negated the possibility of a class movement for change. For the supersession of capitalist relations of production and reproduction, class structure must be constituted by more than economic relations of production. Social relations systemically determined by relations of production continue to produce class as a determinant instance. Thus, they exist only as a category, a position that is structurally determined, and never as subjects of their own making. On the contrary, if social relations are to be seen as more than a system in which actors are determined, social actors must be able to determine their own situation. Thus, social relations are relations in which culture and community are constitutive features. Thus the base/superstructure distinction in which the base is the determinant instance no longer holds. Then knowledge is both a commodity and an autonomous cultural form. The relative autonomy of knowledge and culture is the space in which the possibility exists for class to become more than a "class in itself," to become a "class for itself" and thus become the agent of its own destiny. To hold on to the base/superstructure analysis is to reproduce class as a finite position and to negate the possibility of class as a conscious movement for change.

Our problem then is to demonstrate the centrality of culture within the formation and reproduction of class—to demonstrate that class is a project in the making and not just a form of reification, that this is central to any class analysis in which class is a potential part of the process of the supersession of capitalist domination.

Class relations are social relations. But social relations are not governed by systemic economic relations. They are overdetermined, but not by economics. Class relations are not limited only to the social relations of the labor process; they continue outside the labor process as well. They occur in all aspects of everyday life. They are not just determined instances of the economic base, of the "social laws" of capitalism that in Carchedi's words are "necessary and indispensable."[60] As for Wright and Carchedi, class struggle cannot be reduced to simple class models in which the question of the middle classes can be theorized as contradictory class locations between the asset-rich and the asset-poor, or as collective labor versus unproductive labor. The question is not which side are you on, or with whom you identify; the question is, How do social actors produce class culture, which is necessary and indispensable if class struggle is going to occur?

We referred to Wright's analysis as middle-range Marxism. It was meant as a central component of our critique of Wright's Marxism. Wright's analysis is grounded in Robert Merton's assumptions of an operationalized theory that can be empirically verified. The problem Wright sees in Marx is that there is not a direct correspondence between his theory and empirical research. Marx cannot be operationalized. Thus Wright seems to be making Merton's argument in *Social Theory and Social Structure* that a left-wing functionalism is as possible as a conservative functionalism. Since a middle-range functionalism is a basic requirement for producing a scientific and testable science, "to the extent that the general theoretical orientation provided by Marxism becomes a guide to systematic empirical research, it must do so by developing intermediate special theories."[61] The problem is not that Wright's Mertonian Marxism is mainstream. A mainstream Marxism would be a pleasant remedy to the current dominance of "free-market social science." The problem is that Wright's Marxism, grounded in Mertonian assumptions of what constitutes a social science, is a version of role-set theory in which Merton's own assumptions are grounded in Weber's assumption of "status group." Wright's analysis of middle-class contradictory class locations is closer to role-set and status-group than it is to Marx. It is an analysis of class "as an array of roles" attached to "various social statuses"[62] that are compartmentalized in the bureaucratic institutions of the corporation or the state. Struggle for change is possible only within institutional frameworks, making social movement unlikely. The middle class as a status group, locked into an occupational position, can only be a reified category. In Wright's formulation, class can never be social relations in which people are actively involved in formulating their own existence. Class is a static category; like Tess McGill in *Working Girl*, it has nowhere to go.

"Nowhere to go" also describes the one-dimensional analysis of class of Carchedi and Wright. Wright, though he claims to expand Marx's class analysis, continues a one-dimensional analysis as if struggles over gender, race, and national identity are not mediated by class and as if class is not mediated by these identities. This lack of a multidimensional analysis of class is also a form of class reification. Laclau and Mouffe critique class in these terms, but instead of creating a multidimensional model they read class out of the discourse. For us, class, race, gender, and national identity all determine one another. It is not class versus race, as it is for William Julius Wilson,[63] but class and race as mutually determining. Thus our analysis of Tess is an attempt at an analysis in terms of gender and class without privileging one over the other. Our analysis of *Do the Right Thing* is an attempt to look only at the race/class matrix in the movie in terms of furthering the development of a multidimensional notion of class.

Our intention is not to point a finger at Carchedi or Wright but to argue that both are participants in a 150-year tradition of one-dimensional class analysis. In the 1990s this is no longer tenable. A multidimensional class analysis is a complex task. It must deal with the reality that classes are internally stratified along gender, race, and ethnic lines. For example, the upper class in the United States includes An Wang, David Rockefeller, and Jackie Onassis. They all occupy different positions within the upper class. All of this is mediated by cultural, economic, and political relations under historically specific conditions. For us, this is at the heart of Barbara Ehrenreich's analysis of a middle class in spiritual and material decline.[64] It is here that Alvin Gouldner, in his analysis of the "new class" and its "culture of critical discourse," provides clues for a nonreified and multidimensional analysis of class relations.

Beyond a Reified Analysis of Class Relations

For Gouldner, the "New Class" has become a contender for class power in the world socioeconomic order. The industrial blue-collar worker and the service worker are replaced by automation. The working class as the agency of "universal human emancipation" decreases in importance as knowledge work moves to the center. Knowledge work becomes constitutive of production. It becomes central to both the public and the private world. As a result, the new class becomes less dependent on the corporate world and gains autonomy. This autonomy is founded on their culture of critical discourse. It is the "infrastructure of modern technical languages." It is the culture of critical discourse that constitutes the knowledge that is constitutive of all sectors of society. It is also the "bond between humanistic intellectuals and the technical intelligentsia." It is the basis of their solidarity; they share a "similar education, culture, and language codes."[65]

The new class has traditionally been the servant of corporate capital and the state. Dependent on its special access to cultural capital, this new class was able to produce a space for itself in "public education," a space within which some degree of autonomy could flourish. As new discourses of knowledge become central in education, new-class teachers and professors become crucial. Increasingly, students are socialized into the culture of critical discourse. Family becomes less important as a source of the socialization of children. School becomes the site of a "linguistic conversion," and the new technical languages replace the language of everyday life. Increasingly, the culture of critical discourse becomes the language of everyday life.

Parents are also attached to the cultural discourses of the new class. Just as Tess McGill learns the technical language of the business world, parents have been educated in universities. They are no longer working-class, but are also members of the new class. Their children are nurtured in the culture of critical discourse. And college education becomes the final step of the socialization of their children into the culture of critical discourse.

At the same time that the new class is in formation as an independent and powerful subject, its own self-interest is paramount. Even as its power grows, it is not a unified movement. Unlike Gouldner, we see it as a class that is internally stratified in race and gender terms: certain race and gender groups are excluded, others are valued. This internally stratified group is often conservative in these terms. For Gouldner, this new class is not necessarily a class "in radical chains." Of course, too often it accommodates or defers to the powerful:

> Obsequious professors may teach the advanced course in social cowardice, and specialists transmit narrow skills required by bureaucracies. But Ronald Reagan did not set out to curb the University of California because it was a servant of capitalism. And why the attack on CUNY (City University of New York) if it, too, was only a servant of monopoly?[66]

As Gouldner states, the new class is not only accommodating the centers of power. In its autonomy, intentionally or unintentionally it creates space to confront the contradictions of both postindustrial capitalism and communism. It can dissent and question authority; it can and often does on racial and gender issues. In its domain—the schools, colleges, universities— the culture of critical discourse is no longer totally subordinate to "moneyed capital." And moneyed capital is dependent on the new class, both as the teachers and professors of their own children and because their expertise is required for the production and administration of their own institutions:

> Tertiary education, including the reproduction of the technical intelligentsia, even in capitalist countries, is now less dependent on the private sector and is increasingly dependent on the public sector or state. Some view this as the "socialization" of private industry's research and development costs. . . . This is correct, but it misses the contradictions in the situation. For in "socializing" such costs, the private sector loosens its control over the reproduction of the New Class and increasingly, these become vested in the New Class itself.[67]

As the new class becomes less dependent on the private sector, it increases the space in which it can contend for power. The private sector's grip on the new class decreases, and at its best "public education" becomes the transfer point of cultural capital to a wider and more diverse population of students. Moneyed capital actively increases its attempts to stem the tide of this transformation. Gouldner did see the beginning of this attack, but he did not live to see Reagan become president. As governor of California, Reagan defunded the state university system; as president he was going to do this to the whole country in the name of privatization. Reagan's presidency created an austerity state—an attack on the welfare state and also an attack on the new class.[68] The defunding of education, science, and the arts is an attack on the independence and power of the new class. Attacks on leftists in the university and on "political correctness" are attacks on the dissent, self-interest, and threat to traditional authority that is central to the culture of critical discourse of the new class.

The future of these confrontations, in which knowledge and culture are constitutive features, is crucial to the emergence of the new class. Whether Gouldner's formulation is correct or not, this is a theory that resonates with political, economic, and cultural transformations. Gouldner's theory is not a simple reification of class, as were the theories of Wright and Carchedi.

Much of this book demonstrates the limitations of Gouldner's new class, not the least of which is his own failure to understand the culture of critical discourse in what we have called multidimensional terms. But Gouldner was pointing in the right direction, that is, class relations as disruptions of class as reification. Here we must look at the "subversive side of class . . . that new agents have emerged whose histories are in the process of shifting from subordination to social power."[69] The attack on the new class has resulted in a growing proletarianization. It is not just the result of a global economic and technological restructuring. In fact, the restructuring in part is a result of the felt threat of the new class, the threat of the new class's emerging autonomy and power. Here class must be redefined in terms of these transformations. Class is not just an economic category that expresses itself politically through class struggle. Rather, culture is central to the constitution of class, and it is in the formation of class culture and community that class struggle becomes possible.

Many critical and feminist theorists have decentered class with concerns about race and gender. This has led us to formulate class as multidimensional. Within this multidimensional conception of class, the new conditions of the centrality of knowledge work make possible new politics and new social movements. Thus knowledge/culture embodied in class is still a possibly radical component of the struggle for change.

The stakes of these struggles are defined in terms of the disruptions of the reification of class and a multidimensional class analysis. It is not just that the new technologies in the computer-mediated workplace disrupt the occupational structure, but that knowledge itself is a constitutive feature of the postindustrial, postmodern workplace. The centrality of knowledge is the disruption of occupation and class.

It is not full employment that is the disruption of class relations. No, full employment is a part of the reification that the occupational structure provides employment (independent of a decent income) for everyone.

But the struggle for "zero-work"[70] is a struggle whose central feature is the disruption of both the work ethic and productivity as panacea. Here longshoremen, even though they did not go far enough, exemplify this disruption:

> Brooklyn longshoremen are among the victims of technological development. Thus, automation in the form of containerized cargo and container cranes has displaced tens of thousands of these workers. Yet longshoremen have fared better than most, because they have struggled against automation by opting for pay without work. They won the Guaranteed Annual Income which guaranteed full wages for all workers displaced by technological change on the waterfront.[71]

It is not just the demands for justice by women and racial minorities that are the disruption. But in their demands for "comparable worth," which calls for the overthrow of an obsolete and oppressive occupational and class structure, a new social movement must be built. Class is not a reification when social movement is central. Class as a social movement is the struggle to transform its own cultural representation and formulate a new cultural representation of class, work, and power.[72] It is to these conclusions that we now turn.

Chapter 10

Quantum Measures:
Capital Investment and Job Reduction

High Tech: A Spur to Economic Growth?

"There you are" said Wilcox. "Our one and only CNC machine."

"What?"

"Computer-numerical-controlled machine. See how quickly it changes tools?"

Robyn peered through the Perspex window and watched things moving round and going in and out in sudden spasms, lubricated by spurts of a liquid that looked like milky coffee.

"What's it doing?"

"Machining cylinder heads. Beautiful, isn't it?"

"Not the word I'd choose."

There was something uncanny, almost obscene, to Robyn's eye, about the sudden, violent, yet controlled movements of the machine, darting forward and retreating, like some steely reptile devouring its prey or copulating with a massive mate.

"One day" said Wilcox "there will be lightless factories full of machines like that."

"Why lightless?"

"Machines don't need light. Machines are blind. Once you've built a fully computerized factory, you can take out the lights, shut the door and leave it to make engines or vacuum cleaners or whatever, all on its own in the dark. Twenty-four hours a day."

"What a creepy idea."

"They already have them in the States. Scandinavia."

"And the Managing Director? Will he be a computer too, sitting in a dark office?"

Wilcox considered the question seriously. "No, computers can't think. There'll always be a man, at least one man, deciding how. But, these jobs" he jerked his head round at the row of benches "will no longer exist. This machine here is doing the work that was done last year by twelve men."

"O brave new world" said Robyn "where only managing directors have jobs."

This time Wilcox did not miss her irony. "I don't like making men redundant" he said "but we're caught in a double bind. If we don't modernise we lose competitive edge and have to make men redundant and if we do modernise we have to make men redundant because we don't need 'em anymore."[1]

One of the problems we have faced during the research for, discussions about, and writing of this book is that we have had to face the brutal truth that our perspective is, to put it mildly, out of sync with the assumptions that underlie current debates about work and its discontents and prospects. First, we do not accept the prevailing wisdom that significant levels of investment in plant and equipment and consequent economic growth lead, in the current era, to more permanent jobs. Or, to state this proposition more positively, if the tendency of most investment is labor-saving in comparison to the part played by machinery in production, then the jobs created will be reduced relative to the unit of invested capital. So, from the construction of buildings and the production of machinery, the number of workers—intellectual as well as manual—is reduced by *quantum measures* in computer-mediated labor. Consequently, given the relatively slow rate of global economic growth, and especially the long-term tendency toward sluggish U.S. growth rates, even when new investment creates short-term employment for machine producers, the number of jobs is reduced in the long term.

We used the term *quantum measures* to distinguish computer-mediated production from that of the mechanical era. There is no question that labor productivity, if it is measured as a ratio of a unit of output to the time required for its production, has increased dramatically. To be sure, the part played by labor in the production of the commodity has been reduced more or less progressively since the introduction of pulleys and pumps in the sixteenth and seventeenth centuries. The rationalization of the labor process into smaller units of repetitive tasks further increased the productivity of labor. Later, at the beginning of the twentieth century, the process of labor displacement was accelerated by the application of electromagnetic technologies to engines and motors; the replacement of iron with steel and other metals with oil-based synthetic materials and the production technologies that accompanied this change; and the development of the chemical industry and electrically powered telecommunications. But contrary to the optimistic predictions of early advocates of computer technology—who said that it would proliferate needs and thereby increase employment, especially of specialist professional and technical categories—the long-range effect of the introduction of computer-mediated technologies has been to make possible the utopian (dystopian) dream of the *virtually* automatic factory in which labor is consigned to the role of maintaining and administering a self-reproducing labor process in most decisive sectors, including professions.

As we have seen, this tendency is also present in knowledge occupations such as engineering, medicine, and computer programming, which from technological generation to generation tends to *eliminate* or substantially reduce virtually all time-consuming manual operations from intellectual labor and, with the assistance of programming, to assume what are regarded as "purely" intellectual functions, if this distinction may be made. For example, in hospitals and in group and private practices, physicians rely more and more on sophisticated computer-mediated electronic equipment to make diagnoses. The machines have tended to exceed mere information gathering because of their ability to synthesize data supplied by patient interviews and supply both diagnoses and prescriptions. In this transformation, the physician retains *ultimate* authority over patient care but has relinquished, no less than the engineer working with computer-aided design and drafting, some of the "craft" of the profession. In the name of both efficiency and precision, the affective dimension of medical practice is declining or, more exactly, is being displaced to an underground existence. This is particularly the case in hospitals, where the issue of cost containment for most intents and purposes drives the entire system. In hospitals, this may result in fewer connections between physician and patient; the doctor consults not merely a chart of test results, but also computer-derived proposals for medication. In a period when health care is concerned with "getting rid of patients" in order to achieve cost containment, the physician is subject to the same productivity standards as any other employee.[2] In sum, we work with a sharply different set of perceptions about the implications of the scientific-technological revolution of our time than do those who adopt the prevailing intellectual framework.

Second, our conception of the appropriate *moral economy* corresponding to the contemporary social and historical context within which the debate about work takes place is profoundly at variance with both the liberal and conservative common sense about the significance of paid labor and its future. For example, apart from the current economic stagnation, it seems fairly obvious to us that there are simply not enough jobs of any kind with decent pay in relation to the actual—much less the potential—labor force, which has grown exponentially since 1970, and the long-term tendencies in both manufacturing and in the services that point to the progressive displacement of labor, even during periods of economic growth. Further, as work in the production sectors has diminished, the service jobs created in the 1980s that to some extent took up the slack were lower paying and have begun to disappear. The high rate of business failures is a result not of technological innovation but of long-term economic stagnation and two concomitants: stagnant income levels for many, and falling income for a substantial minority; and the "correction" for 1980s overexpansion in retail and financial services that were propelled largely by debt accumulation.

At the same time, public employment—which still constitutes one-sixth of all jobs and was the big employment-growth industry of the 1960s and 1970s—has barely held its own only because state and local governments have been obliged to assume many of the functions abdicated by the federal government. However, the end of the anticommunist phase of the cold war and the onset of the deficit crisis, stagnation, and decline in defense-related expenditures have reduced military installations and defense contracts, and there is almost no chance that the military sector—which accounted for a large proportion of investment and considerable employment for more than half a century—will resume its growth, much less play a decisive role in the economy.

Yet, as we shall see, liberals as well as conservatives tend to accept, in the wake of these trends, the assumption that in the long run economic growth necessarily entails an expanding *employed* labor force. And as one commentator, Lawrence Mead, contends, social policy depends on what he calls the "competence assumption" that, given a chance, any person who is able and in economic need will actively seek work and accept the best offer for employment even if it is poorly paid. That is, the new consensus thinking about social welfare is that the polity must insist upon work as a *moral obligation*, even the fundamental basis of what we mean by responsible public behavior. Of course, Mead, in concert with other mainstream writers, also makes the point that the best route out of poverty is paid labor, even in the wake of the increasing number of working poor receiving food stamps and "partial checks" in recognition of the fact that despite performing full-time paid labor, they are unable to support themselves and their families without public assistance.[3]

We contend that even when the number of jobs has increased, it has not been sufficient to employ a substantial segment of those who want work or who would be willing to accept jobs other than low-wage, dead-end employment. This disparity does not always show up in the unemployment figures. A slack labor market discourages many from seeking work; first-time job seekers are not included in official figures, and part-time workers are counted by the Bureau of Labor Statistics, the official recorder of employment data, as working. Their "part-time" unemployment is not a category of the official statistical methods. The bureau does not calculate the time they are *not working* as unemployment. And, since the Reagan administration, members of the armed forces are counted as employed, adding to the aggregate labor force and further reducing our collective comprehension of the true dimensions of joblessness. Moreover, we argue here that both the quality and the quantity of paid labor no longer justify—if they ever did—the underlying claim derived from religious sources that has become the basis of contemporary social theory and social policy: the view that paid work should be the core of personal identity.

The economic perception of many mainstream welfare and work experts is that there are plenty of jobs, but they go unfilled because the unemployed have refused them for the unethical reason that they prefer dependence to the more dignified but arduous conditions that might accompany independence: the obligation to engage in dead-end, low-wage labor. In the recent literature on welfare, many experts have refused to accept the explanations given by the unemployed for why they are not working: most of the jobs offer minimum wages and are dead ends; they do not pay enough to permit single parents to purchase day care for their young children; the comparative advantages of dead-end jobs over welfare are simply not great enough in most instances to justify working full time.

Interestingly, advocates of coercive public welfare policy admit that the Reagan administration's policy of reducing benefits substantially below the minimum wage, which for a household of four persons is well below the official poverty line of $14,000 a year, has failed to force recipients to accept dead-end, low-wage jobs. Therefore, they argue, public agencies should adopt a more direct form of coercion—workfare—to force welfare recipients to work, even if it is at hopelessly irrational tasks, as a condition of receiving benefits. The liberals wish to tie job training, expanded day care, and improved health benefits to this requirement. In contrast, consistent with a growing retributive standpoint, the conservatives want to get people off welfare even if funds are not made available for supportive services. In accord with prevailing economic doctrine, they believe that "temporary" economic conditions make jobs scarce, in which case welfare recipients who want to work should not expect to remain on public assistance for more than a short period of time. Nonetheless, both agree with the basic assumption that what Mead calls "nonwork" is at the root a psychological or moral debilitation of the poor and the major "cause" of poverty in our time. Hence, the major objective of the draconian measures that make up the consensus view on welfare policy is to overcome the dependence of, particularly, black men and single mothers on the welfare system.

We begin our discussion of the philosophical, ethical, and economic dimensions of the scientific-technological revolution and the future of work with the issues raised by the contemporary welfare debate because this arena has, more than direct discussions of technological innovation, raised many of the crucial questions that must be addressed in order to evaluate the implications of scientific-technological change: the ethical dimensions of work itself; whether jobs exist for those actively seeking work; and if so whether they are worth filling. Beyond these specific questions, the welfare system has remained the central ideological and institutional context within which the U.S. political system discusses work and its discontents.

On the other hand, outside the welfare debate, politicians, journalists, and most social analysts are content to follow one employment trend, taking the official unemployment statistics as a reasonably accurate measure of economic health and social cohesion. If by this measure less than 7 percent of the workforce is counted as unemployed—the current borderline between safety and danger (thirty years ago it was 3 percent)—the economy ceases to be a hot political issue. And the long-term jobless (unemployed for more than a year) are considered largely responsible for the fact that they have no jobs. Further, if the jobless rate among white men is lower and the unemployed are mostly youth, women, and racial minorities, there is even less to worry about, except of course demonstrations and "riots" grouped under the category of "urban" or "inner-city" discontent. In recent years, these events have been interpreted by conservatives and an increasing number of liberal politicians and policy analysts as principally an issue for criminal justice, social work, and mental health institutions.

The core issue we raise in this book is simply excluded from public debate: What are the implications of technological change—not merely the introduction of new machinery into the workplace but also sweeping organizational changes—for the quantity and quality of paid work? The corollary to this is, If paid work is increasingly "unnecessary" in relation to technological possibilities, should it be encouraged? Assume in addition that in fact there are not enough jobs. Under these circumstances, what is the status of the traditional work ethic (which in actuality always means the compulsion to engage in paid work whatever the compensation)?

In 1989 the average annual factory wage, computed on the basis of a forty-hour week, was $25,500, while the average annual wage in services was $14,000, hovering around the official poverty line. While family income did not decline appreciably in the 1980s, this was almost entirely a result of the mass entrance of adult women into the labor force to take the new service jobs. Many of these jobs were in the fast-food industry, supermarkets, and retail department stores and were offered on a part-time basis.

The great majority of part-time workers enjoyed no health, pension, or vacation benefits.[4]

These statistics tell us that factory labor is almost twice as remunerative as service work, but they do not describe the stratification within the industrial sectors. They do not tell us, for example, that machinists whose work is largely programming numerical controls may earn twenty dollars an hour or $40,000 a year base pay (excluding overtime) in the largest firms. But since the introduction of numerical controls in machining, there are fewer machinists and machine operators in industries using noncomputerized standard machine shop tools such as lathes, milling machines, and drill presses. In small shops, machinists in almost all sections of the country are fortunate to earn fifteen dollars an hour. They tend to work more overtime than workers with similar credentials in larger union shops, so their income is not much different, but they work longer hours than at any other period of their working lives.

Similarly, aggregate statistics concerning wages and employment fail to delineate the massive difference between noncraft labor on an automobile assembly line with a 1992 base pay of about $35,000 a year (subject to compulsory overtime that can increase a worker's gross earnings to $60,000) and operatives of similar qualifications in the textile industry, which is 90 percent nonunion, who earn $15,000 to $20,000 a year and rarely get overtime. Similarly, the retail income figures do not differentiate between the wages of a worker in a fast-food restaurant or a convenience store, which rarely exceed six dollars an hour or $12,500 a year—under the poverty level—and are often offered on a part-time basis and the salaries of full-time department store salespeople whose annual income with commissions ranges from $14,000 to $20,000 a year. The annual income of salaried store managers, buyers, and middle managers in department stores and supermarkets may reach $35,000 in the major cities—roughly equal to the wages of skilled workers in major technologically advanced production sectors, but much higher than the wages of most service workers.

Of course, technological innovation may not be introduced when employers choose to reduce labor costs by an assault on union-negotiated wage and benefits or by layoffs when the economy is in decline. At the beginning of an economic downturn, investment in machines as well as in labor slows to a trickle; wages tend to flatten or move downward, and as the unemployed suffer longer periods of joblessness they are willing to accept lower wages; and union workers are less willing to strike, especially in the United States, where employers enjoy legal sanction to replace them with nonunion labor.

Further, almost none of the many recent commentaries on and studies of work, the evolution of jobs, welfare, and other elements of the labor "market" address the influence of scientifically based technological change on long-term trends in the composition, number, and quality of paid labor. This question is often subsumed under the rubric of labor "productivity," a category that only partially describes the effects of technological change. Roughly, economists measure labor productivity by the value of output, not the number of widgets produced, in current dollars and labor costs. If employers succeed in holding down labor costs by reducing wages—by increasing output in the same amount of time without significantly increasing wages or reducing benefits such as health care and pensions while prices remain either constant or increase—productivity is said to increase. Hence, productivity may rise without the introduction of new technologies.

Or technological change may account for increased productivity without lowering wages and salaries. Computer-aided design and manufacturing and the newer computer-aided software programming permit fewer workers to produce the same or more goods and services, in which case the influence of labor cost on total cost becomes less important. Among these innovations may be organizational changes such as articulating design and production by electronic means or introducing new techniques such as flexible specialization or "just in time" production. As we have already seen, under this "post-Fordist" production regime, large inventories are abandoned for small-batch production that corresponds to specific orders. Under this production system, although some workers become more multivalenced, the workers who formerly were engaged in producing stock parts and other products are displaced.

Just-in-time production reduces labor costs because it gives the employer a mechanism with which to hedge market fluctuations that often leave the company with expensive excess inventories. But it assumes a close relationship between the company and the workers. In the case of the General Motors Lordstown plant, this is an overly optimistic wager. In the summer of 1992, Lordstown's sheet metal plant was struck over issues of job security. Just-in-time methods left assembly plants across the country without vital parts. Within a day, the plant producing the profitable Saturn automobile was forced to suspend production and other plants were preparing to close. Within a week more than 30,000 workers were idled by a strike that initially involved 240 tool and die makers, machinists, and machine operators.

The historical trend is plain to see: industrialization entails technological innovation that reduces, relatively, the part played by direct labor time in the production of goods and services. The economic expression for this historical process is that the capital-output ratio increases. In plain language, after an initial explosion of jobs, over the long term machines tend to replace human labor, which itself contributes to unemployment unless other sectors of the economy are simultaneously expanding, especially those sectors such as government jobs and services, which in the past were labor intensive (the capital-output ratio for services is lower than in most industrial production).

In the 1950s and 1960s, when basic industries such as autos and steel introduced automation, the effects of labor displacement were mitigated on the one hand by a generally buoyant economy and on the other by the fact that most other sectors such as textiles, garments, and metal fabricating were still in the mechanical era of industrial production. In effect, increases in the volume of production and the explosion of consumption were sufficient to overcome, for a time, the labor-displacing impact of technological change; in some industries employment actually grew, fulfilling the most optimistic forecasts that workers' fear of automation was misplaced. Computer and laser technologies are now widely employed in these formerly high-employment industries. Combined with the globalization of textile and garment production, along with autos and steel, technological change has dramatically reduced employment in these once major entry-level industries, especially in the South and on both coasts.

The War against the Poor

In the twentieth century, what many have called Fordism and others have called consumer society (not exactly an equivalent term) absorbed some of the human costs of technologically induced labor displacement. Fordism refers to the introduction of highly rationalized technologies that make possible mass production of consumer goods, which in turn entails the development of a mass consumer credit system and greatly expanded distribution and marketing systems. Even before the 1930s, when the state was mobilized to intervene massively in economic life, Fordism introduced a kind of regulated economy. The expanded buy-now-pay-later credit system produced jobs in construction to build single-family homes on a large scale, more jobs on the assembly line to make cheap cars that many workers could afford, and new jobs in banking, insurance, and retail services. At the same time, owing to the decision of the federal government to take responsibility for broad areas of economic activity as well as to provide limited "transfer payments" to retirees, to those unable to work, to the temporarily unemployed and the underemployed (the working poor), the size of the combined federal, state, and local government bureaucracies doubled during the Depression to 3 million and multiplied by a factor of five after the war. Despite the general view that the New Deal was antibusiness, it introduced a new system of regulation that benefited large corporations. One of the major purposes of regulation was to make labor a predictable cost of production as well as a known factor in the political system. No New Deal reform was as important as the National Labor Relations Act of 1935; in return for relative labor peace, insurgent unions negotiated progressive wage and salary increases for a substantial minority of the labor force. In turn, in an economy marked by long-term net exports, corporations in key oligopolistic sectors of the economy were relatively free to calibrate prices to these increases.

From 1935 to 1970, collective bargaining drove industrial relations but also provided much of the fuel for mass consumption, which in turn helped cushion the blows of technological change, capital flight, and recessions. During this period the provision of income to the long-term as well as the short-term jobless, to the aged and disabled was justified by economists allied to the New Deal-Fair Deal coalition as a "floor" under consumption that, sufficiently funded, would retard recessionary tendencies in the economy. This claimed effect of transfer payments has been sharply disputed by both conservative and radical economists. Radicals argue that the welfare state functions to legitimate the capitalist state and, more broadly, the capitalist system. Conservatives oppose transfer payments on various grounds, perhaps most importantly (except for defense) the view that government intervention reduces private investment incentives by both raising the general wage levels and thereby reducing profits and unfairly competing with private business.

Since politics and policy concern *perception* rather than what may be called the "objective" case, income transfers were supported for their contribution to the economic system as much as their benefit to individuals. When conservative economic and political ideology regained hegemony over large segments of the population, and especially politicians, journalists, and policy intellectuals, it was not a result of the superiority of their arguments; rather, it was largely abetted by the claim that economic stagnation was caused by the large public debt and that government spending, especially welfare benefits, lay at the root of the debt crisis. Conservatives argued that this spending, combined with the Federal Reserve policy of supporting low interest rates to facilitate consumer loans, produced inflation. At the same time, "big government" was fueled by exorbitantly high taxes that squeezed the middle class. Hence the prescription: cut big government by cutting welfare and other "entitlements," and privatize investment so that economic growth can resume.

It is difficult to pinpoint the historical moment when the relatively widespread economic perception that welfare was both necessary in the wake of economic vicissitudes and beneficial to the economy was no longer broadly accepted. A simple yet powerful explanation is that perceptions began to change when the era of capitalist regulation began to be taken apart in the late 1960s. Certainly the crisis of the welfare state was intensified by the taxpayers' revolt that first broke out in 1976 in California with the Jarvis-Gann amendment, itself a premonition of the end of long-term economic growth. In the late 1970s it became evident that postwar prosperity had been the necessary condition of broad support for antipoverty programs such as job training and expanded educational opportunities for the poor, especially minorities. Feeling the pinch, many white working-class and middle-class people jettisoned modern liberal ideology in favor of a doctrine of social and cultural retribution. Now, the absent black family (the neoliberal explanation), the amorality of black youth (a more conservative version, but not necessarily inconsistent with the lack-of-family-values thesis), and the opportunism of single black mothers (another variant of displacement to the victims) were held responsible for their increasing poverty.

These local perceptions were brought together in a new version of the so-called culture of poverty thesis first advanced in the affluent 1960s by social theorists who followed Oscar Lewis's conclusion, based on ethnographic work in Puerto Rico and Mexico, that the poor were "structurally" separated from the working class not only in income but also in distinct cultural characteristics. The basic idea was that blacks and Latinos failed to enter the mainstream of economic opportunity because they had become caught in a "cycle" of poverty and had developed their own culture. According to Lewis and his intellectual progeny, the poor remained at the bottom in this period primarily as a result of *cultural* factors. Lewis posited the distinction between the "underclass" poor and the working poor. Tacitly, he "found" that in the midst of unprecedented economic development in Puerto Rico after World War II, an underclass that did not work regardless of whether there was paid work available had come into existence.[5]

In the United States as well as Puerto Rico, this cycle may have been produced in the past by economic circumstances beyond its victims' control—the chaotic development of U.S. capitalism, for example, and the agricultural crisis of the South after World War I. But the reproduction of intergenerational poverty after World War II—when at least 20 percent of the population remained poor in the midst of the great leaps of economic growth, in the United States as well as in Puerto Rico—could no longer be explained, according to some social analysts, by racism, or even by the Marxist conception that growth entails the existence of a relatively large "industrial reserve army" that can be mobilized to take paid labor in periods of expansion. Rather, they tried to establish a correlation between illiteracy, poverty, and a culture in which the work ethic played virtually no role. The nonworking poor could be found in industries such as drugs and prostitution and swelled the welfare rolls on a more or less permanent basis.

Daniel Patrick Moynihan and Nathan Glazer argued, from a generally liberal viewpoint, that the disintegration of the black family and other forms of social disorganization were the main culprits in apparently intractable black poverty. Without a male head of household and the internalization of "values" such as the importance of work and family cohesion as vital components of personal identity, even civil rights laws, education, and training would not enable blacks to transcend the conditions of their historical oppression.[6] To be sure, these views were already components of the Great Society programs of the Kennedy and Johnson administrations—programs such as Head Start (organized to give black children preschool experiences that could inculcate values counter to those of their own poverty culture), job training, and day care that might, together with proper indoctrination, succeed in breaking the cycle of black poverty. Job creation was a subordinate and poorly funded aspect of the antipoverty crusade because the working assumption was that the problem was not a lack of jobs but a mismatch between available jobs at all levels of skill and credentials and the qualifications of potential (black) applicants.

By 1969, when Moynihan became Richard Nixon's domestic policy adviser, there was some indication that he was tacitly revising his own analysis in the wake of the first signs that, even in the midst of the Vietnam War, economic growth was slowing to a crawl and unemployment was growing among white workers. Although Moynihan did not entirely abandon his social-psychological orientation, he concluded that poverty had a structural dimension as well as a basis in social disorganization among blacks; as the number of white poor grew, he was prepared to modify his earlier views. He proposed a guaranteed income plan, which, although the amount was below the poverty line, was shot down by conservatives: such ideas violated the precepts of their free-market ideology. For the next fifteen years, as conservative hegemony in social policy gradually assumed the status of common sense, we saw the steady rise, in different registers, of the culture of poverty thesis, which eclipsed the once accepted view that even after the Great Depression poverty had become an endemic feature of the economic system.

The new expert perception of the role of income transfers amounts to a sea change among politicians and social analysts concerning the causes of poverty. There is almost no mention of the economic function of these payments; nor, indeed, do many erstwhile liberals argue that poverty is either chiefly or partly the result of failures of the economy. We have definitively entered an era in which social welfare is considered a negative for its recipients as well as, for society, inflationary, a contributor to the massive debt, and morally unacceptable. The muddled but unmistakable implication of these judgments is that people in need might be better off without public assistance. In effect, there is an often unstated but by no means marginal view that the New Deal social justice program—its labor reforms as well as its income transfer programs—might as well be repealed.

This perception is based almost entirely on a rather traditional cultural ideal: the proposition, first advanced in England in the eighteenth century, that moral character is built by economic independence and destroyed by dependence. While no "responsible" policy intellectual proposes (yet) the return of the debtor's prison or strengthening vagrancy laws to deal with homelessness, there is an increasing tendency to address poverty and homelessness as "public safety" issues (*public* here refers to everyone except the unemployed poor).

As we have noted, the "post-Fordist" era, which roughly corresponds to the emergence of a truly global economy and the relative decline of the dominance of the United States, has shaken the foundations of the old welfare-state arrangements. In an era of reduced expectations, the welfare state is now perceived as an economic deficit rather than a depression-fighting tool. Leisure, one of the more pleasurable aspects of the American cultural ideal, is transmuted into self-indulgence and hedonism—evils that both "left" and "right" critics would replace with the fundamental values and moral obligations of work, family, and community.

By the 1970s a new intellectual "movement" of liberals and leftist intellectuals emerged, calling itself "communitarianism." Among its early landmarks were Christopher Lasch's books *Haven in a Heartless World*, a paean to the importance of family and family values in building individual character and healthy social relationships, and *The Culture of Narcissism*. Together they became virtual manifestos of liberal and "left" social conservatism because they boldly articulated the discomfort felt by many who in the previous decade had easily dismissed Moynihan and Glazer's work if not as racist, then as an extreme example of the tendency to blame the victim.

But Lasch reached down into the political unconscious of many intellectuals who became genuinely disturbed by what they perceived to be a massive black "refusal" to work. In the 1960s, Marcuse's advocacy of the Great Refusal to participate in the emerging society of total administration had excited anarchist and revolutionary sentiments among many young intellectuals who at the same time were far from sympathetic to the counterculture. The fragmentation and decline of the civil rights and antiwar movements and the simultaneous rise of radical feminism abetted the return of the repressed. They (we) were accused of collectively oppressing women by, among other things, taking their homes as their castles. Typically, men have refused to share housework equally, to assume *any* of the tasks of child rearing, or to recognize and encourage the professional, intellectual, and political freedom of their spouses.

Lasch was the first theorist with strong leftist credentials to articulate a new traditionalism; rather than addressing these charges directly, he argued a general theory of American moral turpitude masked in such feminist staples as the attack on the family, the cult of "excess" as opposed to discipline and responsibility, and overweening individualism. In sum, Lasch launched what has become emblematic of all social conservatism, but with special focus on its liberal and left variants: he combined the defense of the family with an attack on freedom and pleasure. He then deployed the eternal values of family and community to characterize feminism and the counterculture as willful, self-indulgent upper-middle-class heresies. This deft concatenation of class issues with cultural issues mobilized a substantial segment of the intellectual "left" in a way ordinary social conservatism of the right could not have done.

In the nearly twenty years since the appearance of this doctrine, we have witnessed a spate of commentary stemming from an underlying repudiation of the excesses of the 1960s radical movements, especially the counterculture—a section of the New Left for whom unwork was not a violation of social responsibility but a rational response to the scientific-technological revolution and to the pervasive alienation of labor from itself—and the feminist movement that had rejected the traditional bourgeois cultural system that it believed kept women in bondage. Jean Elshtain and others have focused on the essential irresponsibility of feminism's critique of the family, and Elizabeth Fox-Genovese has recalled Lasch's attack on individualism and the irresponsibility of sexual radicalism.[7] Robert Bellah and Ann Swidler have cast their sociological studies in these terms.[8]

There are clear elective affinities between the efforts to reinstate social conservatism on the left and the sharp divergence of many policy intellectuals from liberal social welfare assumptions. At the level of socioeconomic theory, the key innovation was the "discovery" of an underclass that could no longer be considered an industrial reserve army because it was not available for work or was otherwise excluded from the labor markets even in periods of economic expansion. The question is how to interpret such exclusions. According to what is perhaps the dominant trend in contemporary social policy, the key to an explanation is the psychological and social factors leading to poverty. While, as is generally acknowledged among postliberal writers, there may have been economic and political reasons for dependence, its intergenerational reproduction must be ascribed to psychological and social factors that together constitute a culture. Among the psychological factors is the perception, widely shared among poor black men and single mothers, that work is not a redeeming social value.

We think that the relationship between the theory of the underclass and the culture of poverty thesis is more than coincidental. Although some, notably William Julius Wilson, trace the genealogy of the urban underclass to political and economic developments and are not at all in the postliberal camp on social welfare issues, there is in the *hegemonic* policy discourse a close identification of the two concepts.[9] Thus, there is no longer a necessary connection between the perception that millions are routinely excluded from the labor market and a determinate explanation for this phenomenon.

Work without End

From a political perspective, perhaps the most important consequence of the end of the guarantee for most workers, union and nonunion alike, of progressively rising living standards, measured largely by consumption levels, was the three-sided whammy suffered by many if not most of them. In the late 1970s and the 1980s, almost 6 million factory workers lost the best-paying jobs they will ever have. In high-wage industries such as autos, steel, oil, mining, and metal fabricating and in lower-wage industries such as textiles, shoes, and garments, plants closed because employers were unwilling to make the investments needed to modernize them. In working plants and offices, labor forces were reduced by labor-saving technologies. The end of U.S. global dominance—not only because of Japanese competition, but also competition from Germany, France, and, in advanced technological sectors, Italy—reduced the size of both U.S. export and home markets and consequently the size of the labor force. Henceforth, the crucial exports were agricultural products, for which only the French provided genuine competition, and the products of the knowledge industries, which were contested after the late 1970s by the Japanese, Germans, and Italians. New jobs were being created in knowledge-based industries such as bioengineering, computers, and health, but factory workers were not qualified for them, and the number of these fairly well paid new jobs was smaller than the number of well-paying jobs lost in industrial production. For example, during an extensive public relations campaign that heralded computers as the "solution" to virtually all economic, educational, and health problems, after almost thirty years of the spread of computer-mediated labor in factories, offices, and homes, there are about 1.5 million jobs in the production, distribution, programming, servicing, marketing, and selling of computer hardware and software. There are, of course, many more production and clerical jobs where the computer displaced older technologies. We are referring here to the degree to which computerization created new jobs as opposed to the millions of jobs that were abolished and the millions of jobs that required workers to operate computers rather than, say, levers, typewriters, and pencils. IBM, Hewlett Packard, Xerox, Apple, and other large producers of computer and electronic equipment pay competitive wages to their industrial workers, their professionals, and their managers. But many computer production jobs such as making chips are poorly paid.

In the past two decades, however, jobs for workers classified as skilled and semiskilled did not increase, even as these companies expanded, because they were subject to labor-saving technological innovation. Companies like IBM and Xerox maintained large industrial labor forces, but the growth occupations in the 1980s required scientific and technological *knowledge*. To qualify for employment, programmers, managers, and sales, service, and repair people must have either specialized training or credentials. In the 1990s, as we shall see, many industrial workers are leaving high-technology industries, but there is considerable attrition among professionals and managers as well.

Although the health services industry grew in this period by more than a million and a half jobs to become the largest industry in terms of both value of output and employment in the United States, occupations that require specialized training and credentials grew faster than jobs that do not. Nonprofessional workers in hospitals and other medical facilities—technicians, for example—must have credentials or formal training but are frequently paid much less than ordinary industrial labor in large plants and occupations such as trucking, for which the worker requires only on-the-job training or a relatively short training program. In New York, Philadelphia, Los Angeles, and other cities where hospital workers are unionized, nonprofessionals in patient care, housekeeping, and dietary departments enjoy the benefits of a union contract and earn the same as or more than ordinary factory labor in the largest companies, but unions represent only 25 percent of the nonprofessional workers in the health industries. In professional and technical categories, the proportion of unionized employees is much smaller.

The jobs offered to displaced factory workers and to women who entered the labor force (in part as a result of loss of partners) were for the most part of three varieties: full-time nonunion construction; trucking or factory jobs that paid about half of the factory workers' former wages and provided few if any benefits; and part-time and full-time employment in the services that were part of the last period of the postwar Fordist expansion. These service jobs pay between $14,000 and $25,000 a year, most of them about three-fifths of the 1990 average industrial annual wage of $26,000. Many below the supervisory level earn much less. When fringe benefits are calculated as part of income, the proportion dips below 50 percent. An increasing number of part-time jobs, now about 12 million or 13 percent of the total, are at or close to the minimum wage.

Moreover, as we indicated in chapter 2, clerical jobs are being displaced by technological change at a rate that rivals displacement of factory labor. By the mid-1980s, after a steep increase in clerical jobs in large private corporations such as banks and insurance companies, government employment, and the retail service industries that lasted almost twenty years, growth had come to a standstill. By the end of the decade, the once-secure clerical job was a memory; even when their pay was considerably lower than that for more volatile factory jobs, clerical workers became subject to market viscissitudes, thereby erasing the traditional comparative advantage of this occupation: protection from market fluctuations. And as computers replaced typewriters and other conventional business machines, employers replaced full-time employees with part-time word processors under the sign of flextime. The old ethic of corporate loyalty was dead; now employees, many of them working from their homes, were encouraged to take the money and run.

During the long-term economic recession that has afflicted the global economy, tendencies toward technological displacement are of course somewhat obscured by the overwhelming impact of stagnation and decline. Employers have adopted all of the major strategies to cut labor costs and maintain profit levels: they have introduced new technologies across all major industries and effected large-scale layoffs and introduced wage reductions (in real or monetary terms, including cuts in health and pension benefits) and increased the workloads of the workers who remain on the job.

As Juliet Schor has argued, contrary to the attribution by social conservatives of both the right and the left of lagging productivity to worker sloth, we are in a time of "the overworked American."[10] Working hours have lengthened even as unemployment remains high and stubbornly resists (slow) economic growth. The "double shift" has become a routine part of the lives of many women.[11] For many, paid labor and unpaid household work together absorb twelve to fifteen hours a day, and there is little weekend respite. Millions of men and women hold two or more jobs in order to earn what one good industrial job yielded a decade earlier. Professional labor of all sorts continues to betray the norm of the forty-hour week.

In fact, because their productivity cannot be easily measured in terms of units of physical or monetary output, those engaged in the production and dissemination of knowledge experience no time-bound definition of their work. Put another way, the boundary between work and leisure is forever blurred; even sleep may provide little respite. As we have shown, this lack of distinction may or may not be experienced by the professional as a blessing. Those happily at the cutting edge of creative activity—artists and those producing new "scientific" or critical knowledge, whatever the uses of their labor—are a small minority whose privilege is increasingly threatened by the general decline in living standards. For many professionals and managers, even routine tasks have no limitation on hours, especially if the work involves interacting with patients, clients, or other people. Many of these employees routinely take work home in the form of reports, memos, and technical and professional journals that provide information needed to keep up with new knowledge in their fields. Among the more ambiguous aspects of contemporary professional and technical work is the degree to which it requires that those who perform it be constantly "wired" by means of a beeper or telephone.

The character played by Tony Roberts in *Annie Hall* provides a grotesque of this phenomenon. Roberts is a manager. Wherever he goes, he frantically commandeers a telephone to call his office to get messages. The movie, released in the 1970s, prefigured the fate of an entire cohort of professionals and managers. In the 1990s, a latter-day Roberts can return calls from a car, a golf course, a tennis court, or a restaurant. The beeper has become a necessary business tool, but it is experienced by many professionals and middle managers in the corporate world as an oppressive sign of constant surveillance, a chain of engagement that fosters an odd kind of dependence. We offer this picture to counter the valid but overemphasized characterization of such people as workaholics. What is popularly seen as personal pathology may signify instead the growing social pathology of overwork that is attendant on survival and advancement in the workplace for many managers and professionals. (A recent *New York Times* story reported that professionals and managers who sleep a normal eight hours are far less productive than those who can manage on three or four hours a night. The "sleepers" risk losing opportunities for career advancement in proportion to their insistence on "normal" sleep habits.)

In addition, the end of the era of the relative "full employment" economy (in which reported joblessness was less than 5 percent) was marked by increasing maldistribution of income and wealth. The rich got richer and the poor were poorer, and the working "middle class"—that indefinite majority category of the American population—got squeezed. While an important minority of professionals, managers, and entrepreneurs in the growth industries such as computers, financial services, and health gained social power in addition to high salaries, real wages—even for most technical and professional employees outside the growth industries—slipped during the 1980s. Salaries and wages declined, in large measure a result of the steady retreat of collective bargaining and of unions, which after the war had tied their fate to what may be termed contract unionism.

The number of single-parent households increased dramatically in proportion to a rising divorce rate. Perhaps more salient for the hard times suffered by such families was the inability or refusal of the men to provide support. Many men suffered reduced incomes, in addition to their perception that they had lost the privileges associated with family life. The decision of many couples to seek separation and divorce, initiated more frequently by women after the rise of the feminist movement, in many instances provoked male resentment that patriarchial entitlements had been destroyed.

Working-class family income rose, though only marginally, but this may be ascribed almost entirely to the entrance of women into the labor force. The two-paycheck family has become the norm but does not signify a higher living standard for most working-class families. Until the late 1980s most were able to keep up with mortgage, appliance, and car payments and were able to send their children to college. Since the recession of 1989-90, however, this uneasy equilibrium has been profoundly disturbed: with mounting layoffs, foreclosures and auto repossessions have skyrocketed, and even as the costs of higher education are rising, many families have been forced to reduce or cease contributions to their kids' tuition and living expenses. Working-class and middle-class students who remain in school often must take out huge loans to defray the costs of schooling, and they are working more hours than ever before. Instead of finding their own apartments, they are frequently obliged to live at home. Moreover, the common expectation that a degree amounts to a guaranteed entry-level technical, professional, or line management job has been shattered at all levels. Nor will most of them be able to afford a home of their own.

The conjunction of technological unemployment and underemployment, disinvestment, and long-term economic stagnation has meant that a significant portion of formerly stable working-class families has sunk into the working poor; a smaller, although growing, segment has been condemned to nonwork. This is particularly evident in small industrial communities where a single large employer has either closed the plant, the mine, or the office or laid off a substantial portion of the labor force. In such areas there are few alternatives to working for that employer, and the remaining jobs typically pay six dollars an hour or less.

We can observe this situation in upstate New York towns where employers in erstwhile cutting-edge knowledge industries—IBM, Xerox, Eastman Kodak—remain the chief private-sector employers within a large area. In the past decade they have reduced their labor forces by one-third, in part as a result of the eclipse of their near-monopolies over technological innovation and therefore over the high-tech markets. Consistent with the no-layoff tradition adopted by IBM and Kodak during the period after World War II when unions persistently attempted to organize their employees, the first workforce reductions were relatively painless for those with long service: they were offered early retirement, often on extremely favorable terms. At the same time, less senior employees were retained on the assumption that economic conditions would soon improve.

As the slump in computers stretched out into the early 1990s, IBM's traditional no-layoff policy was strained to the breaking point. It now became necessary to find ways to get rid of those not eligible for early retirement without giving the appearance of betrayal. In the new arrangement, employees with a specified minimum service were offered early retirement with an enhanced benefits package. Younger employees were "bought out" with lump sum payments, which, although they were frequently substantial, did not provide more than temporary respite, since the job market has become increasingly bleak. These measures were intended to preserve the no-layoff policy. But for others—especially at Kodak, which reversed its no-layoff policy—and equally profoundly for the young people in the area for whom an IBM or Kodak job represented the pinnacle of occupational aspiration, the effects have been devastating. Some workers are able to get jobs in the leisure industry (motels and hotels and other retail outlets), in construction, and in the burgeoning state and local prison systems where the pay is perhaps twice that offered in near-minimum-wage retail trades. But where in the 1980s both family partners were able to find jobs, the recession has in many cases resulted in layoffs for one or both of them.

This pattern has with some variations been reproduced in other areas, especially where military-related employment was the basis of the local economy for the past fifty years. The recent closings of army, navy, and air bases as well as cutbacks in government contracts for various weapons has seriously impaired local economies in Texas, Florida, Georgia, and California. Layoffs at General Dynamics and other major defense contractors, combined with the depression in oil prices that has virtually wiped out domestic drilling, has plunged Texas into an unprecedented decline. Similarly, the cutbacks in aerospace and shipbuilding contracts and the closing of auto assembly plants and metalworking companies, notably Kaiser Steel, reversed the apparently boundless upward spiral of the Southern California economy.

Lawrence Mead, speaking from the postliberal conservative perspective that stubbornly holds to the truly American view that there are always jobs available for those willing to work, acknowledges but ultimately discounts the degree to which many of these jobs were the best jobs that manual workers with no formal craft or education qualifications can get in the United States. More egregiously, he ignores the symbolic significance of technological change and deindustrialization, preferring to minimize the impact of the more than 5.5 million lost jobs in manufacturing. He derides the claims of economists such as Barry Bluestone and Bennett Harrison who first called attention to the potentially devastating consequences of capital flight by invoking not the information that demonstrates that comparable opportunities are now or soon will become available, but the experience of the 1980s boom to show that many jobs remained unfilled in low-level service occupations.

The jobs associated with the mechanical era of industrial production were the bedrock—for the workers, their families, and their communities—of the ability of generations of immigrants and native-born workers of rural backgrounds to achieve a higher standard of living and also a significant measure of their industrial citizenship. Whatever the tendencies toward bureaucratization that developed within the American labor movement and the often controversial consequences of consumer society associated with the era of Fordist regulation, these developments represented a period during which a significant segment of workers lifted their heads from the subordination to which they were subjected in the nonunion era of industrialization. Wages in key sectors of production and transportation industries were, for the first time in history, equivalent to a family wage. That is, many women who in earlier years were *obliged* to seek and perform wage labor could stay at home, and children were no longer forced by economic conditions to work alongside their parents in the "satanic mills" or with their fathers in the dungeonlike coal mines.

To be sure, laws and custom that "retired" women to housework were not without their ironies. For some women it was a welcome liberation from the double shift since male working-class culture abjured shared housework, regarding much of it, especially child rearing, as a violation of manhood. The dark side of high, largely male wages is that it disempowered women and made them economically and sexually dependent on men by removing them from paid work. Where in the early decades of industrialization women worked in the mines and iron mills as well as in printing and other crafts, by the early twentieth century a mythology had arisen that identified "heavy" work in basic American industries with male strength and encouraged legislation to "protect" women from the oppression of certain industrial work and crafts. This ideology was decisively repudiated by the large influx of women into key production industries during World War II but was, remarkably, reinstated after the war as it became evident that despite unparalleled economic expansion, there was not enough room in the industrial workforce for both men and women. Hence the enormous popularity of the doctrine that women "fulfilled" themselves biologically and emotionally (often conflated) by having and raising children. "Homemaking" was elevated into a vocation in the Christian sense of the term and the print and the electronic media issued a relentless barrage of information, advice, and entertainment directed to the majority of women who worked at home.

318

The industrial culture has all but disappeared in dozens of Midwestern, Southern, and Northeastern communities that have experienced massive capital flight and technological labor displacement since the late 1970s. In some cases, the community itself was dismembered. Textile and steel mill towns, mining communities, lumber and shoe areas were left to nature as hundreds, sometimes thousands, of longtime residents were forced to migrate from the so-called frost belt to the sun belt, where, according to contemporary economic wisdom, opportunity was once more unlimited.

Pittsburgh and its neighboring steel and electrical towns were utterly transformed into a commercial, education, and technical center. For the scientific and technological professionals who built new knowledge industries, Pittsburgh became a technological cornucopia. But tens of thousands of steelworkers and their families watched the mills shut down, and many were forced to leave the area or accept jobs at much lower pay. Although he admits that there are millions of working poor who cannot seem to get ahead of poverty, Mead believes they are better off than those on welfare, especially in the 1970s and 1980s when deindustrialization revised their economic expectations downward.

But, as Mead argues, many managed to maintain their living standards:

> Income per capita has risen more quickly than earnings per worker because more members of families have gone to work to keep up with rising prices, while the size of families has fallen. Especially wives and other women have taken jobs in record numbers allowing the majority of families to keep ahead of inflation.[12]

The basis of this relatively successful adjustment to the new hard times that Mead and others admit has occurred because women have reentered the labor force may be endemic to America's economic future. The social identity of these families was ineluctably tied to one of the cardinal elements of working-class culture: to stay independent of the welfare system, even if to do so meant accepting dead-end, low-wage jobs or giving up homes or possessions purchased on credit. To which we would add that the stigma associated with this specific type of public assistance since the end of the Depression has for many been a strong incentive to remain within the wage labor system at any cost.

But we must also calculate the effect of public policies since the war. First, public housing, once regarded as a significant public gain, became stigmatized as a holding pen for the poor. Federal policy encouraged families to purchase single-family homes by providing guaranteed low-interest loans.

As the one-family home became the core of the new cultural ideal, the higher stratum of the working class was effectively split from the poor, even the working poor, who in the environment of the great American celebration that followed the war became invisible. Second, the panoply of income transfer programs from unemployment compensation to public assistance is infused with the fiction that ours is a full-employment economy and joblessness is a temporary condition. Therefore, we have never enacted a guaranteed-income program or even raised short-term jobless benefits that approximate a living wage. Beyond the obvious market argument that keeping jobless benefits low encourages workers to reenter the paid labor force, all of these programs are suffused with Christian moralism that regards the long-term poor as essentially irresponsible, a concept that was never far from the lips of the Democratic presidential candidate in the 1992 election.

Finally, public policy since the 1960s has been oriented toward systematically reducing welfare benefits in proportion to the lowest-paying jobs. Since the minimum wage has been declining in real terms for almost two decades, welfare payments are constantly adjusted downward, eligibility requirements are stiffened, and amenities are reduced to the vanishing point. Thus, there is a rational basis for staying off welfare. In keeping with the view that work incentives had to be maintained through public policies, the Reagan administration made aggressive and largely successful efforts to hold welfare costs down by reducing the rolls and drastically cutting payments to those who remained on them. Still, under Reagan, despite buoyant economic growth in the last half of the 1980s, welfare spending grew by 3 percent. This was a result of features of the system that respond automatically to movements of the labor market. Interestingly, welfare expenditures grew by 25 percent in the first years of the Bush presidency. Since Bush's policies did not foster draconian cuts in welfare or drastically change eligibility but instead tried to contain spending by freezing benefits and new programs, spending rose because the number of newly poor expanded dramatically between 1989 and 1992. In summer 1992 nearly 27 million people were receiving food stamps—more than 10 percent of the U.S. population. At the same time, the number of people on AFDC (Aid to Families with Dependent Children) grew by 2 million to 13 million.

There is considerable evidence that white working-class families are having increasing difficulty staying off welfare. While many may still be able to avoid AFDC, the dramatic explosion of the food stamp program signifies an increase in the working poor not only in coal mining, textile, garment, and other industrial towns of the South, the border states, and the Northeast, but also in California and other West Coast states. Some are finding it difficult to stay off direct-income payments because it is no longer as easy even to get jobs paying the minimum wage or slightly more in retail establishments such as fast-food restaurants and department stores in a period of bankruptcies and store closings. When these positions are available, they are mostly part time and offer no benefits.

Until the late 1960s, during what has commonly been termed the "progressive era," unemployment and its country cousin, poverty, became public issues that invoked various movements for redistributive government policies: job creation, job training and education, higher unemployment compensation rates, and improved welfare benefits for the poor. Mead points out that the competence assumption was the unstated ethical foundation for such policies.[13] In the current political and ideological climate, however, even though there is a widespread perception of economic crisis or at least serious trouble for the U.S. economy, the traditional progressive solutions are not even on the social policy agenda.

Nonetheless, there is little doubt that many single parents, white as well as black, and black men resist the proposition that any job is better than the alternatives. In this respect, Mead is right to point to what he calls psychological barriers to the competence assumption in some sectors. But he never entertains the possibility that the refusal of work may have a rational ground, even if not his, that there are reasons that cannot be ascribed to deficits in the moral economy of the nonworking poor that account for their unwork. Implicitly, his argument for a new politics of welfare based on a revival—by education and training, but also coercion—of the work ethic does not rest on grim economic realities because his perception is that the white workers who are part of two-parent families are for the most part successfully integrated by the assumptions of the work imperative. Beyond the disputable morality that work at any price is better than welfare lies the dubious assumption that there are jobs out there for the willing.

Indeed, the 1988 and 1992 elections were fought in part on these assumptions. According to the presidential candidates of both major parties, "old solutions" such as the expansion of the social welfare state must be abandoned in favor of fiscal policies to stimulate economic growth in the private sector, or public-sector investment policies to upgrade deteriorating infrastructures such as roads and rails. In a more affluent period, Presidents Kennedy and Johnson responded to what was generally viewed as an unacceptable rise of poverty amid general affluence in the 1950s and 1960s. Every presidential election and national administration since 1964 has witnessed the consolidation of a broad perception that the progressive era is decisively over and if government is to act to end poverty it must recognize that the main strategy is to help the *deserving* poor, that is, only those who agree to help themselves and thereby end the new era of dependence that accompanied the Kennedy-Johnson War on Poverty. The chief weapon in this effort, it is agreed by both "sides," is to encourage the poor to get off the welfare rolls and go to work.

Mead, who speaks from the standpoint of a chastened liberal Democrat, is one of the more articulate proponents of this view. He deplores "nonwork" among the poor and ascribes it to social disorganization as well as the resistance of blacks to internalizing the value of paid labor as a moral precept. Contrasting the response of the white middle class and working class to that of the black poor, he argues that the whites will take lower-paying jobs because they have accepted the proposition that the only way to combat their own poverty is to engage in paid labor (which he mistakenly calls "work"). In the face of stagnant income in the 1980s and early 1990s that did not keep up with inflation—in part as a result of the decline of unions—white women entered the labor force and, together with their husbands, found themselves working harder than ever before to maintain their living standards.

According to Mead, the number of workers displaced by deindustrialization is far fewer than left critics of contemporary economic policy have claimed.[14] But Mead's alternating moral and quantitative approach to poverty cause him to seriously underestimate the significance of these reductions. The loss of almost 6 million factory jobs (his figure is 5.1 million as of 1989-90, but 1 million more were lost in construction, transportation, and factories during the early 1990s) represents almost a quarter of all such jobs.

In addition to the decline of steel employment by two-thirds in the 1970s and 1980s and the cuts in auto employment by one-third, there were reductions of tens of thousands of over-the-road truck drivers, rail workers, longshoremen, miners, and construction workers, especially those in the "trowel" trades: the work of bricklayers and plasterers was replaced by pre-fabricated materials. Strip mining and other technological innovations reduced the number of coal miners from nearly 500,000, mostly organized into unions in the mid-1960s, to barely a quarter million, half of them non-union, by 1990.

A major technological innovation, containerization, eliminated huge quantities of the labor of longshoremen; the labor force on any given day in East and West Coast ports was reduced by 90 percent. The San Francisco waterfront, once the heart of West Coast shipping, has been transformed into a tourist attraction. The jobs have moved to Oakland and especially to Los Angeles and Puget Sound, but because of containerization the total number of West Coast longshoremen does not equal the number that once worked in San Francisco alone. Similarly, before the widespread introduction of containers on the Brooklyn waterfront, there were 40,000 members of the union and 30,000 worked on any given day. In 1992, there are no more than 3,000 union members, and most do not work regularly.

Employers and unions long ago recognized that the introduction of labor-saving technologies in the auto, steel, printing, telecommunications, and transportation industries would permanently eliminate most jobs of the mechanical era. In the late 1950s unions began to demand that new collective agreements reflect the tremendous productivity gains that would permanently wipe out millions of jobs. As early as the 1950s, the auto and steel unions abandoned the historic approach of dealing with economic shifts and technological changes by work sharing. They gave up the fundamental demand for shorter hours in favor of plans that would guarantee employment and income. Benjamin Kline Hunnicut's study of the history of the eclipse of the shorter hours movement found that:

> Higher wages, collective bargaining rights and fringe benefits . . . have virtually eclipsed any new calls for shorter hours. For over forty-five years, labor has done nothing comparable to its nineteenth and early twentieth century initiatives and successes. Moreover, since the depression, no major party has made shorter hours a political issue; no resolutions have been passed, and no convention platform has been constructed with reference to this traditional reform.[15]

Even though contract provisions that required employers to pay a steep premium for overtime work had been adopted in the late Depression years in order to discourage this practice, after the war overtime had become an essential component of industrial operations and the income of many workers. Consequently, especially in the early 1970s when cost-cutting became the watchword in American production, compulsory overtime was quietly or turbulently instituted in the auto industry and many other workplaces. Cars and single-family homes had become necessities rather than luxuries or privileges in the wake of the suburbanization of America, especially the decentralization of industrial location. The historic argument, based on principles of solidarity, that overtime put workers out of a job was laid to rest.

Among the immediately affected sectors, the labor movement's initial response to automation was deep skepticism concerning the assurances of economists, employers, and political leaders that workers had little to worry about, that technological change would lead to more, not fewer, good jobs. In our view, this skepticism was more than justified by both the long-term and short-term results of the massive introduction of new technologies into the clerical and industrial workplaces. But from the mid-1950s to the late 1970s, America's dominant world economic position seemed unimpeachable, and despite three recessions during these years, employment in nearly all sectors either held its own or kept growing. This helped to buttress the mood for more work that held a strong grip on many in labor's ranks who had accumulated extensive personal debt.

Equally powerful was the emerging mood of neocorporatism that was taking hold in the industrial union leadership. In 1955, under the tutelage of its progressive president Walter Reuther, the auto workers' union signed an unprecedented five-year no-strike agreement with major car manufacturers that provided for substantial wage increases, cost of living adjustments, and income improvements based on productivity gains. The union tied itself to the progress of the industry, abandoning the old assumption that unions sought to prevent employers from squeezing out more work by methods such as speedup and overtime. Today, in nearly all basic auto plants and a considerable portion of other manufacturing plants as well, overtime is compulsory. In fact, in spring 1993, as bank and government economists were proclaiming that the economy was slowly recovering, employers were solving production problems with overtime rather than by hiring new workers. Citing the exorbitant costs of health insurance and other benefits, employers preferred lengthening the work day to ten hours. However, economic uncertainty appeared to be the greater reason for hesitating to create new jobs in the wake of rising orders.[16]

On the assumption that the pace of economic growth would successfully absorb displaced workers, the "automation" agreements with the Big Three auto companies provided for supplementary unemployment benefits for laid-off workers for up to one year; with further technological innovations and a sluggish market in the 1970s the union negotiated a "thirty and out" retirement provision that gave workers with thirty years seniority their full pension. More recently, the union has won limited guaranteed income and job security provisions for senior employees, in return for which the union leadership was inclined to permit the Big Three auto corporations to close plants and shed tens of thousands of workers.

Other labor agreements, notably in longshore and printing, pioneered the guaranteed-income approach. In effect, these agreements institutionalized nonwork, but as a condition for receiving their pay workers were required to report regularly for work on the assumption that they were ready and willing to do the job if they were needed. In practice they actually worked a few weeks every year. The basic presupposition of the arrangement is the willingness of workers to grant employers the right to introduce labor-saving technologies in return for guaranteed wages whether they work or not. After a prolonged period of conflict during which skilled *New York Times* typesetters tried to block the computerization of their jobs, they finally entered an agreement that provides for guaranteed income, but in lieu of reporting to work employees may apply for an extended "sabbatical" during which they may do research or other approved activities. Despite these much-heralded agreements, guaranteed-income plans through collective bargaining, which may be viewed as a shorter work year agreement, have proven to be the exception rather than the rule.

The communications industry, for example, is among the most techsnologically advanced workplaces. The telephone companies, most prominently AT&T, have through several generations of technological displacements all but eliminated direct labor in placing calls. Operator-assisted calls are by far the exception rather than the rule, and most employees are now in administration and maintenance. In the last decade, the era of computer-mediated communications and organizational change, redundancy has become the crucial issue for workers, professionals, and managers in the industry. However, in contrast to the auto and transportation industries where technological change was introduced by mutual agreement, labor-minimizing technologies are part of the constitution of the telephone industry and predated unionization by decades.

Therefore, the power of the communications union, never as formidable as that of the auto, steel, and rubber workers, or transportation unions, has been increasingly constrained by computer technologies that permit management and clerks to operate the system for considerable periods during strikes. To be sure, the strike weapon has not entirely disappeared from the union's bargaining strategy, but in an industry that has been marked by long and turbulent strikes in the past, the likelihood of strikes in the future is next to zero. For this reason, in recent negotiations with management over labor-force reductions there has never been a question of a successful effort to reduce working time or to achieve guaranteed employment or income. Instead, the union has negotiated early retirement and buyout deals that resemble the basic approach of the rest of the communications industry, especially IBM, in order to mitigate the effects of the rapidly accelerating pace of layoffs. The pattern can also be observed at Western Union, where, in recent years, a Teamsters local has agreed to modest buyouts and early retirement for an overwhelming majority of the company's employees.

Professional Redundancy

The economic slowdown that began in the late 1980s witnessed a unique and startling new development in the United States that for many constituted nothing short of a profound psychological as well as social trauma. In contrast to all previous postwar recessions, when factory workers and other manual categories bore the brunt of unemployment, during the recessions of 1982-84 and the longer one of the late 1980s and early 1990s, for the first time since the 1930s professionals and managers experienced the dread as well as the reality of redundancy. In the 1990s professionals and managers have suffered periods of long-term unemployment lasting between six months and two years. For those over fifty, losing a job may signal the termination of a "career" and its replacement by a "job," sometimes outside the field. A large portion of these displaced employees are able, for the time being, to find other positions within their profession, but because of the widespread layoffs among these categories in the recent past, often at a substantial reduction in pay. Many others have been forced to take working-class jobs as, for example, parking lot attendants, cab drivers, and salespeople in retail trades. Even some lawyers, teachers, and social workers are having trouble finding jobs, especially at prevailing salaries. Still others, especially those over fifty, find that they are unable to find any employment at all.

In the prior half-century, although some professionals and managers, especially in defense-related industries, suffered "frictional," that is, temporary, unemployment when their employers lost contracts, the competition for their services was too intense in most sectors to permit employers to lay them off. Companies preferred to keep them on the payroll rather than risk being unprepared when business picked up, as it almost inevitably did. Thus, the very idea of unemployment was simply not in the vocabulary of professional, technical, and managerial employees. After 1989, however, many companies, already awash in debt accumulated during fierce buyout fights or during long-term recessionary conditions, became desperate to cut overhead expenditures. They cut their production and clerical staffs, but in many cases these reductions had been so steep that they had reached their limit. Their only recourse, short of going out of business, was to address cost problems by turning on those who had themselves been responsible for implementing policies of introducing labor-saving technologies and maintaining the labor process. Technologies such as computer-aided design and drafting and computer-aided software programming are meant to reduce the amount of labor required for design and program writing and, as a recent Electronic Data Systems ad argued, eliminate these high-priced categories.[17]

Middle managers have been especially hard hit, along with some engineers and technical categories. The Big Three automakers have recently initiated attrition programs for these employees that resemble those in the electronics and communications industries; thousands have lost their jobs, albeit with some remuneration, in many computer, electronic equipment, and optical companies; the aircraft industry, partially in response to defense cuts and the depression in airline travel, is hemorrhaging administrative employees as well as factory workers; and in summer 1992 the postmaster general announced an attrition plan for 40,000 managers in an effort to fully bring management practices of the quasi-public Postal Service in line with those of the private sector.

Beyond economic conditions, it became evident that "software" is itself a manager of the labor process; no less than that of the craft machinist or the engineer, the manager's knowledge has been transferred to the machine. Computers can issue instructions and provide both information and informed production and administrative decisions. In the mechanical era, every new technical innovation in the direction of further rationalization signaled the appearance of an army of managers to make sure that the quantity and quality of the product and the flow of materials were not disrupted in the wake of the increasingly detailed character of the labor; now much of the work of administration on the shop and office floor is inherent in the computer program.

Just as large amounts of clerical labor were displaced by computerization and by devices such as the answering machine, voice mail, and the personal computer, a significant quantity of managerial labor can now be displaced by the computer's capacity to do the work of coordination. The irony of this development is that the computer and other electronic devices that have displaced manual and clerical labor can also be turned on those whose scientific and technical knowledge promulgated the displacements. Management and computer consulting firms are widely employed to develop software programs and organizational proposals designed to eliminate knowledge producers. The displacer has been displaced.

We have offered a broad historical perspective for the argument that there are simply not enough jobs for those who are ready and willing to work. There are no signs to justify forecasts of a reversal in the elements of this judgment: in Europe as well as the United States, unemployment has stabilized for the foreseeable future at more than 10 percent if the *real extent of unwork is calculated*; with the exception of the German labor movement that has fought and won a shorter workweek, despite a relatively unfavorable world economic climate, we anticipate no sharp alteration, short of a new era of militant labor struggles, in the United States or most of the rest of the world, of labor's policies of exchanging job security for a minority of the labor force for the solidaristic demand for shorter hours; the unchallenged introduction of labor-saving technologies displacing both intellectual as well as manual labor; and long-term economic stagnation, especially in the post-cold war era.

Chapter 11

The Jobless Future?

Is Work a Need?

"Men like to work. It's a funny thing, but they do. They may moan about it every Monday morning and they may agitate for shorter hours and longer holidays, but they need to work for their self-respect."

"That's just conditioning. People can get used to life without work."

"Could you? I thought you enjoyed your work."

"That's different."

"Why?"

"Well, it's nice work. It's meaningful. It's rewarding. I don't mean in money terms. It would be worth doing even if one wasn't paid anything at all. And the conditions are decent—not like this."[1]

This conversation between Vic Wilcox and Robyn Penrose, a lecturer in English at a thinly disguised University of Birmingham, is from David Lodge's novel *Nice Work*. Penrose has been assigned to "shadow" Wilcox, the managing director of Pringle, a medium-sized, diversified, metalworking company in the increasingly deindustrialized Midlands region of England. In this dialogue, Vic summarizes one of the crucial elements of what has commonly been designated as the "work ethic." It is not merely the comparative economic advantage of paid labor over an increasingly inferior "dole" that motivates "men" to take jobs. Nor does the meaning of work derive from its intrinsic interest; in principle, technology eliminates the workers' skills and, finally, the workers themselves. Instead, we are driven by the fact that the "self" is constituted, at least for most of us, by member-ship in the labor force, as a member of either the job bourgeoisie—the "pro-fessions"—or the working class.

Thus paid work is, in Penrose's amendment, a socially and psycholog-ically constructed "need" shared by those who have been successfully habit-uated to think that the link between holding a job and having "dignity" is a given. Put bluntly, in this view the self is identical to its place in the paid labor force. No job, no (secure) self. Individuals under retirement age (an increasingly indeterminate threshold) are motivated, indeed urged, to seek and hold paid labor when, for whatever reason, they experience a need that does not depend entirely on rational considerations such as how much a job pays or fear of sinking into penury—better to take pride in the fact that, as workers, they are able to provide for self and family without state aid or charity.

Even when one-third of the U.S. labor force was officially unemployed throughout the 1930s, and many workers were on short-time schedules, they still still blamed themselves for their joblessness. There was no dignity for those who could not find jobs; the conventional wisdom, shaken for more than a decade but not displaced, was that there was "always" plenty of work for those who wanted it. This homily derives from the larger American ideology according to which there cannot, by *definition,* be a dis-junction between broad economic growth and jobs. Individuals, not the eco-nomic and social system, are ultimately responsible for their fate; the market adjusts itself at a level approximating full employment, and any joblessness is "frictional"—that is, temporary—for responsible and able-bodied indivi-duals. This key precept of the dominant ideology resumed its virtually un-contested hegemony after World War II, when official statistics recorded jobless rates of less than 6 percent until the early 1980s.

There are, of course, exceptions to the universal principle of paid labor as the sole path to male (and, increasingly, female) dignity, but these turn out to be only variations on the theme that work is a "need." One may retain "dignity" if income has been "earned" through past usury or owner-ship of business. Unwork becomes dignified only if income is derived from retirement or disability. The implicit assumption is that the retired and the disabled would have remained in the paid labor force if they were able-bodied or younger. Retirement is still considered a reward for a lifetime of faithful paid work, although some research has contended that relatively few retirees in the United States prosper unless they have income acquired through labor or property in addition to their Social Security benefits. From the standpoint of the conventional ethic, paid labor is considered optional for women.

Inherited income is ambiguous. Even when heirs do not need to take paid work in order to live, it is always implied that they should be subject to the work ethic. Even in higher circles the "playboy" and "playgirl" are morally condemned; heirs who live on income that derives from trust funds and other repositories of the past labor of others but are unwilling to engage in socially useful activities face censure from peers. The socially responsible heir seeks redemption through performing good "works" in charities or other civic activities or may become a patron of the arts, science, or other types of knowledge.

Although Robyn Penrose objects to the notion of paid work and its surrogates as a "need," she invokes another criterion for working: it is "meaningful" in nonmonetary terms. Intellectual work yields its own rewards when it fulfills two principal criteria: it is done under decent conditions and it produces new knowledge that, by implication, transcends its intrinsic pleasures for the producer. The work is produced not only for subjective gratification or survival, but has a larger meaning.

Much of David Lodge's novel concerns Robyn's struggle to reconcile her desire to hold a permanent teaching job at a time of contraction for academic institutions with her belief that intellectual work is intrinsically meaningful. She is oppressively aware that her chances to land a job are enhanced if she writes and publishes a second book. That she is interested in what she wants to write about is plainly secondary to the requirement that it be written and published.

Robyn must secure paid labor in order to live, but she cannot envision life other than teaching and writing. Her "self" is clearly intertwined with being an intellectual, which generally entails holding an academic post. As the narrative unfolds, we are made aware that her pursuit of a job is indistinguishable from the production of a suitable text, the contents of which, although putatively significant, are subordinate to its role in enhancing the commodity value of her labor.

Vic, alternately her antagonist and her lover (he is an archetypal anti-intellectual), also lives to work. Despite the apparent gulf that separates their interests, he derives as much pleasure from playing the power game, as much "meaning" from building the company, as Robyn does from writing a book or delivering a successful lecture. He suffers from the inevitable petty intrigues in his situation, as does Robyn in hers: she must curry favor and lobby intellectual "inferiors" in the quest for a job. What unites them is the satisfaction they get from the *nature* of their work rather than merely its monetary rewards. Yet their identities are impelled in different ways by the requirement that they occupy a definite position within their respective job hierarchies and can thereby manage a determined, if unequal, living standard. The product the workers at Lodge's fictional company, Pringle, make is far less important than how much of it they make and how cheaply.

Craftspeople and common laborers work for the sake of making a living, not to fulfill themselves. Vic, like Hegel's master, is driven by the hope that, via mastery, he can achieve recognition and, perhaps more to the point, a dwelling place through the labor of "his" workers. He deplores the present orientation of managers and employers in much of British industry; he takes considerable care to make sure that "his" firm recovers from years of unprofitability and is prepared by what he calls "rationalization"—new-product development, technological innovation, and organizational effi-ciency— a program that implies that the firm would be willing to abandon its diversified character and become more specialized. Despite his success in achieving these goals, which he was hired to pursue, Vic loses his job when the firm is sold to its archcompetitor by the holding company that owns it; his interest in Robyn and the intellectual life has made him suspect among his peers. In the end, he is a victim of the creed by which he lives: rational-ization. A manager is a manager and an intellectual is just that. Robyn fares better, but she learns that the university is subject to the same bottom-line rationality.

A Distinction between Labor and Work?

Hannah Arendt has insisted on the distinction between labor and work; she calls work a reified activity—that is, it has an objective existence indepen-dent of the immediate needs of the producer or, for that matter, of its role in consumption.[2] In fact, Arendt argues, proceeding from Marx but not from the same standpoint, labor and consumption are part of the same system. Both function to reproduce human labor power and, while they are neces-sary human activities, they have a profoundly different significance from that of work.

Arendt sees the products of labor as transitory; all that counts is their reproductive function. In contrast, work ensures the "durability of the world":

It is this durability which gives the things of the world their relative indepen-dence from men who produce and use them, their "objectivity" which makes them withstand, "stand against" and endure, at least for a time, the vora-cious needs and wants of their living makers and users. From this viewpoint, the things of the world have the function of stabilizing human life, and their objectivity lies in the fact that—in contradiction to the Heraclitean saying that the same man cannot enter the same stream—men, their ever-changing nature notwithstanding—can retrieve their sameness, that is, their identity, by being related to the same chair and the same table. In other words, against the subjectivity of men stands the objectivity of the man-made world rather than the sublime indifference of an untouched nature.[3]

Arendt does not accept the Platonic claim for the ontological status of pure "form"; she recognizes the relativity of durability. Tables and chairs disintegrate, if they are neglected or worn out, to their state as wood (but not as trees; their artificial character is irreversible). They require "care in preservation" just as the "tilled soil, if it is to remain cultivated, needs to be labored upon time and again." There is no permanence, no objectivity that can resist the sands of time. But just as the artwork, whatever its satisfactions from the point of view of consumption, remains independent of the labor/consumption system, so the "works" of science and technology and of craft can be separated from their incorporation into the commodity-form.

Far from constituting a manifestation of the alienation of labor, the independence of fabricated things from the conditions of their production and consumption is that they remain objects that retrieve the self, which is ineluctably lost in the never ending life process. When, in Arendt's terms, Marx "joyfully" announced the primacy of labor as the historical and ethical equivalent of the production and reproduction of "life," he also proclaimed the subordination of the individual to the social character of life's flow.

But Arendt discovers a contradiction in Marx's thought. On the one hand, he shows that labor is the means by which humans and nature constitute and reproduce the metabolism by which nature is humanized and humans are naturalized. In this perspective humans are part of nature: *Homo sapiens* are the latest evolutionary stage of natural history. Accordingly, the reproduction of labor power is equivalent to the reproduction not only of human life, but of life itself. On the other hand, Marx never tires of restating his claim that communism consists not merely of the emancipation of the laborer from exploitation, but of the emancipation of labor itself. Communism develops the productive forces that are brought into existence by capitalism but remain "fettered" by the class system. With the full development, through scientifically based technology of human powers *over* nature, it becomes possible to liberate the worker from labor and to reinstate work, which, however, is neither subsumed under the commodity-form nor free to become a form of play.

Arendt argues that with the advent of these productive forces the link between humans and nature is severed. In the age of automation (and, one might add, cybernetics), nature becomes a playground to indulge the new leisure needs produced by massive labor displacement. For Marx and Engels, the passage from the realm of necessity to that of freedom signifies the abolition not only of the specific form of wage labor, but also of the close identification of the head with the hand, of the constitution of the self with its role in making the object world. Perhaps only the designers of machines that make things—the architects of automatic production—retain anything like the sense of a self that is identified with fabrication, with self-objectification. Against Marx, Arendt wants to separate means from ends. In her view, historical materialism does not get us very far from being laboring animals for whom reproduction is the sole end. Relegating art and other works to the realm of play only compounds the problem, for we have no reliable means of acheiving a specifically *human* identity without our ability to attain a "free disposition of tools"—free from the division of labor and the rhythmically repetitive labor process in which the individual is subordinated to the requirements of organizational precision. Arendt writes that "what dominates the labor process and all work processes which are performed in the mode of laboring is neither man's purposeful effort nor the product he may desire, but the motion of the process itself and the rhythm it imposes on the laborers."[4]

The incorporation of machines into the labor process in order to make the activity of laboring easier has failed to restore laborers to their humanity. Instead it has further subordinated workers to the machine and to the forces of nature by imposing a regime in which the process of re(production) mimics the physical processes of animal existence and dominates life. As Marx himself remarked, under capitalism we engage in production for the sake of production. The age of automation, according to Arendt, has been so organized that humans do not produce in order to provide themselves with a "dwelling place" but to further the efficiency and operational capacity of the machines:

> For the society of laborers, the world of machines has become a substitute for the real world, even though this pseudo world cannot fulfill the most important task of the human artifice, which is to offer mortals a dwelling place more permanent and more stable than themselves.[5]

Rather, machines are seen, as Marshall McLuhan noted, as "extensions" of our biological selves. In turn, we are attached to machines as to ourselves.

Arendt rejects the collapse of work into labor, and also what she considers Marx's error of relegating work to "play," his protestations to the contrary notwithstanding.[6] In the final accounting, for Arendt and for many intellectual critics of technology, the artwork remains the best repository of that dwelling place where thought and action are fused, ends are not subordinated to means, and, perhaps most important, *Homo faber* and *vita activa* (fabricating "man" and active life),[7] may create a durable world. Arendt deplores the degree to which virtually all human purposes and relations that once found their apogee in the political realm have been transformed into a process of production of goods that may be exchanged in the marketplace. Since for Arendt social production becomes an extension of biological reproduction aimed at the reproduction of labor power, people are not really "doing something" that can be distinguished from the labor of animals, *work* as opposed to labor consists in the fusion of thought and making in activity that contributes to that which really constitutes freedom —politics, the one sphere in which specifically *human* interaction is possible. Yet when politics is not possible because all activity has been subsumed under labor or, more precisely, science, which may be the most powerful force of modernity, she tragically but triumphantly concludes that it is better to do nothing, to be alone.

Of course, Arendt's privileging of politics over labor is by no means unique in the history of social and political thought. This view, most powerfully expressed in Plato's *Dialogues* and Aristotle's *Ethics*, and whose most recent manifestation is Habermas's theory of communicative action, accepts the instrumental character of the relation of humans to nature.[8] According to this school of thought, work, however necessary, is ultimately no measure of man. The repository of truly human action is language. In Habermas's version of "discursive" ethics, speech is directed toward rational decision making outside the consensual sphere of production.[9]

This perspective explicitly denies that humans negotiate their relationship to nature through labor and thus form themselves. For this reason, there is little consideration in Arendt's and Habermas's theoretical discourse of the ethical dimensions and implications of ecological questions; labor and nature having been relegated to a nether world, the "social" is removed from its natural referent. In making this separation, Arendt and Habermas follow a tendency that embraces a wide spectrum of contemporary thought. The question we wish to explore here is its *defensibility*.

The Decline of "Work"

Both Arendt's discourse on the significance of work as opposed to labor and David Lodge's skeptical response to this view raise the fundamental question of whether there is work worth doing in a regime in which production is merely a means to make profits and, for the rest, to reproduce life; in which apart from the important qualification that under current conditions labor displacement results in economic pain, workers, engineers, and managers have no reason to defend labor itself and are right to be indifferent to both the process and the product of labor. In fact, when they are given the opportunity, workers—skilled and unskilled alike—are pleased to be relieved of participation in the labor process provided they are guaranteed an income adequate to the current "decent" standard of living.

With the decisive passing of craft, except in the crevices of the modern labor system, the main value of having a job (besides its economic function for individuals and households) is that it once provided a "community," which for many men replaced the traditional agricultural family. With the partial breakdown of the urban family, many women found the workplace a source of social solidarity as well. Moreover, contrary to popular depiction of craft labor as intrinsically satisfying in comparison to stupefying mass production, historical and ethnographic evidence demonstrates that skilled workers were no less eager to be liberated from the workshop than assemblers and laborers. Much of Arendt's invocation of the "reified" object as the permanent testament to humanized work turns out to be as mythological as the supposed beneficent effects of the slave system on its subjects, for it was not the object that provided satisfaction for the craftspersons, but the opportunity for a richer human interaction. This benefit of collective labor was not generally available to assembly-line and other less-qualified laborers. Production workers found after-hours conviviality in bars and social clubs.

But the culture of the factory and the large office is dying; for most workers, even those classified as "skilled," the old bonds are considerably loosened, even when they have not completely disappeared, for a very good reason: most craftspersons—in construction, in factories, car and instrument mechanics—know that the division of labor and the computer rationalize craft nearly as much as they do manual labor. Most auto and instrument mechanics, for example, rely on computer-mediated electronic tests for diagnostic purposes and work with parts that are engineered as modules. When Vic tells Robyn that the men like to work, what he forgets to mention is the reason for this affection. It certainly does not lie in work "satisfaction," that ambiguous sociological and psychological category that might issue from the substantial aspects of the tasks, but in the reality of the shop as a "dwelling place," a home that has little to do with the end product of their collective labors.

Achieving a home in the traditional industrial culture did not entail crafting a reified object that represented suprabiological dignity, as Arendt and so many romantic accounts of the artisanal era have claimed. The "bonds" of labor derive from the immanence of laborers' collective power and shared recognition of their collective subordination. The markers of these bonds are less "tools," instruments of fabrication, than a shared discursive universe replete with rituals, linguistic codes, jokes, and worldviews—in a word, a culture. The shop or office may be regarded as a universe that visits exhaustion and frustration upon its inhabitants but provides, at least for some workers, an irreplaceable network of relationships and, taken in its multiple significations, a discourse, which together constitute the class culture of the factory.

Contrary to the ideologically conditioned theory shared by sociologists, psychologists, and policy analysts that "nonwork" produces, and is produced by, social disorganization and is symbolic of irresponsibility and personal dysfunctionality, recipients of guaranteed annual income who are relieved of most obligations to engage in labor do not fall apart. The incidence of alchoholism, divorce, and other social ills associated with conditions of dysfunctionality does not increase among men who are not working. Nor do they tend to experience higher rates of mortality than those of comparable age who are engaged in full-time work. Given the opportunity to engage in active nonwork, they choose this option virtually every time.[10]

For example, East Coast longshoremen who are not working but receive adequate income find many things to occupy their time. Many spend more time with their families, some engage in side businesses, and others take up hobbies or fix up the house. They retain their industrial community and much of its culture. Most important, they are happier because they do not have to labor every day at a hard, often life-threatening job where the dangers associated with loading and unloading cargo are compounded by the need to handle materials that are frequently hazardous to their health. Because of the pleasures of nonwork—work in the specific sense used here, paid labor under a hierarchical management system—the men are not pleased to be called to put in a day's labor.

Most of all, they have regained "free" time. This freedom, perhaps more than the activities in which they become absorbed as an alternative to paid labor, fulfills the premier promise of technological displacement that in its earlier ideological expressions was heralded by the labor movement and intellectuals as the main historical benefit of industrialization. An alarming number of workers, both intellectual and manual, surrender nearly all of their waking and even dreaming time to labor. The by now ancient slogan of the movement for shorter hours—"eight hours work, eight hours sleep, and eight hours to do with what we will"—has been abandoned. The notion of free time is as distant from most people's everyday experience as open space. Labor has been dispersed into all corners of the social world, eating space and time, crowding out any remnants of civil society that remained after the advent of consumer society, and colonizing the life world. We are able neither to work nor to play; unlike the older industrial model where labor was experienced as an imposition from above, the dispersal of work makes the enemy invisible because labor is now experienced as a compulsion dictated by economic anxiety more than by the "need" to work.

Under current economic and social conditions, the major casualties of technological changes on the waterfront and, increasingly, in the auto, electronics, and communications industries, are the children and grandchildren who will never have the chance to work on the docks or in the factories and accumulate enough time to achieve dignified nonwork. The time of the new generations of never-to-be industrial workers is not free even though they are relieved of paid labor. Instead, it is suffused with anxiety that they may never again enter the cycle of labor and consumption that defined working lives in the Fordist era, or they displace the anxiety of nonwork without income in lives of petty crime (in which case they need not apply for public assistance). Whereas before containerization, as late as 1960, sons of longshoremen certainly would have followed their fathers onto the waterfront, in the postwork society, life for the children of dockers is in most respects harder than it was for their parents. Many among the next generation who have been unable to accumulate the requisite cultural capital to qualify for employment in one of the knowledge industries or have not had the luck to find a job in one of them are, lacking guaranteed income, reduced to undignified nonwork—or worse, are driven to seek low-wage dead-end jobs because they are suited neither for unemployment nor for lives as drug dealers or petty thieves.

The remaining repositories of "work" within the wage-labor system are, despite the ruthless transformations of virtually all of its products and producers into commodities, the diminishing instances of petty craft production (occupations that are frequently suffused with the uncertainties connected with self-employment), art, and the products of the relatively small proportion of those who produce knowledge of all kinds, even those, like teachers and writers, who through transmission (or translation) re-produce knowledge. This work retains its objectivity, depending on neither a knowing subject nor the immediacy of the labor process.

It may be objected that this argument seems Eurocentric; it applies at best to the fate of work in so-called "advanced" industrial countries. As industrial production moves away from the United States and Europe, especially to Latin America and Asia, some have envisioned the rebirth of the industrial proletariat. And presumably, because of this shift, few of these issues such as joblessness and class decomposition apply. According to this view, the old slogans of labor solidarity have not disappeared; they have merely been deterritorialized. This thesis, however, ignores the fact that industrialization in formerly agricultural regions is occurring at an accelerating rate under the new scientific-technological regimes, which are by no means local. Computer-mediated labor processes are the standard against which global labor is measured, not merely labor in the traditionally industrialized countries. For example, some of the *maquiladoras* on the Mexico-U.S. border are often more technologically developed than older U.S. plants that produce the same products. They make auto parts, computers, and other high-tech commodities as well as furniture and textiles, which, as we have seen, are increasingly produced with computer-mediated and laser technologies. Thus, whereas earlier capital migrations relied on low and intermediate technologies because the advantages of employing low-wage labor outweighed the costs of introducing advanced machine processes, we may now observe new forms of capital migration that tend to make the labor process—if not (yet) wages and working conditions—uniform.

Workers on the Mexican side of the border may earn at most about eighty dollars a week, but they are often paid much less. And even though Mexico has some protective factory and environmental legislation, it is observed more in the breach than by enforcement. Moreover, as workers organize in Mexican factories, employers have not hesitated to steal away in the night to other areas in the country where wages are even lower and workers are less prepared to form independent unions.[11]

The North American Free Trade Agreement (NAFTA) may be viewed as merely the conclusion of the first chapter in a long process of overcoming some aspects of the traditional unequal division of labor between north and south. In the next decade, U.S. wages and living standards are likely to continue to deteriorate. If labor organization emerges in Mexico and other parts of Latin America, wages there will rise, but not by enough to deter migration of U.S. plants, at least during the 1990s. In the near future, Texans and Californians will cross the border in greater numbers every day to work in Mexico and Mexican workers will continue to migrate to certain jobs in the United States, approximating the situation at the already blurred U.S.-Canadian border. At the same time, as in Canada, Mexican industry is increasingly subject to U.S. investment; this will set a pattern for transnational investment in other countries of Latin America, particularly Brazil, which, along with Mexico, had before the current economic crisis succeeded to some extent in developing its own industrial base.

Deterritorialized production applies also to knowledge. By the early 1990s, for example, China and India were offering U.S., Japanese, and European capital access to highly qualified scientific and technical labor. U.S. computer corporations began to let contracts to software corporations in India. Du Pont and other chemical corporations were building petrochemical complexes in Shanghai, employing Chinese engineers and chemists at eighty dollars a month. The fairly well developed Mexican bioengineering sector is actively negotiating with U.S. corporations to "share" discoveries and technical achievements.

Even in science and technology, whose products are situated in their own historical and institutional contexts and as often as not are appropriated for socially dubious purposes, the product never entirely "disappears" into consumption but is incorporated into the common built environment. Yet although the work of some of those engaged in the production of arts and science retains excitement, challenge, and end products that possess genuine durability, few have the good fortune to be custom cabinetmakers, theoretical physicists, literary critics, social scientists, molecular biologists, or computer engineers.

Just as the scientific-technological revolution has utterly transformed the workplace in all categories of labor, we are obliged to examine its consequences for the conception of work that undergirds cultural identity, the self, and our collective understanding of the norms by which the moral order imposes a mode of conduct upon us. In his notebooks, Marx wrote in 1857–58:

> The free development of individualities, and hence not the reduction of necessary labour time so as to posit surplus labor, but rather the general reduction of the necessary labor of society to a minimum, which then corresponds to the artistic, scientific etc. development of the individuals in the time set free, and with the means created, for all of them. Capital itself is the moving contradiction [in] that it presses to reduce labour time to a minimum, while it posits labor time, on the other side, as the sole measure and source of wealth.[12]

It may be argued that the history of capitalism during the last hundred years may be recounted in terms of this contradiction. This transformation in industrial production has stunningly fulfilled the tendencies that were prefigured in Marx's description: once based chiefly on the practical knowledge handed down to succeeding generations by craft traditions, production is now based on abstractions of organization and on science.

The promise of this movement, however, has been subsumed almost entirely under the sign of capital reproduction. Capital fears its own moving spirit. Vast quantities of labor are set free from the labor process, but rather than fostering full individual development, production and reproduction penetrate all corners of the life world, transforming it into a commodity world not merely as consumption but also in the most intimate processes of human interaction. Intellectual labor, its ideology of professional autonomy in tatters as a result of its subordination to technoscience and organization, becomes a form of human capital the components of which are specialized knowledge and differentially accumulated cultural capital determined mainly by hierarchically arranged credentials. Most professionals, let alone "liberated" manual workers, enjoy little free time for artistic and scientific development, either of their individuality or indeed of the productive forces. To the contrary, we live in a time when not only are individuals thwarted, but the political economy of late capitalism appears —at least in one crucial area, research and development—to fetter the new productive force: knowledge.

On June 30, 1993, the *New York Times* reported that U.S. companies are cutting funds for scientific and technological research:

> Scientific research by private industry, the traditional powerhouse of innovation and technological leadership in the United States, is suffering deeper financial woes than previously disclosed, suggesting that America is slipping in the international race for discoveries that form the basis of new goods and services. The National Science Foundation reported in February that industrial research on research and development had begun to shrink after decades of growth.[13]

Of course, much of the previous growth was military, and was therefore driven by and dependent on public funds. But with recession, the tapering off of the cold war, and the enormous deficits accumulated by government and by corporations caught up in the swirl of the leveraged buyout mania of the 1980s, funds have dried up. For example, as we noted earlier, the National Institutes of Health, which formerly funded a third of the research proposals submitted to the agency, supported only 10 percent in 1991. More to the point, the priorities of the federal scientific and technical bureaucracies, which are increasingly tied to the requirements of corporations, have restricted the *kind* of research they are willing to support. Consequently, there is almost no hope that biomedical projects that fall outside the purview of molecular biology and biophysics will be funded. And, as we have seen, research scientists are feeling pressure to make arrangements with private corporations in order to obtain desperately needed research funds. In short, the commodification of basic science, combined with its increasingly technical character and declining funds, may in the future all but seal the fate of the United States as a major economic power.

For the plain truth is that overfunding and "useless" knowledge is the key to discovery. From the discoveries of Galileo to the "idle" ruminations of Frege, Gödel, Einstein, and Bohr, patronage, whether public or private, permitted unbounded dreaming that led to new ways of seeing and ultimately— but only ultimately—new modes of producing. When government and corporate policy makers insist on "dedicated" research as a condition of support, they announce that they have opted for failure rather than long-range innovation. This blatant act of research shooting itself in the foot is by no means intentional. Rather, it is the result of the logic of technoscience and the human capital paradigm according to which unsubordinated knowledge is perceived to threaten the social order either by draining economic resources or by proposing unpalatable jolts to the imagination. Moreover, it signals a profound failure of nerve, a refusal to take the risk that some knowledge can never be translated into technology and will remain outside the framework of accepted science and that some knowledge might even subvert cherished beliefs within the prevailing social order. For the social sciences and the humanities, cost reductions exacted a steep toll on research, but during the Reagan-Bush era many projects were rejected by conservative leaders of the National Endowment for the Arts and the National Endowment for the Humanities on political grounds, a manifestation of the conservative ideological attack on postmodern cultural expression.

The crisis in research of course has serious consequences for the U.S. national economy, but it augurs equally badly for hope that intellectual work will be possible for more than a tiny fraction of scientists and artists in the future. Its effects are even more far-reaching. For, in a higher education system already incurring severe criticism for the low number of U.S.-born scientific and technical majors and graduates at the undergraduate and graduate levels, the decline in basic research constitutes a disincentive for young people to enter the sciences. At leading universities, many if not most advanced-level physics, mathematics, and chemistry students are foreign born.

The irony of this situation is that the completion of the process by which science is almost entirely subsumed under capital and which, concomitantly, transformed intellectual work into human capital, is by no means in the system's interest. For just as the emergence of knowledge as a productive force "solves" the problem of productivity while at the same time intensifying the problem of how capital valorizes itself, so the subordination of knowledge to the imperative of technical innovation undermines one of the central presuppositions of innovation: *unfettered* free time for knowledge producers.

In recent years this contradiction has been at play in universities, even in first-tier institutions, which place increasing administrative burdens on faculty; the second and third tiers impose, in addition, heavier teaching loads. Under the impact of economic constraints we have entered a new era of academic cost cutting and of surveillance whose intended as well as unintended effects are to discourage independent intellectual work. For a society that trumpets the growth imperative as the key to its survival, and for which knowledge is the acknowledged economic spur, such measures are, of course, self-defeating.

Clearly, the promise that the scientific and technological revolution will usher in an epoch in which the full development of the individual is finally fulfilled is thwarted as long as Bentham's Panopticon dominates the political unconscious of established authorities. The current tendency is to "resolve" the contradictions prompted by the emergence of knowledge as both the salvation of capitalism and its nemesis by transforming the intellect, as it did craft, into human capital.

Toward a New Labor Policy

Ours is a time when questions are much more easily posed than answered, when the fragmentation of economic and political perception among both experts and what may be termed the "public" prevents solutions that have a chance to garner wide support. Further, the question What is to be done? has become tainted because it points to calumnies committed in the name of traditional ideologies of emancipation. Indeed, to speak of liberation, of the emancipation of labor, appears utopian, in the bad sense; in a period of the transformation of global politics by the breakup of the Soviet Union, the intellectual discrediting of Marxism, even by some of its erstwhile practitioners, and, on a larger canvas the steep decline of socialism recalling these words conjures images of betrayal.

Nevertheless, without bold alternatives, we condemn ourselves to the present state of affairs. This is the fundamental conundrum of the Frankfurt school, of some French philosophers and social theorists, and of others who have, with some reason, concluded that ours is the epoch of the death of the subject. If this means the subject as God's surrogate, solitary consciousness, or human dominion over nature, we have little to quarrel about. What is often meant by this phrase, however, is the possibility of agency, of opposition, or even of resistance. To posit the end of this kind of agency is a kind of reconciliation with the established order.

This book situates itself in the discourses of human emancipation, freedom, and hope because we align ourselves with the agents of opposition and alternative: the "new" social movements—feminism, ecology, sexual self-determination—and elements of the "old" social movements, particularly in black, Latino, and Asian freedom movements and the labor movement. Needless to say, in arguing for a perspective on the future of work that takes into account the scientific-technological revolution of our time, we make no claim to be doing more than suggesting some pathways. As intellectuals we speak to, not for, other intellectuals and those in the social movements who might find our viewpoint useful.

Our proposals are based on the presuppositions of this study: that economic growth grounded in technological innovation does not necessarily increase employment unless there is a sharp reduction in working hours, and even then may not be sufficient to sustain a level approaching full employment; and that since a considerable number of recently created jobs are part-time, poorly paid dead ends, there is a powerful argument that we have reached the moment when less work is entirely justified. In addition, our proposals assume the goal of assuring the *possibility* of the full development of individual and social capacities.

These statements further imply that—if our assertions that the world economy will not sustain full employment in the coming decades and the social safety net will remain full of holes are correct—we need to reconsider the pace of technological change and the effects of corporate reorganizations that have shed tens of thousands of employees in the past several years. Until measures such as a substantial reduction of working hours, a guaranteed income plan, a genuine national health scheme, and the revitalization of the progressive tax system have been introduced into law and union contracts, job-destroying technologies and mergers and acquisitions should be rigorously *evaluated* in terms of their implications for the well-being of communities and workers. In an era of uncontrolled growth amid economic stagnation, corporate efforts to make workers and communities pay the costs of falling profits are exacting heavy tolls and should be stopped.

We have used the general concept "evaluation" to connote the urgent need for social controls, perhaps in the form of re-regulation of business, over untrammeled labor-saving technological change and mergers that result in permanent reductions of labor forces. Needless to say, this proposal directly opposes the dominant strategy of U.S., European, and Japanese corporations and would assume a political situation in which national states were independent of these corporate interests. Unfortunately, for the most part this is the case among neither conservative nor social-democratic and social-liberal regimes. Free enterprise and free market ideology enjoy global hegemony in the current political and economic environment. Thus, even the suggestion that technological change and mergers may not be in the public interest flies in the face of the prevailing common sense.

Needless to say, we do not support technophobic perspectives on technological transformation. As our critique of Arendt and the earlier discussion of work and skill show, we do not mourn the passing of craft. Given guaranteed income, shorter hours, and work sharing, we welcome the coming of a postwork society and have tried to refute the sociological and psychological "wisdom" that labor is an intrinsic need beyond survival. In fact, we have claimed that, as a mode of life, its historicity has been demonstrated by the nature and the spread of cybernetic technologies.

We wish to point out, however, that deregulation has been most consistently applied to corporate prerogatives: to reduce labor in production; to relocate, at will, factories and professional services; to eliminate workers

343

through consolidation; to put labor in competition with itself by breaking and otherwise reducing the traditional protections provided by union contracts for decent wages and against working conditions that threaten health and safety; and to weaken employer- or government-financed health and pension benefits.

When it comes to regulating the poor, there is no absence of programs: workfare; more prisons for convicted drug dealers and users; armed guards in urban high schools. The largest corporations have never insisted upon competition for government contracts, nor have they hesitated to support tariff and trade restrictions when their particular interests are at stake. Nor have conservatives failed to temper their opposition to open borders to Latin American, Caribbean, and Asian immigrants, demanding draconian measures to regulate their flow.

Recall the words of economic historian Karl Polanyi: "It should need no elaboration that a process of undirected change, the pace of which is deemed too fast, should be slowed down, if possible, so as to safeguard the welfare of the community. Such household truths of traditional statesmanship . . . were in the nineteenth century erased from the thoughts of the educated by the corrosive of a crude utilitarianism combined with an uncritical reliance on the alleged self-healing virtues of unconscious growth."[14]

Finally, social ecology, which has emerged as a major paradigm of social and economic theory as much as it is a significant social movement, has taught us that untrammeled growth is by no means an unmixed blessing. At the most fundamental level, ecological thought is a powerful counterweight to the Western idea of progress. We have learned that technologically driven growth has had disastrous consequences that can no longer be ignored. Hazardous waste, industrial pollution and its consequent global warming, life-threatening power sources (alternating-current electricity, nuclear energy), and increased radiation resulting from high-powered computer technologies are some of the most visible results of the rapid expansion of industrial and consumer societies.

To be sure, a theoretical model according to which human survival depends on maintaining a sustainable biosphere and stable ecosystems suggests that there may be enormous costs to uncontrolled economic growth, but no consensus has emerged, even among the most insistent critics of uncontrolled growth, concerning possible solutions. At one end of the spectrum are those who warn that unless growth is arrested, even reversed, the ecosystems that sustain life are in imminent jeopardy. This view proceeds from the indisputable fact that "development," one of the cherished names for capital accumulation and urbanization, has exacerbated what are called "natural" disasters—soil erosion, floods, global warming, the cancer epidemic that afflicts nearly a third of all people in industrial societies (by the year 2000, the figure may rise to 40 percent). Cancer, which many biologists argue is directly linked to living and working conditions, is rapidly becoming the major disease of industrial societies. Beyond industrial and commercial pollution, it is linked to the spreading contamination of food and water. This position argues for elimination of entire sectors of industry—especially nuclear energy, many branches of chemical production, the use of most fossil fuels—and conversion of the highly centralized electric power industry to locally based water, wind, and solar energy. Decentralizing power production suggests bioregional economies in which communities produce and distribute their own basic foods and many other everyday products. In this economic arrangement, the scale of production is reduced. This regime would not entirely eliminate the division of labor and commodity exchanges, but would limit these activities in order to minimize disturbances to the ecosystem.

At the other end are the proponents of environmental protection through state and voluntary regulation by industrial corporations and developers. This group includes many social liberal governments and their professional retainers, who insist that the political and economic climate is permanently unfavorable to draconian ecological regulations. Growth can be selectively moderated by conservation measures such as creating national parks and wilderness areas, limiting the use of fossil fuels, encouraging industrial and consumer recycling, and requiring business to clean up after itself. In the United States, a federal law imposes heavier taxes on polluters but relies on market mechanisms for remedial action. Large employers, most public policy professionals, and some trade unionists are largely hostile to proposals to restrict automobile travel and expand mass transit; to declare a moratorium on many types of industrial and residential development, especially in rain forests and rural areas; and to restructure industrial production to eliminate or sharply curtail carcinogenic processes and products. Faced with declining profits, corporate capital resists innovations that add to the cost of doing business. Lacking an alternative to jobs, many trade unions and their members have opposed ecological protections. For example, most union labor sided with timber companies in Oregon against environmentalists' demands for substantial restrictions on timber production and in the fights over nuclear power during the 1970s and 1980s almost invariably ignored warnings by ecologists and public interest groups that reactors endangered the safety and health of workers and of neighboring communities. Nor have some unions been willing to insist on strict health protections lest plants be shut down and jobs lost. There are exceptions, notably the Oil, Chemical and Atomic Workers, which played a crucial role in the passage of the Occupational Safety and Health Act in 1971.

In light of the mounting evidence of ecological crisis, the idea that economic policy can no longer fail to incorporate fairly sweeping ecological perspectives seems to us to be incontrovertible. However, if this argument is accepted, we can no longer rely on growth to address problems such as technologically induced unemployment and to improve living standards. Yet, in a remarkable example of failure of political imagination and will, uncontrolled growth remains the basis of world and national efforts to resolve long-term economic woes in nearly every major country. The ideological hegemony of growth economics, combined with the powerful threat of globalization, has virtually eliminated from public debate the characteristic industrial-era imperatives of social justice and equality. Since the social-justice left and the labor movement have largely dropped their traditional demands for redistributive justice in favor of growth, the political environment seems more unfavorable than at any time in recent memory for addressing ecological issues. In addition, what might be called the spread of *virulent nationalism* in nearly all industrialized countries as well as in Eastern Europe militates against international efforts to deal with ecological disaster. The breakup of communism, which from a democratic perspective may prove to be an important milestone in human history, has done nothing to improve the chance to revitalize the world's ecosystems. On the contrary, having embraced the doctrine of the free market, many governments and political forces in Eastern Europe seem resigned to accepting the trade-off of economic vitality for ecological disaster, an approach that was consistent with the policies of the former communist regimes.

Given ecological and economic crisis and world economic and political restructuring, there is an urgent need for thinking that refuses to remain mired in the impossibilities of the present. For to insist in advance that possibilities are limited to the givens of the social and political world leads to the conclusion that no genuine transformation is possible, which in turn gives rise to the dark conservatism that holds that change is not desirable, and even is evil. Since it can be easily demonstrated that international "competition" is only one and perhaps a minor feature of the current economic situation but that we have entered an unprecedented period of central power over most economic decisons on a global scale, we have no compunctions in offering a practical and necessary discussion of how and why the new era of postwork may be addressed.

The Need to Reduce Working Hours

There has been no significant reduction in working hours since the implementation of the eight-hour day through collective bargaining and the 1938 enactment of the federal wage and hour law. Since then, we have witnessed a slow increase of working time despite the most profoundly labor-displacing era of technological change since the industrial revolution. People are laboring their lives away, which, perhaps as much as unemployment and poverty, has resulted in many serious family and health problems. In turn, the lengthening of working hours has contributed to unemployment and poverty among those excluded from the labor system.

Therefore, there is an urgent need for a sharp reduction in the work-week from its current forty hours—a reduction of, *initially*, at least ten hours. The thirty-hour week at *no reduction in pay* would create new jobs only if overtime was eliminated for most categories of labor. And, although some people may prefer flexible working arrangements that are more compatible with child-rearing needs or personal preference, the basic workday should, to begin with, be reduced to six hours, both as a health and safety measure and in order to provide more freedom from labor in everyday life. Finally, we envision a progressive reduction of working hours as technological transformation and the elimination of what might be termed make-work in both private and public employment reduces the amount of labor necessary for the production of goods and services. That is, productivity gains would not necessarily, as in the past, be shared between employers and employees in the form of increased income, but would result first in fewer laboring hours.

Obviously, restricting laboring hours raises some important questions: How do families maintain their living standards if income is substantially reduced by restricting overtime and other work-sharing arrangements? Will people use free time to develop their capacities or will time be absorbed destructively? Who will pay for work-sharing? Is it feasible in a global economy where capital moves freely in search of cheap labor? We will address the last question first because although it is politically agonizing, it poses fewer conceptual problems.

The experience of the German labor movement is instructive in this regard. In 1985, the Metalworkers Union (IG Metall), which represents auto, steel, and metal fabricating workers, struck for reduced hours. After a relatively short walkout involving millions of workers in the most technologically advanced sectors of the economy, employers yielded to the demand for a thirty-five-hour workweek, to be implemented in stages over five years. Gradually, other sectors have adopted the shorter workweek, but there is no federal law because the labor-supported Social Democrats are out of power. The competitive position of German industries is not suffering because of this innovation, in part because of the tremendous productivity of German workers made possible by cutting-edge technologies that have been widely introduced in production. Moreover, in countries such as Germany where the social wage includes substantial government-administered health benefits and guaranteed income and pensions, labor costs to employers may be lower than in the United States, which does not have these state-sponsored provisions, even when wages are higher. In the United States, employers have shouldered much of the burden of the welfare state, spending as much as 40 percent of wages on fringe benefits.

While notions of solidarity have suffered in Europe in the past several decades, particularly in the wake of a major influx of immigrants from the Middle East and Africa, Italian and German labor movements nevertheless retain considerable ideological loyalty to concepts such as class unity. The victory of the German metalworkers—and a parallel struggle by public employees—attests to the power of discursive and ideological influences in determining the shape of the politics of work. Although the German economy has suffered during the recession of the past decade and there is considerable xenophobia throughout German society against immigrants during a period of high joblessness, the discourse of social justice has not disappeared because the labor movement insists that employers share the pain of economic woes. Moreover, the unions have insisted that the promise of pay equity between East and West Germany be fulfilled.

This is not the case in the United States, where public discourse is dominated by demands that business be protected at all costs. The privatization of welfare and antediluvian social policies, especially the lack of national health care and a strong pension system, places onerous burdens on enterprise labor costs to provide these social wages. Of course, to require industries to be "competitive" presumably entails a sharp curtailment of these company-paid benefits. But if workers agree that they have a "responsibility" to help their own companies, current conditions demand that they accept wage and benefit reductions and suspension of hard-won work rules that protect their health and their jobs. Clearly, U.S. workers and their unions gave back many of their previously won gains in the 1980s, but these concessions failed to reverse capital flight. Although movements of capital, especially from north to south, are often meant to reduce labor costs, there are other factors that motivate such shifts. One of them is historical. After World War II, strong U.S. unions were able to increase wages and benefits in the private sector even as social wages remained static. But technological innovations developed within this country were on the whole translated into reinvestment in a wide range of intermediate (that is, mechanical), technology industries such as steel, machine tools, and metal fabrication. As a result, U.S. industries remained stuck in earlier technological regimes while other countries, through computerization, were transforming their labor processes. In the United States, advanced technologies were introduced mainly in military-related industries.

Regulating Capital

In some countries, capital may not freely export jobs without consultation with unions and the government. Clearly, reducing working hours without simultaneously addressing the issue of capital flight is unthinkable. In 1988 the U.S. Congress passed modest plant-closing legislation requiring employers only to notify employees and the community of their plans to close a facility. This law could be strengthened to compel collective bargaining with unions and local governments over the conditions of capital flight, including the extent of compensation and effects on the community. To discourage plant closings, employers could be required to pay substantial compensation to displaced workers and to communities, and they could be required to offer transfer rights to their employees. Unions have sought to protect jobs by persuading Congress to pass the so-called domestic content bill according to which a percentage of the components of commodities (autos and garments, for example) sold in the United States would have to be produced by U.S. workers. This provision has been incorporated into the North American Free Trade Agreement (NAFTA) for some items; it could be extended to become a basis of plant-closing legislation.

The most important issue raised by our proposals is international coordination of labor demands. It is evident that the purely national framework within which labor movements operate is for many purposes archaic. But although there are some instances of genuine coordination of strikes, bargaining, and even legislation, labor movements are often at loggerheads over their own position in the international division of labor. In the face of global competition, it is nothing short of suicidal for labor to remain in competition with itself. Unless these issues are addressed, discussions of the need for shorter hours can never advance beyond the proposal stage.

The question of living standards strikes at the heart of the cultural dimension of this issue. For millions of Americans, working almost all the time is the only way they can maintain their homes and provide for the care and education of their children. Here we offer three suggestions. First, single-family, privately owned homes should not be the most important source of new housing. Publicly financed, affordable multiple-dwelling rental housing would lift an enormous burden from the shoulders of working people. The value of their homes—whether they were cooperatively owned or rented—would no longer depend on market fluctuations that have in recent years severely reduced equity in millions of homes and, perhaps more egregiously, spurred lender-provoked evictions. Second, we need free, publicly provided child care services like those in many European countries. Since mortgage payments or rent plus child care absorb as much as 50 percent of the income of many households, they bear on laboring practices.

Third, the United States could adopt the European system of treating post-secondary education as a public resource therefore a public expense.

At the same time, we would propose that higher education be a right rather than a privilege reserved for a minority of the population, as it is in most of Europe and the countries of the Americas. Here we can observe considerable differences between the United States, Western Europe, and developing countries. In most of the world, all education is paid for by the state, but access to education is severely regulated. In Europe, a relatively small percentage of students enter postsecondary programs, including technical institutes. In most of the less developed regions of the world, most people are denied a decent elementary and secondary education, much less opportunities for university degrees.

Since the 1960s, U.S. colleges and universities have been more accessible to students than they used to be. Some 50 percent of high school students enter some kind of postsecondary education program; about half of them go to community colleges and technical schools. Dropout rates, however, are enormous, and sometimes as high as 70 percent. Plainly, if the revolution in scientific and technical knowledge has occurred, fairly high levels of educational achievement are now a necessity for larger numbers of young people. Just as secondary education became a right at the beginning of the twentieth century, so higher education must become a right at the turn of the twenty-first.

Paradoxically, just at the historical juncture when knowledge work becomes more important, U.S. colleges and universities have entered into a period of downsizing due to budget constraints. Public universities have been especially affected by massive cutbacks. In 1992 and 1993 the University of California suffered a budget cut of nearly 10 percent; plans to build new campuses were shelved; graduate admissions were restricted; and professors were obliged to accept a 5 percent salary reduction. New York's City University, the country's largest urban university whose student body is mostly black and Latino, sustained four consecutive years of cutbacks totaling 20 percent. The huge California State University system was similarly beleaguered by reductions, and parallel developments occurred in New Jersey and Massachusetts, among many other states. In the midst of these cuts, enrollments were still rising, placing a huge burden on an already depleted faculty. William Honan, writing in the *New York Times,* cited Robert Zemsky of the University of Pennsylvania's Institute for Research in Higher Education, who argues that if faculty members do not adjust to the new "realities" by accepting the new austerity, changes will be imposed from without: "The 'without' are the brute economic realities for private institutions and further cutbacks by city, state and federal legislatures for the public institutions."[15]

Zemsky, echoing the views of university administrators, cannot envision an alternative to the reality of austerity, a regime that has already forced administrators to consider measures such as eliminating programs and departments, laying off faculty, and eliminating a wide range of services such as private telephones, faculty access to duplicating facilities, and paid sabbaticals. Many universities have encouraged senior faculty to take early retirement, but have not offered to replace them on anything near a one-to-one basis, even with assistant professors.

All over the United States, faculty have said no to cutbacks, but have failed to propose alternative schemes to preserve democratic access to higher education. Rather, they have been cast, in Honan's article and elsewhere, as staunch defenders of the status quo. To be sure, some faculty unions have lobbied legislatures against budget reductions, and faculty senates have refused to accept the elimination of departments, especially in the traditional disciplines. But the trend toward restoring the concept of higher education as a *privilege* rather than a *right* is pervasive.

At a basic level, our proposals involve much more than an effective legislative struggle. They also require a significant effort to pose alternatives to the values that have propelled American cultural ideals since the end of World War I. The persistence of the old values, many of them crucially tied to the period of American economic expansion and world dominance, has constituted one of the most significant tools in the arsenal of insurgent conservatism. The conservatives have been able to mobilize working-class and professional constituencies with a populism that is based on resisting the implications of change.

We do not want to be interpreted as falling in line with the belt-tightening, antipleasure ideologies of the communitarians. To the contrary, we are arguing that the only chance to maintain and advance our living standards is by means of a bold, intelligent reassertion of the values of more equality and more high-quality public services as the basis of social policy. The fifteen-year bipartisan experiment in deregulation has failed, miserably, to reach any of its major goals, except that of lining the pockets of a small cabal of business interests. In the wake of deregulation many small businesses have failed, workers have lost their jobs, and services have deteriorated. Moreover, the private sector has failed to provide moderate-cost housing, day care, and education while the public sector has been ruthlessly gutted in an orgy of cost-cutting measures.

There Is Still Work to Do

Despite labor-saving technologies, there is still much work to do. Our roads, bridges, water systems, schools, and cities need rebuilding and repair. We need a mass-transit system to counteract traffic jams and the deleterious effects of auto and truck emissions. With a growing population, we require more garbage collection, cleaner streets, and refurbished parks, forests, and other public spaces. And, as always, there is an urgent need to reclaim "wilderness" areas that have been subjected to industrial and real-estate development. Surely we require a new, balanced development policy, since the long-awaited era of the post-paper, steel, and fossil fuels society seems still far in the future. We could spend vastly more funds to research, develop, and produce solar, windmill and water power.

This work is frequently labor-intensive and physically hard, but, because it improves the quality of life, it is worth doing. And because much of it is onerous but necessary, pay should be higher than for many other occupations. Workers should be paid on a principle of what might be called *reverse renumeration,* that is, paying more for jobs that are more unpleasant but enhance "public goods": manual, routine, or dirty tasks such as cleaning the streets and parks and collecting garbage; heavy work and routine mass production tasks; caring for children, older people, and the sick. If we are serious about the arrival of the so-called post-cold war era, paying for these services should pose few additional burdens on ordinary incomes because the military budget still hovers around $300 billion. Even a 50 percent reduction in the military budget and transfer of funds would result in a net increase in jobs, since much current military spending is extremely capital intensive. At the same time, a long-range commitment to expanding public services such as mass transit is expensive.

We should replace the current tax system, which favors the rich, and reintroduce the progressive tax, dropping the fiction that tax incentives are a major impetus to new investment within the U.S. economy. As the experience of the 1980s—when the Reagan administration presided over not only a tax giveaway but also one of the sharpest redistributive tax measures in history—amply demonstrates, putting more money in the pockets of the rich does not guarantee new investment within the borders of the United States or, indeed, better-paying jobs.

If we were committed to abolishing the hierarchical division of intellectual and manual labor, such tasks could be shared through a program of universal public service. A new commitment to universal public education to prepare the multivalenced worker would replace the current focus on specialization. Many if not all tasks in what is conventionally regarded as "mental" work could be shared among a wider portion of the labor force. In this regime of task equalization, every person would be obliged to perform some of the least desirable tasks, regardless of accumulated credentials; these jobs would not be permanently assigned to any class of people.

This is a long-term perspective, but in the wake of the objective possibilities inherent in new technologies for eliminating vast quantities of manual and clerical labor in both "advanced" and developing areas of the world, the question before us is whether the polity is prepared to tolerate *permanent mass unwork* or whether share-work values and programs will begin to bridge the gulf between knowledge-based labor and manual and clerical work.

A Guaranteed Income

Accordingly, if unwork is fated to be no longer the exception to the rule of nearly full employment, we need an entirely new approach to the social wage and, more generally, "welfare" policy. If there is work to be done, everyone should do some of it; additional remuneration would depend on the kind of work an individual performs. But shorter work days, longer vacations, and earlier retirement imply that most of us should never work anything like "full time" as measured by the standards of the industrializing era. We need a political and social commitment to a national guaranteed income that is equal to the historical level of material culture. That is, everyone would be guaranteed a standard of living that meets basic nutritional, housing, and recreational requirements. Everyone would assume the responsibilities of producing and maintaining public goods, so no able citizen would be freed of the obligation to work. This would place a large burden on the private production sectors to induce people to engage in routine labor, presumably at wage rates higher than the guaranteed income and equal to tasks in the public sector. These rates would constitute a further incentive to invest in labor-saving technologies, which would free people from routine tasks without plunging them into a state of penury.

There would be no welfare system because the distinction between workers and "idlers" would disappear. Services such as health care (including counseling and therapy), education, and social work would expand and be paid for through general tax levies, but, assuming a new perspective on "jobs" and the division of labor, would shift their emphasis from work toward solving problems, exploring possibilities, and finding new ethical and social meaning.

School curricula, for example, could concentrate on broadening students' cultural purview: music, athletics, art, and science would assume a more central place in the curriculum and there would be a renewed emphasis on the aesthetic as well as the vocational aspects of traditional crafts. We suppose this would lead to a revival of what has become known as "leisure

353

studies": psychologists and sociologists would study, prescriptively as well as analytically, what people do with their time, no longer described in precisely the same terms as it was thirty years ago. Concomitantly, space and time themselves become objects both of knowledge and, in the more conventional science fiction sense, of personal and social exploration. Consequently, lifelong learning, travel, avocations, small business, and artisanship take on new significance as they become possible for all people, not just the middle and upper classes. Some may choose to participate in the technoculture as a crucial component of the exercise of the right to *pleasure* as well as work. Others may avoid the technological construction of social and personal meaning.

Only when social policy has been transformed can the conditions that have produced the ecological crisis—consumerism, for example—be redirected. In a society in which the preferred route to pleasure is buying commodities, to propose a new asceticism in the name of ethical renewal, as Lasch and others have done, merely perpetuates the repressive cultural regime of our era. Concomitantly, those who will not address issues of social justice and economic equality should keep silent about ecological disaster, for to expect that the vast majority of people will sacrifice their living standards to preserve the spotted owl without provision for the means of life is either naive or blatantly class biased.

Clearly, there is an urgent need for a new ethic that addresses the proliferation of waste in our communities. No observer of urban politics can fail to notice the struggles in black, Asian, and Latino communities against government plans to build incinerators to deal with the mountains of garbage that have accumulated in overcrowded cities and suburbs. Yet it is not enough to resist the most polluting methods of garbage disposal, nor will alternatives that do not ask the hard questions about how we have constructed our lives around consumption be adequate.

A New Research Agenda

In the immediate as well as the medium-range future, we need a renewed public commitment to scientific and technological research. Research would not be confined to developing new products or motivated exclusively by considerations such as enhancing the national economy in an era of global competition. Fighting disease, protecting the environment, and finding new ways to construe time and space in both "wilderness" and urban settings would absorb considerable resources.

Expert opinion, echoing similar conclusions after World War II, reached a virtual consensus that the Reagan and Bush administrations lacked science, energy, and technology policies beyond those required by the military. In fact, the Bush administration had an active science policy: using the pared-down federal research budget to encourage collaboration between government agencies, universities, and private corporations in the pursuit of growth-directed research. Federal Courts, for example, ruled that private corporations could "own" patents on new life forms; the federal government reduced its direct participation in scientific research and instead encouraged collaborative arrangements between universities and biotechnology companies. As we have noted, funding for basic research was reduced by almost two-thirds, even as European countries were increasing their public funds for these purposes. Production and exploration of domestic oil, gas, solar power, synthetic fuels, and wind power were actively discouraged in favor of increasing oil imports. Research on alternative fuel sources was reduced to the vanishing point as the United States became increasingly dependent on imported fossil fuels.

Clearly, ecological considerations impel seeking alternatives to reliance on oil as our primary energy resource and on plastics as the major packaging material and as a substitute for natural fibers, among other nefarious uses. We should reexamine the almost universal use of alternating current for electricity in the wake of strong suspicion that large, centralized generators may contribute to cancer. Direct current and solar and wind energy might not be as dangerous. In any case, this is a vital part of the research agenda. In addition, a concerted effort to curtail and in some instances ban automobile use, especially in cities, is long overdue. Mass transit and new patterns of settlement in multifamily dwellings would create jobs and at the same time address ecological concerns.

It is by now evident that among the costs of a relentless pursuit of industrialization was pollution of our water and air. Congress has established a national cleanup fund and imposed some regulation on industrial polluters, but so far has not been willing to reexamine the historical costs of industrial enterprise. More than forty years ago, R. William Kapp took the first steps when he argued that we have failed to calculate the "social costs of private enterprise":[16] when a coal mine or a metal plant is abandoned, even if the employer is required to "clean up" the surrounding area, people from miles around have already suffered the deleterious effects of air and water pollution. To be sure, public agencies have made our drinking water safer; the Clean Air Act imposed regulations on employers but provided extremely small inspection and enforcement teams.

Even if they were vigorously enforced, these measures are all after the fact. They take for granted the historical regime of industrial production that requires huge quantities of fossil fuels, employs large-scale power plants and disposes of waste in large dumps that pollute the water bed. We erect regimes of treatment to counter the cancer- and heart disease-causing effects of industrialization and spend billions of dollars to construct hospitals and other medical facilities; we produce pharmaceuticals as our primary life-preserving weapon against the identification of industrialization with civilization itself. We take for granted that capital provides jobs, and the job culture becomes the new religion of advanced industrial societies.

We do not yet fully possess the knowledge required to fundamentally and radically shift the basis of our production regimes: wind, solar, and hydro energy are still in their infancy; we have only scant experience with radically decentralized small-scale production methods, except in agriculture; and alternatives to the life-threatening (as well as life-enhancing) effects of medical treatment require more work (some alternative methods, based on traditional cultures and organicist worldviews, have proven effective against certain diseases, while others remain hypothetical).

There is little doubt that the two main killers—cancer and heart disease—and another key cause of premature death—automobile accidents—can be reduced only if we decide to reverse the blind, compulsive march of industrial culture. In the interest of averting the health crisis as well as the ecological crisis—not for saving labor costs—we need to deploy cybernetic and other labor-saving technologies in conjunction with the development of small-scale, ecologically sound production regimes. Undoubtedly we will find that some of these technologies, especially those that use large quantities of energy, are ecologically problematic; others will certainly prove to be ecologically beneficial.

One of the major issues in our emerging ecological crisis is how to reduce the amount of nonbiodegradable synthetic materials that have replaced cotton, wool, and wood products in packaging, furniture, clothing, cars, houses, and appliances. Despite claims that many plastics are biodegradable, some have argued that they seep into and pollute the water supply. Obviously, for "convenience" and cost savings, we have agreed to the trade-off, just as Americans have become addicted to their cars and single-family homes.

Clearly, the inevitability of both the jobless future and the ecological crisis demands a conclusive cultural shift, for we cannot simply legislate a change of this scope. But this shift cannot be achieved without a national and international effort to reduce the degree to which people would be required to give up some components of the "good life" associated with consumer goods. For just as we do not advocate nonwork without adequate income, we see the need to mobilize the scientific and technological revolution to meet ecological and health needs. In the last instance, this becomes the basic goal of science and technology policy.

Ending Endless Work

Since the democratic revolutions of the eighteenth and nineteenth centuries, images of the Athenian city-state have suffused the work of political philosophers. Hannah Arendt's polemic against Marx, for example, is directed not chiefly to the glorification of work over labor (although this remains a significant aspect of her picture of an ideal civilization), but to the notion of the primacy of politics. For Arendt as well as other political theorists, among the more egregious consequences of the Enlightenment was the displacement of politics by social relations and, indirectly, our relations with nature. Accordingly, Marx's critique of the relation of power to domination erased the very idea of the polity since it could be shown that "citizenship" was an ideology that masked what should have been evident: that the individual as the subject of politics had not yet emerged—indeed, could not emerge until the relations of production were decisively transformed. For the people as a self-governing body presupposed, as in ancient Athens, rough economic equality, at least for a fraction of the nonslave population.

Like many who have come before us, we believe that among the crucial tools of domination is the practice of "work without end," which chains workers to machines and especially to the authority of those who own and control them—capital and its managerial retainers. To be sure, labor did not enter these relations of domination without thereby gaining some benefit. In the Fordist era, as Hunnicut has brilliantly shown, organized labor exchanged work for consumption and abandoned its historical claim of the right to be lazy, as Paul Lafargue put it.[17] Here, within limits, we affirm that right but confess another: the freedom of people emancipated from labor to become social agents.

Needless to say, we reject the idea that liberal democratic states have already conferred citizenship and that apathy is the crucial barrier preventing many from participating in decision making. Such optimism, unfortunately promulgated by many intellectuals of the left as well as the right, blithely ignores the social conditions that produce "apathy," especially the structural determinants of disempowerment, among them endless work. Nor are we prepared to designate the economic sphere, including the shop-floor "rational-purposive" activity that on the whole has been effectively depoliticized and functions only in terms of the perimeters of instrumental, technical rationality.[18] Management's control over the workplace is an activity of politics. There are winners and losers in the labor process. To render the workplace rational entails a transformation of what we mean by rationality in production, including our conception of skill and its implied "other," unskill; a transformation of what we mean by mental as opposed to physical labor and our judgment of who has the capacity to make decisions under regimes of advanced technologies.

Politics as rational discourse—as opposed to a naked struggle for power—awaits social and economic emancipation. Among the constitutive elements of freedom is *self-managed time.* Our argument in this book is that there are for the first time in human history the material preconditions for the emergence of the individual and, potentially, for a popular politics. The core material precondition is that labor need no longer occupy a central place in our collective lives, nor in our imagination. We do not advocate the emancipation from labor as a purely negative freedom. Its positive content is that, unlike the regime of work without end, it stages the objective possibility of citizenship.

Under these circumstances, we envision civil society as the privileged site for the development of individuals who really are free to participate in a public sphere of their own making. In such a civil society, politics consists not so much in the ritual act of selection, through voting, of one elite over another, but in popular assemblies that could, given sufficient space and time, be both the legislative and the administrative organs. The scope of popular governance would extend from the workplace to the neighborhood. For as Ernest Mandel has argued, there is no possibility of worker self-management, much less the self-management of society, without ample time for decision making.[19] Thus, in order to realize a program of democratization, we must create a new civil society in which freedom consists in the first place (but only in the first place) in the liberation of time from the external constraints imposed by nature and other persons on the individual.

The development of the individual—not economic growth, cost cutting, or profits—must be the fundamental goal of scientific and technological innovation. The crucial obstacle to the achievement of this democratic objective is the persistence of the dogma of work, which increasingly appears, in its religious-ethical and instrumental-rational modalities, as an obvious instrument of domination.

Notes

Introduction

1. Peter Kilborn, "Solid Jobs Seem to Vanish Despite Signs of Recovery," *New York Times,* December 26, 1992.

2. As quoted in Peter Kilborn, "New Jobs Lack the Old Security in a Time of 'Disposable Workers,'" *New York Times,* March 15, 1993.

3. "All Things Considered," National Public Radio, March 5, 1993.

4. Barry Bluestone and Bennett Harrison, *The Deindustrialization of America* (New York: Basic Books, 1982), 25–26.

5. Wassily Leontief and Faye Duchin, *The Future Impact of Automation on Workers* (New York: Oxford University Press, 1986); for statistics on unemployment in the steel industry see David Bensman and Roberta Lynch, *Rusted Dreams: Hard Times in a Steel Community* (Berkeley: University of California Press, 1987), 3.

6. "America Rushes to High-Tech for Growth," *Business Week,* March 28, 1983.

7. Karen Pennar, "The Productivity Paradox: Why the Payoff from Automation Is Still So Elusive—And What Corporate America Can Do about It," *Business Week,* June 6, 1988.

8. Bureau of Labor Statistics, *Monthly Labor Review* 108, no. 11 (November 1985).

9. Michael J. Piore and Charles F. Sabel, *The Second Industrial Divide: Possibilities for Prosperity* (New York: Basic Books, 1984), 251–80.

1. The New Knowledge Work

1. Steve Lohn, "Top IBM Issue: How, Not Who," *New York Times,* March 25, 1993.

2. The concept of human capital in its classic enunciation is in Gary Becker, *Human Capital: A Theoretical and Empirical Analysis with Special Reference to Education* (New York: Columbia University Press, 1964). Building on the work of T. W. Schultz, Becker argues that the growth of physical capital (machinery, buildings, and so forth) accounts for "a relatively small part of the growth of income" when compared to "education and skills." Hence his argument that education, which, presumably, upgrades skills and knowledge, is crucial for growth policies.

3. The words used by the secretary of labor after the Bureau of Labor Statistics reported an increase of 365,000 jobs in February 1993 (*New York Times,* March 24, 1993).

4. William DiFazio, *Longshoremen: Community and Resistance on the Brooklyn Waterfront* (South Hadley, Mass.: Bergin and Garvey, 1985).

5. See chapter 10 for a fuller discussion of this point.

6. Phillip K. Dick, *The Three Stigmata of Palmer Eldritch* (London: Jonathan Cape and Granada Books, 1978). First published in 1964, Dick's novel foreshadows the development of virtual reality technology, linking it to a future when most people can no longer live on Earth but are afforded the means to simulate a life on this planet from a position somewhere in the galaxy.

7. Immanuel Wallerstein, *The Modern World System* (New York: Academic Press, 1974). Building on the work of "dependency" theorists such as Giovanni Arrighi, Cardozo and Faetto, Andre Gunder Frank, and, especially, the Annales school of French historiography (Braudel, Lucien Febvre), Wallerstein demonstrates that, since the sixteenth century, capitalism has been a global system, albeit one of unequal exchange.

8. In 1992, Indian engineers and computer scientists emerged as world-class players in high-tech design. American and European corporations began letting contracts to Bombay- and Delhi-based software firms.

9. Karl Marx, *Grundrisse* (New York, Vintage, 1973), 701–5.

10. In the United States, nearly 70 percent of women had entered the labor force by 1990. In recent years, many have been able to obtain only part-time jobs.

11. Stanley Aronowitz, *False Promises: The Shaping of American Working Class Consciousness,* with a new introduction and epilogue (Durham, N.C.: Duke University Press, 1992 [1973]); *Working Class Hero* (New York: Pilgrim, 1983).

12. Beth Sims, *Workers of the World Undermined American Labor's Role in U.S. Foreign Policy* (Boston: South End Press, 1992). For an earlier study of this question, see also Ronald Radosh, *U.S. Labor and American Foreign Policy* (New York: Random House, 1971).

13. *Wall Street Journal,* March 9, 1992.

14. Martin Sklar, "On the Proletarian Revolution and the End of Political-Economic Society," *Radical America,* June 1969. This remarkable article provides a theory, from the perspective of political economy, of the end of real capital accumulation. According to Sklar, the major tendency of contemporary advanced capitalist societies was toward an overaccumulation of capital; thus, the task of investment is to get rid of this surfeit, to disaccumulate capital. Hence advertising, the production of waste in the form of planned obsolescence, the proliferation of "services," and, of course, in the United States, massive military expenditures.

15. Harry Braverman, *Labor and Monopoly Capital* (New York: Monthly Review Press, 1974).

16. E. P. Thompson, *The Making of the English Working Class* (New York: Knopf, 1963).

17. Frederick Winslow Taylor, *Principles of Scientific Management* (New York: Norton, 1967 [1911]); Lillian Gilbreth, *The Homemaker and Her Job* (New York: Appleton, 1927). Gilbreth's book is an application to "women's work," more specifically the work of the "mother" who must provide a "place of rest" by making invisible the myriad household tasks, of the principles of scientific management developed with her husband, Frank, and Taylor.

18. Christian Palloix, "From Fordism to Neo-Fordism," in *The Labour Process and Class Strategies* (London: Conference of Socialist Economists, 1976); Bernard Doray, *From Taylorism to Fordism,* trans. David Macey (London: Free Association Books, 1988).

19. Michel Aglietta, *A Theory of Capitalist Regulation,* trans. David Fernbach (London: New Left Books, 1976).

20. Sidney Pollard, *The Genesis of Modern Management* (Cambridge, Mass.: Harvard

University Press, 1965); Daniel Nelson, *Managers and Workers* (Madison: University of Wisconsin Press, 1975).

21. Alvin Toffler, *The Third Wave* (London: William Collins, 1980); Jacques Attali, *Millennium: Winners and Losers in the Coming World Order* (New York: Random House, 1991).

22. Mark Poster, *The Mode of Information* (Chicago, University of Chicago Press, 1990).

23. Martin Heidegger, *The Question Concerning Technology and Other Essays* (New York: Harper & Row, 1979).

24. For an extraordinary account of the sociology of industrial accidents, see Tom Dwyer, *Life and Death at Work: Industrial Accidents as a Case of Socially Produced Error* (New York: Plenum, 1991). For the transformation of American industrial relations, see Kochan, Katz, and McKersie, *The Transformation of American Industrial Relations* (New York: Basic Books, 1986).

25. Anthony Hyman, *Charles Babbage: Pioneer of the Computer* (Princeton, N.J.: Princeton University Press, 1982).

26. Joseph Agassi, *Faraday as a Natural Philosopher* (Chicago: University of Chicago Press, 1971).

27. David F. Noble, *America by Design: Science, Technology and the Rise of Corporate Capitalism* (New York: Knopf, 1977).

28. Martin Kenney, *Bio-Technology: The University-Industrial Complex* (New Haven, Conn.: Yale University Press, 1986).

29. Alfred Sohn-Rethel, *Intellectual and Manual Labour* (London: Macmillan, 1977).

30. Marx, *Grundrisse*, 699.

31. John Dewey, *The Public and Its Problems,* in *Collected Works of John Dewey,* vol. 1, *Later Works* (Carbondale: Southern Illinois University Press, 1981).

32. Walter Lippmann, *Public Opinion* (1919).

33. Lewis Corey, *The Crisis of the Middle Class* (New York: Covici Friede, 1936); James Burnham, *The Managerial Revolution: What's Happening in the World* (New York: John Day, 1940); A. A. Berle and Gardiner Means, *The Modern Corporation and Private Property* (New York: Macmillan, 1932).

34. Steve J. Heims, *John Von Neumann and Norbert Weiner* (Cambridge, Mass: MIT Press, 1980), 179.

35. Ibid., 180.

36. C.Wright Mills, *White Collar* (New York: Oxford University Press, 1951).

37. C. Wright Mills, *The Power Elite* (New York: Oxford University Press, 1956); Mills, *The New Men of Power* (New York: Harcourt, Brace, 1948).

38. Daniel Bell, *The Coming of Post-Industrial Society* (New York: Basic Books, 1973).

39. Alain Touraine, *Post-Industrial Society* (New York: Random House, 1971); Serge Mallet, *The New Working Class* (London: Spokesman, 1979); Andre Gorz, *A Strategy for Labor* (Boston: Beacon, 1966).

40. Rudolph Hilferding, *Finance Capital* (London: Routledge and Kegan Paul, 1981).

41. Seymour Martin Lipset, *Political Man* (New York: Doubleday, 1961); Ralf Dahrendorf, *Class and Class Conflict in Industrial Society* (Palo Alto, Calif.: Stanford University Press, 1957).

42. Harvey Swados, "The Myth of the Happy Worker," *Nation,* April 26, 1957.

43. Daniel Bell, *The Cultural Contradictions of Capitalism* (New York: Basic Books, 1978).

44. Talcott Parsons, *The Social System* (Glencoe, Ill.: Free Press, 1951).

45. Paul Willis, *Learning to Labor* (New York: Columbia University Press, 1981).

2. Technoculture and the Future of Work

1. Richard Rhodes, *The Making of the Atomic Bomb* (New York: Simon & Schuster, 1986), 143.

2. David Dickson, *The New Politics of Science* (New York: Pantheon, 1984), 25.

3. Fritz Macchlup, *Knowledge: Its Creation, Distribution and Economic Significance*, vol. 1, *Knowledge and Knowledge Production* (Princeton, N.J.: Princeton University Press, 1980).

4. John McDermott, "Technology: The Opiate of the Intellectuals," in Albert Teich, ed., *Technology and Man's Future*, 2d ed. (New York: St. Martin's, 1977).

5. Barbara Katz Rothman, *Recreating Motherhood* (New York: Norton, 1989).

6. Karl Marx, "Economic and Philosophical Manuscripts," in David Fernbach, ed., *Karl Marx: Early Writings* (New York: Vintage, 1982).

7. Herbert Marcuse, *One Dimensional Man* (Boston: Beacon, 1964).

8. Max Horkheimer and Theodor Adorno, *Dialectic of the Enlightenment* (New York: Continuum, 1972).

9. Gilles Deleuze and Félix Guattari, *Anti-Oedipus* (New York: Viking, 1977).

10. "It is moot whether, without restoring the category of the sacred, the category most thoroughly destroyed by the scientific enlightenment, we can have an ethics able to cope with the extreme powers which we possess today and constantly increase and are almost compelled to wield" (Hans Jonas, *The Imperative of Responsibility in Search of an Ethics for the Technological Age* [Chicago and London: University of Chicago Press, 1984], 23).

11. This conclusion was reinforced for one of the authors during a recent conference of North and Latin American intellectuals in Mexico City on issues of knowledge. A substantial number of the papers were profoundly informed by many of the ideas of the Frankfurt school and similar views by other tendencies of European critiques of science and technology.

12. Rachel Carson, *Silent Spring* (Boston: Houghton Mifflin, 1962); Barry Commoner, *The Closing Circle* (New York: Pantheon, 1969); Rene Dubos, *Reason Awake: Science for Man* (New York: Columbia University Press, 1970).

13. Walter Lippmann, *The Phantom Public* (1922).

14. The beneficent effects of a strong military and nuclear arsenal were reinforced by the fact that many students were able to complete their undergraduate degrees only with the help of National Defense Education Act grants and the GI Bill of Rights, which also provided funds for graduate research in science. In fact, it may be argued that the rubric of "defense" became the key umbrella under which federal aid to education and to science and technological research were promulgated in the United States. Otherwise, given strong states' rights sentiments, especially in education, Congress has been unwilling to put the federal government in the business of education and research.

15. Robert Merton, *The Sociology of Science* (Chicago: University of Chicago Press, 1973), 270.

16. Ibid., 278.

17. Ibid., 264.

18. Martin Kenney, *Bio-Technology: The University-Industrial Complex* (New Haven, Conn.: Yale University Press, 1986).

19. Andrew Pickering, *Constructing Quarks* (Chicago: University of Chicago Press, 1988).

20. For a fuller discussion of this point, see chapters 3 and 4 of this book.

21. McDermott, *Opiate,* 189.

22. Ibid., 198.

23. Ibid., 201. Emphasis added.

24. Marcuse, *One Dimensional Man,* 257.

25. Ibid., 256.

26. Ibid., 230–31.

27. McDermott, *Opiate*, 198.

28. Ibid., 196.

29. Ibid., 194.

30. Ibid., 195.

31. Ibid.

32. See Sheldon Krimsky, *Genetic Alchemy: A Social History of the Recombinant DNA Controversy* (Cambridge, Mass.: MIT Press, 1982), for a detailed treatment of the Cambridge citizens' movement against bioengineering; the ACT UP/NY Women & AIDS Book Group, *Women, AIDS and Activism* (Boston: South End Press, 1990). See also Gilbert Elbaz, "ACT-UP," Ph.D. dissertation, City University of New York Graduate School, 1992, perhaps the most comprehensive treatment of the successful exercise of scientific citizenship in recent U.S. history.

33. The literature on the social and cultural context of scientific work is vast. See especially Bruno Latour and Steve Woolgar, *Laboratory Life* (Los Angeles and London: Sage, 1979); Bruno Latour, *Science in Action* (Cambridge, Mass.: Harvard University Press, 1987); Stanley Aronowitz, *Science as Power* (Minneapolis: University of Minnesota Press, 1988).

34. Constance Penley and Andrew Ross, eds., *Technoculture* (Minneapolis: University of Minnesota Press, 1991).

35. Herbert Dreyfus, *What Computers Can't Do* (New York: Harper & Row, 1979); John Broughton, "The Surrender of Control: Computer Literacy as Political Socialization of the Child," in Douglas Sloan, ed., *The Computer in Education: A Critical Perspective* (New York: Teachers College Press, 1984).

36. Martin Heidegger, *The Question Concerning Technology and Other Essays,* trans. and with an introduction by William Leavitt (New York: Harper & Row, 1977).

37. Howard Rheingold, *Virtual Reality* (New York: Summit, 1991), 345.

38. Ibid., 346.

3. The End of Skill?

1. The increasingly large literature on technophilia includes the following: from a cultural perspective, Constance Penley and Andrew Ross, eds., *Technoculture* (Minneapolis: University of Minnesota Press, 1991); from a historical perspective, Howard P. Segal, *Technological Utopianism in American Culture* (Chicago: University of Chicago Press, 1985); and from a sociological/management perspective, Shoshana Zuboff, *In the Age of the Smart Machine: The Future of Work and Power* (New York: Basic Books, 1988).

2. There are many versions of this story. John Barnard tells it this way: "When Reuther toured a new highly automated Ford engine plant in Cleveland a company engineer taunted him with this remark: 'You know not one of those machines pays dues to the United Automobile Workers.' Reuther shot back: 'And not one of them buys new Ford cars either.'" John Barnard, *Walter Reuther and the Rise of the Auto Workers* (Boston: Little, Brown, 1983), 154.

3. Robert H. Zeiger, *American Workers, American Unions, 1920–1985* (Baltimore: Johns Hopkins University Press, 1986). See chapter 5, "Affluent Workers, Stable Unions: Labor in the Postwar Decades."

4. Karl Marx, *Grundrisse* (New York: Vintage, 1973), 692–93, 699. Emphasis added.

5. Ibid., 700.

6. Michael J. Piore and Charles F. Sabel, *The Second Industrial Divide: Possibilities for Prosperity* (New York: Basic Books, 1984).

7. Harry Braverman, *Labor and Monopoly Capital* (New York: Monthly Review Press, 1974).

8. Frederick Winslow Taylor, *The Principles of Scientific Management* (New York: Norton, 1967).

9. Mike Cooley, *Architect or Bee? The Human-Technology Relationship* (Boston: South End Press, 1980), 36.

10. Ibid., 129.

11. Three books that deal with Braverman's position but take a different approach are Michael Burawoy, *The Politics of Production* (London: Verso, 1985); Larry Hirschhorn, *Beyond Mechanization: Work and Technology in a Postindustrial Age* (Cambridge, Mass.: MIT Press, 1984); and Paul Thompson, *The Nature of Work: An Introduction to Debates on the Labor Process* (London: Macmillan, 1983).

12. Paul Adler, "Technology and Us," *Socialist Review*, no. 85 (January–February 1986): 82.

13. Ibid., 83.

14. Paul S. Adler, "Automation, Skill and the Future of Capitalism," *Berkeley Journal of Sociology: A Critical Review* 33 (1988): 3.

15. Ibid., 2.

16. Ibid., 33.

17. Stanley Aronowitz, "Marx, Braverman, and the Logic of Capital," in *The Politics of Identity: Class, Culture, Social Movements* (New York: Routledge, 1992). William DiFazio, *Longshoremen: Community and Resistance on the Brooklyn Waterfront* (South Hadley, Mass.: Bergin & Garvey, 1985).

18. Paul S. Adler, "When Knowledge Is the Critical Resource, Knowledge Management is the Critical Task," *IEEE Transactions on Engineering Management* 36, 2 (May 1989): 92.

19. Adler, "Automation, Skill and the Future of Capitalism," 33.

20. "Study of New Jobs Since 79 Says Half Pay Poverty Wage," *New York Times*, September 27, 1988. Frank Levy states on current income stagnation, "By 1975 median family income had fallen by $1,700. It gained most of this back by the end of 1979, but fell sharply in the 1980–82 recession and stood at $26,433 in 1984. This sudden break in trend—twenty-six years of income growth followed by twelve years of income stagnation—is the major economic story of the postwar period" (*Dollars and Dreams: The Changing American Income Distribution* [New York: Norton, 1988], 17).

21. Bennett Harrison and Barry Bluestone, *The Great U-Turn: Corporate Restructuring and the Polarizing of America* (New York: Basic Books, 1990), 37–38.

22. Manuel Castells, *The Informational City* (Oxford: Blackwell, 1989), 342.

23. Harold Salzman, "Computer-Aided Design: Limitations in Automating Design and Drafting," *IEEE Transactions on Engineering Management* 36, no. 4 (November 1989).

24. Ibid., 255.

25. Adler, "Automation, Skill and the Future of Capitalism," 3.

26. Feminist Majority Foundation, "Empowering Women in Business" (Arlington, Va.: Feminist Majority Foundation, 1991).

27. John Rule, "The Property of Skill in the Period of Manufacture," in Patrick Joyce, ed., *The Historical Meanings of Work* (Cambridge: Cambridge University Press, 1987), 108. See also Maxine Berg, *The Age of Manufactures, 1700–1820* (London: Fontana, 1985).

28. Paul Starr, *The Social Transformation of American Medicine* (New York: Basic Books, 1982), 391.

29. Robert Gray, "The Languages of Factory Reform in Britain, c. 1830–1860," in *The Historical Meanings of Work*, 150.

30. Stanley Aronowitz, *Science As Power* (Minneapolis: University of Minnesota Press, 1988), 298.

31. Charles F. Sabel and Jonathan Zeitlin, "Historical Alternatives to Mass Production," *Past and Present* 108 (August 1985).

32. Piore and Sabel, *Second Industrial Divide,* 269.

33. Ibid., 261.

34. Ibid., 270.

35. Ibid., 272.

36. Erle Norton, "Future Factories: Small Flexible Plants May Play Crucial Role in U.S. Manufacturing," *Wall Street Journal,* January 13, 1993.

37. Ibid.

38. Ibid.

39. Fred Block, *Postindustrial Possibilities: A Critique of Economic Discourse* (Berkeley: University of California Press, 1990), 94–103.

40. Tom Forester, *High-Tech Society* (Cambridge, Mass.: MIT Press, 1987), 172.

41. Block, *Postindustrial Possibilities,* 101; Ramchandran Jaikumar, "Postindustrial Manufacturing," *Harvard Business Review* 64, no. 6 (November–December 1986).

42. Block, *Postindustrial Possibilities,* 103.

43. Ibid., 94.

44. David Harvey, "Flexibility: Threat or Opportunity?" *Socialist Review* 21, no. 1 (January–March 1991).

45. Ibid., 73. Emphasis in original.

46. Sabel and Zeitlin, Piore and Sabel.

47. Block, *Postindustrial Possibilities,* 96.

48. Zuboff, *In the Age of the Smart Machine,* 61.

49. Ibid., 397.

50. Reinterpretation of Karl Marx's ironic use of "The Law of the Tendential Fall in the Rate of Profit" in vol. 3 of *Capital* (New York: Vintage, 1975), 315.

4. The Computerized Engineer and Architect

1. Magali Sarfatti Larson, *The Rise of Professionalism: A Sociological Analysis* (Berkeley: University of California Press, 1977). The "professional" engineer who possesses not only advanced degrees but also certification from leading professional associations is the exception, not the rule, in most industrial design workplaces. Since the United States has experienced severe shortages for the past decade, the labor force in engineering, as in medicine, includes an increasing number of foreign-born professionals who are not considered qualified by U.S. associations.

2. Antonio Negri, *The Politics of Subversion,* trans. James Newell (London: Polity Press, 1989).

3. Karl Marx, *Grundrisse* (New York: Vintage, 1973), 705.

4. Stanley Aronowitz, Patricia D'Audrade, William DiFazio, et al., *Time for Decision: Computer-Aided Design and the Future of the Civil Service Professional in New York City* (New York: Institute on Labor and Community, August 1984), section 6, 3. Computer-aided design (CAD) is the term in general use throughout the profession. Computer-Aided Design and Drafting (CADD) was the title of the section where we began our studies of the computerization of the design work of engineers and architects in the Department of Environmental Protection in New York City. In design, traditionalists have insisted on the division between drafting and design, thus CADD. But computer-aided design has made this division unrealistic; drafting has been combined with design, making drafting redundant.

5. *Business Week,* August 12, 1991, 61.

6. Tom Forester, *High-Tech Society* (Cambridge, Mass.: MIT Press, 1988), 210.

7. Jan Forslin, Britt Marie Thulestedt, and Sven Andersson, "Computer-Aided Design: A

Case Strategy in Implementing a New Technology," *IEEE Transactions on Engineering Management* 36, no. 3 (August 1989): 192.

8. Paul S. Adler, "CAD/CAM: Managerial Challenges and Research Issues," *IEEE Transactions on Engineering Management* 36, no. 3 (August 1989): 205–6.

9. Wassily Leontief and Faye Duchin, *The Future Impact of Automation on Workers* (New York: Oxford University Press, 1986); Harley Shaiken, *Work Transformed* (New York: Holt, Rinehart and Winston, 1984); Mike Cooley, *Architect or Bee? The Human-Technology Relationship* (Boston: South End Press, 1980).

10. Stanley Aronowitz, William DiFazio, and Eric Lichten, *Contracting Out, Computer Aided Design, Engineering and the New Jersey Department of Transportation* (New York: Public Technology Study Group, 1985), section 2, 10.

11. Shaiken, *Work Transformed*, 219–20.

12. Harold Salzman, "Computer-Aided Design: Limitations in Automating Design and Drafting," *IEEE Transactions on Engineering and Management* 36, no. 4 (November 1989): 255.

13. Ibid., 255.

14. Aronowitz et al., *Time for Decision*, section 1, 1–2.

15. Ibid., section 6, 3.

16. Michael T. Cetera, "Conference Proceedings," GDS User Conference (Cambridge, England: King's College, September 1988), 48–49.

17. Computer Aided Drafting and Design Committee, "Pilot Project Evaluation," prepared for the Department of Transportation, City of New York, December 1988, 31.

18. Douglas F. Stoker, "The Disappearance of CAD in the '90s," *Architectural & Engineering Systems* 6, no. 1 (January 1990): 34.

19. Adler, "CAD/CAM: Managerial Challenges," 214.

20. Alvin W. Gouldner, *The Future of Intellectuals and the Rise of the New Class* (New York: Oxford University Press, 1979).

21. Eric Lichten, *Class, Power and Austerity* (South Hadley, Mass.: Bergin & Garvey, 1986).

22. Stanley Aronowitz, William DiFazio, and Eric Lichten, "Unions, Technology and Computer Aided Design," in Pamela Wilson, ed., *Here Comes Tomorrow: Technological Change and Its Effects on Professional, Technical and Office Employment* (Washington, D.C.: Department of Professional Employees, AFL-CIO, 1988).

23. Ibid., 3.

24. Ibid., 5.

25. Ibid., 8–9.

26. Stephen Levy, *Hackers: Heroes of the Computer Revolution* (New York: Dell, 1984).

27. Seymour Papert, *Mindstorms: Children, Computers and Powerful Ideas* (New York: Basic Books, 1980).

28. For a thorough critique of computers in education, see especially John M. Broughton, "The Surrender of Control: Computer Literacy as Political Socialization of the Child," in Douglas Sloan, ed., *Computers in Education: A Critical Perspective* (New York: Teachers College Press, 1984). Also see Edmund Sullivan, "Computers, Cultures and Educational Futures: A Critical Reflective Meditation on Papert's *Mindstorms*," Toronto, *Strategic Planning Documents on Computers in Education*, 1983.

29. As quoted in Hubert L. Dreyfus and Stuart E. Dreyfus, "Putting Computers in Their Proper Place: Analysis Versus Intuition in the Classroom," in *Computers in Education*, 62.

30. Walter Benjamin, "A Small History of Photography," *One Way Street* (London: New Left Books, 1979).

5. The Professionalized Scientist

1. By *elite* Magali Larson means monopolistic control of a professional marketplace based on expert knowledge. Magali Sarfatti Larson, *The Rise of Professionalism: A Sociological Analysis* (Berkeley: University of California Press, 1977).

2. Bruno Latour, *Science in Action* (Cambridge, Mass.: Harvard University Press, 1987).

3. Ibid., 131.

4. Ibid., 68–69.

5. National Science Board, *Science & Engineering Indicators: 1991,* 10th ed. (Washington, D.C.: U.S. Government Printing Office, 1991), 75, 290.

6. Latour, *Science in Action,* chapter 6.

7. National Science Board, Committee on Industrial Support for R&D, *The Competitive Strength of U.S. Industrial Science and Technology: Strategic Issues* (Washington, D.C.: National Science Foundation, August 1992), ii.

8. *Science & Engineering Indicators: 1991,* 93.

9. Alain Touraine, *Return of the Actor* (Minneapolis: University of Minnesota Press, 1988), xxv.

10. Brahmins in Spencer Klaw's sense: "The basic researcher is free, in principle, to work only on what interests him—a freedom he shares with artists, poets, and people who inherit money. Basic research also has the fascination of being a game in which victory—the discovery of a new relationship or the formulation of a new law or concept—confers on the victor the gratifying sense of having changed the universe." Spencer Klaw, *The New Brahmins: Scientific Life in America* (New York: Morrow, 1968), 45–46.

11. Mandarins as defined by Charles Derber et al.: "In the world's greatest ancient civilization, bureaucratic scholar-officials or 'mandarins' ruled China for over 1,000 years. They contended that, according to the laws of nature, 'there should be two kinds of people: the educated who ruled and the uneducated who were ruled'" (Dun Li, "The Four Classes," in Molly Joel Coye, Jon Livingston, and Jean Highlands, ed., *China* [New York: Bantam, 1984], 48, quoted in Charles Derber, William A. Schwartz, and Yale Magrass, *Power in the Highest Degree: Professionals and the Rise of a New Mandarin Order* [New York: Oxford University Press, 1990], 4–5). The mandarins created a formal class hierarchy based on Confucian credentials conferred by exams.

12. Alfred Sohn-Rethel, *Intellectual and Manual Labor: A Critique of Epistemology* (Atlantic Highlands, N.J.: Humanities Press, 1978).

13. Andrew Pickering, *Constructing Quarks: A Sociological History of Particle Physics* (Chicago: University of Chicago Press, 1984), 136.

14. Stanley Aronowitz, *Science as Power* (Minneapolis: University of Minnesota Press, 1988).

15. Plato, *Timaeus and Critias* (London: Penguin, 1977). Aristotle, *Metaphysics* (New York: Columbia University Press, 1952).

16. Aronowitz, *Science as Power,* 331.

17. John Dewey, *The Quest for Certainty* (New York: Putnam, 1929), 21.

18. Robert Weinberg quoted in Jerry E. Bishop and Michael Waldholz, *Genome* (New York: Touchstone, 1990), 21.

19. Karl Mannheim, *Ideology and Utopia* (New York: Harcourt Brace Jovanovich, 1936), 157.

20. Quoted in Philip J. Hilts, "Congress Urged to Lift Ban on Fetal-Tissue Research," *New York Times,* May 27, 1992.

21. Warren Leary, "Gene Altered Food Held by the FDA to Pose Little Risk," *New York Times,* May 26, 1992.

22. Warren Leary, "Cornucopia of New Foods Is Seen as Policy on Engineering Is Eased," *New York Times,* May 27, 1992.

23. Aronowitz, *Science as Power,* 337.

24. Andre Gorz, "Technology, Technician and Class Struggle," in Andre Gorz, ed., *The Division of Labor* (Sussex: Harvester, 1976).

25. Stanley Aronowitz, "On Intellectuals," in Bruce Robbins, ed., *Intellectuals* (Minneapolis: University of Minnesota Press, 1990), 4.

26. Russell Jacoby, *The Last Intellectuals: American Culture in the Age of Academe* (New York: Basic Books, 1987).

27. Jürgen Habermas, *The Structural Transformation of the Public Sphere* (Cambridge, Mass.: MIT Press, 1989).

28. G. W. F. Hegel, *Phenomenology of Spirit* (Oxford: Oxford University Press, 1977), 7. Emphasis in original.

29. Ibid., 29–30.

30. Willard Van Orman Quine, "Two Dogmas of Empiricism," in *From a Logical Point of View* (Cambridge, Mass.: Harvard University Press, 1953).

31. Aronowitz, *Science as Power,* ix.

32. Aronowitz, "On Intellectuals," 6.

33. Mary Hesse, "The Explanatory Function of Metaphor," in *Revolutions and Reconstructions in the Philosophy of Science* (Bloomington: Indiana University Press, 1980).

34. Sharon Traweek, *Beamtimes and Lifetimes: The World of High Energy Physics* (Cambridge, Mass.: Harvard University Press, 1988), 103.

35. Bettina Aptheker, *Tapestries of Life: Women's Work, Women's Consciousness and the Meaning of Daily Life* (Amherst: University of Massachusetts Press, 1989), 39.

36. Sandra Harding, *Whose Science? Whose Knowledge? Thinking from Women's Lives* (Ithaca, N.Y.: Cornell University Press, 1991), 126.

37. Ibid., 133.

6. Contradictions of the Knowledge Class: Power, Proletarianization, and Intellectuals

1. Alvin Gouldner, *The Future of Intellectuals and the Rise of the New Class* (New York: Seabury, 1979), 28–29.

2. Michel Foucault, *Power/Knowledge* (New York: Pantheon, 1980).

3. Ernesto Laclau, "The Impossibility of the Social," paper presented at the "Marxism and the Interpretation of Culture" conference, University of Illinois, Champaign-Urbana, July 1983. This essay does not appear in the volume of the same name edited by Cary Nelson and Larry Grossberg.

4. Stanley Aronowitz, *Science as Power* (Minneapolis: University of Minnesota Press, 1988).

5. For a good account of this process, see Andrew Pickering, *Constructing Quarks* (Chicago: University of Chicago Press, 1984).

6. Foucault, *Power/Knowledge.*

7. We can see this process vividly expressed in the career of Russian President Boris Yeltsin. Unlike his Czech counterpart, Václav Havel, Yeltsin not only led the social movement that toppled the Soviet state (even if much of its apparatus remains to be dismantled), but also has led the process of partial privatization, having survived numerous attempts to dislodge him from power. Still, as of this writing (January 1994), his hold on power is ever tenuous, especially since his bold move to dissolve Parliament in 1993.

8. But, as illustrated by Václav Havel's fall from presidential power, the cultural intellectuals have had a more difficult time of it than political intellectuals, who tend to have a firmer

connection to the old state bureaucracy and, perhaps more importantly, are prepared to bow to nationalism in order to remain in power.

9. J. P. Nettl, "Intellectuals," in Phillip Reiff, ed., *Intellectuals* (New York: Doubleday Anchor, 1968).

10. Sylvia Scribner, "Everyday Cognition," in Jean Lave, ed., *Everyday Cognition* (Berkeley: University of California Press, 1988).

11. Richard Hofstadter, *Anti-Intellectualism in American Life* (New York: Vintage, 1963).

12. William James, "The Will to Believe," in *Essays in Pragmatism* (New York: Hafner, 1948).

13. Michael Weinstein, *The Wilderness and the City* (Boston: University of Massachusetts Press, 1985).

14. Charles Sanders Peirce, "The Fixation of Belief," in Justus Buchler, ed., *Selected Writings* (New York: Dover, 1954).

15. In a personal conversation with Aronowitz in spring 1976, Wynter used this term to designate the oxymoronic state of holding a salaried job and, simultaneously, being part of the bourgeoisie, which, as is well known, is constituted as a class of owners.

16. Serge Mallet, *The New Working Class* (Bristol: Spokesman, 1975); Andre Gorz, *Strategy for Labor: A Radical Proposal* (Boston: Beacon, 1967).

17. Barbara and John Ehrenreich, "The Professional/Managerial Class," in Pat Walker, ed., *Between Labor and Capital* (Boston: South End Press, 1976).

18. The use of the term *majority* may at first appear to be excessive, but more than half of all physicians and attorneys, at least 90 percent of nurses and engineers, and virtually all teachers and social workers hold salaried jobs rather than being self-employed. The largest professions are, indeed, proletarianized, at least in comparison to the beginning of the twentieth century, when most professionals were still self-employed.

19. Arlie Hochschild with Anne Machung, *The Second Shift: Working Parents and the Revolution at Home* (New York: Viking, 1989).

20. Fritz Machlup, *The Production and Distribution of Knowledge in the United States* (New York: Columbia University Press, 1962, 1969, 1983).

21. Shoshana Zuboff, *In the Age of the Smart Machines* (New York: Basic Books, 1988).

22. J. David Bolter, *Turing's Man* (Chapel Hill: University of North Carolina Press, 1989).

7. Unions and the Future of Professional Work

1. William E. Leuchtenburg, *Franklin D. Roosevelt and the New Deal* (New York: Harper Torchbooks, 1963), 71.

2. Daniel Nelson, *Managers and Workers* (Madison: University of Wisconsin Press, 1981); Sidney Pollard, *Genesis of Modern Management* (Cambridge, Mass.: Harvard University Press, 1965).

3. Alfred E. Sloan, *My Years with General Motors* (New York: Prentice-Hall, 1965).

4. Frederick Winslow Taylor, *Principles of Scientific Management* (New York: Norton, 1967).

5. Thorstein Veblen, *The Engineers and the Price System* (New York: Harcourt, Brace & World, 1963 [1921]).

6. Compare salaries of these professionals to those of auto workers in 1992. Fully employed auto workers earn $50,000 a year (albeit for a fifty-hour week) while teacher salaries average about $40,000 (slightly higher in metropolitan areas) and engineers average $48,000. Social workers who are not managers earn about the same as teachers but, like teachers, start at salaries in the high twenties. There are, of course, much higher paid engineers, especially those employed by large corporations in the military and aerospace sectors and those who work as design engineers.

7. H. J. Habakkuk, *American and British Technology in the 19th Century* (Cambridge: Cambridge University Press, 1962, 1967). Habakkuk cites the following: "Until well into the 20th century, Americans have been content to let most of the basic discoveries in science and technology originate in Europe, while they themselves have followed a policy of adapt, improve and apply" (J. B. Rae, "The 'Know-How' Tradition: Technology in American History," *Technology and Culture* 1, no. 2 [1960]: 141, quoted in Habakkuk, 202). Habakkuk invokes examples: the early dynamo improvements were the work of Gramme, a Frenchman, and Von Hefner Alteneck, a German, but were developed by Edison and Bush in the United States; "Garz and Company of Budapest were the first to perfect the transformer, and it was there that Westinghouse got his ideas. . . . The internal combustion engine was another borrowed idea" (ibid.). And, we may add, the U.S. development of nuclear weapons and energy was largely based on the work of Albert Einstein, Enrico Fermi, Leo Szilard, and other Europeans.

8. Politically generated because, although it began as a program geared to the special needs of the physically handicapped, it evolved into a way of segregating "hyperactive" or "disturbed" kids and, finally, into a hidden tracking device to separate some "slower" learners from the "normal" kids. Special education often brings in funds school systems would not otherwise get from states and municipalities, so it is in the interest of the school system to have extensive special education programs and many students in them.

9. For a fuller discussion of the rise of public-employee and white-collar unionism, see Stanley Aronowitz, *Working-Class Hero: A New Strategy for Labor* (New York: Pilgrim, 1983).

10. For a sympathetic account of the rise of teacher unions, see Phillip A. Taft, *United They Teach* (Ithaca, N.Y.: ILR Press, 1974); for a hostile account, see Robert J. Braun, *Teachers and Power* (New York: Simon & Schuster, 1972). Clearly, the often abrasive personality of Albert Shanker has produced considerable controversy among critics of the labor movement. In fact, Shanker was instrumental in transforming many organizations of professionals from associations into unions. The example of the 1964 strike in New York reverberated to the massive National Education Association, currently the largest union in the United States, but also to associations of nurses, social workers, and, and more recently, attorneys and physicians. At the same time, Shanker and the teachers proved a loyal ally of the most conservative forces in U.S. foreign policy, were generally found in the centrist wing of the Democratic party, and, until the last ten years, were notable for their craft union approach to educational unionism.

11. Leon Fink and Brian Greenberg, *Upheaval in the Quiet Zone: A History of the Hospital Workers Union Local 1199* (Urbana and Chicago: University of Illinois Press, 1989). The authors render the later split within the union with notable dispassion and considerable skill, although their affections are clearly with the founders and organizers, Leon Davis and Elliot Godoff.

12. For a discussion of university teachers and their problems, see chapter 8 of this book.

13. H. L. Nieburg, *In the Name of Science* (New York: Quadrangle, 1965); Martin Kenney, *The University-Industrial Complex* (New Haven, Conn.: Yale University Press, 1986); Dorothy Nelkin, "Intellectual Property: The Control of Scientific Information," *Science* 216 (May 14, 1982).

14. Charles Derber, William A. Schwartz, and Yale Magrass, *Power in the Highest Degree* (New York: Oxford University Press, 1990).

15. George Konrad and Ivan Szelenyi, *Intellectuals on the Road to Class Power* (New York: Harcourt Brace Jovanovich, 1981); Alvin Gouldner, *The Future of Intellectuals and the Rise of the New Class* (New York: Seabury, 1979).

8. A Taxonomy of Teacher Work

1. Lucius Outlaw, "Toward a Critical Theory of 'Race,'" in David Theo Goldberg, ed., *Anatomy of Racism* (Minneapolis: University of Minnesota Press, 1990).

2. See especially books by bell hooks, for example, *Ain't I a Woman? Black Women and Feminism* (Boston: South End Press, 1985).

3. Joan Scott, "Comment," *October,* no. 61 (Summer 1992).

4. "Minorities in Science," *Science,* November 22, 1992.

5. Some may dispute this judgment, but in comparison to professors in teaching colleges, especially the private four-year institutions, high school and elementary school teachers, and almost all other professionals, the tenured professor in a research university with a four-course yearly teaching load has much more time to do work that is not prescribed by the institution.

6. Others include City University of New York, some of the campuses of the State University of New York, and some campuses of the California State University system.

7. In December 1992 a commission appointed by the chancellor of the City University of New York and headed by President Leon Goldstein of Kingsborough Community College issued a report calling for elimination of ninety-seven programs throughout the university and consolidation of some key areas of the liberal arts in a few campuses. Within four months, the faculty senates of nearly all of the campuses rejected these recommendations by wide margins. The main source of dissent was faculty belief that the autonomy of the colleges that constitute the university would be lost to the central administration.

8. "Workloads of Faculty in Institutions of Higher Education" (Washington, D.C.: State Higher Education Executive Officers, 1992).

9. "U.S. Colleges Are Forced to Restrict Access to Students," *New York Times,* November 11, 1992.

10. Steven Brint and Jerome Karabel, *The Diverted Dream: Community Colleges and the Promise of Educational Opportunity in America 1900–1985* (New York: Oxford University Press, 1989).

11. Emily Abel, *Terminal Degrees* (New York: Praeger, 1986).

12. *Lingua Franca,* March 13, 1993.

13. Carnegie Foundation Report on Teaching in Higher Education, 1991.

9. The Cultural Construction of Class: Knowledge and the Labor Process

1. Karl Marx, *Critique of Hegel's Philosophy of Right* (New York: Cambridge University Press, 1970), 141.

2. Andrew Feenberg, *Lukács, Marx and the Sources of Critical Theory* (New York: Oxford University Press, 1981), 19.

3. Karl Marx, *The Poverty of Philosophy* (New York: International Publishers, 1963), 173.

4. Spike Lee with Lisa Jones, *Do the Right Thing* (New York: Simon & Schuster, 1989), 140.

5. Ibid., 142.

6. Jean-Paul Sartre, *The Critique of Dialectical Reason* (London: Verso, 1976). Sartre defines *practico-inert* as "matter in which past praxis is embodied" (829).

7. Lawrence Mishel and David M. Frankel, *The State of Working America: 1990–91* (New York: Sharpe, 1991), 221.

8. Harold Garfinkel, *Studies in Ethnomethodology* (Englewood Cliffs, N.J.: Prentice Hall, 1967), 146.

9. Pierre Bourdieu, *In Other Words* (Stanford, Calif.: Stanford University Press, 1990), 137–38.

10. Pierre Bourdieu, "Outline of a Theory of Art Perception," *International Journal of Social Science* 2 (1968): 598.

11. Ronnie Steinberg, "Debate on Comparable Worth," *New Politics* 1, no. 1 (Summer 1986): 111.

12. Donald J. Treiman and Heidi Hartman, *Women, Work and Wages: Equal Pay for Jobs of Equal Value* (Washington, D.C.: National Academy Press, 1981), 28, 66–67.

13. Linda M. Blum, *Between Feminism and Labor: The Significance of the Comparable Worth Movement* (Berkeley: University of California Press, 1991). "Simply put, whereas affirmative action aims to move women into men's work, comparable worth aims to raise the value of women's work" (4). She says of affirmative action: "The policy aims primarily to ensure equality of opportunity in the labor market; it strikes down barriers assigned according to ascriptive traits and gives preference to members of underrepresented groups in order to equalize competition. For feminism, it has represented a job integration strategy, one that has succeeded in demonstrating that women's capacities are indeed the same as men's" (18–19).

14. Karl Marx, *The 18th Brumaire of Louis Bonaparte* (New York: International Publishers, 1963), 124.

15. Karl Marx, *Grundrisse* (New York: Vintage, 1973), 706.

16. Jürgen Habermas, *Knowledge and Human Interest* (Boston: Beacon, 1968), 208.

17. Ellen Meiksins Wood, *The Retreat from Class: A New "True" Socialism* (New York: Verso, 1986), 4.

18. Ernesto Laclau and Chantal Mouffe, *Hegemony and Socialist Strategy: Towards a Radical Democratic Politics* (London: Verso, 1985), 154.

19. Ibid., 177.

20. Ibid., 191–92.

21. Nicos Poulantzas, *Political Power and Social Classes* (London: Verso, 1978), 77.

22. Nicos Poulantzas, *State, Power, Socialism* (London: Verso, 1980), 132.

23. Ibid., 247.

24. Robert K. Merton, "On Sociological Theories of the Middle Range," in *Social Theory and Social Structure* (New York: Free Press, 1968), 39–72.

25. Nicos Poulantzas presented a detailed analysis of "new petty bourgeoisie" in *Classes in Contemporary Capitalism* (London: Verso, 1974). In "The New Petty Bourgeoisie," *Insurgent Sociologist* 9, no. 1 (Summer 1978): 60, he makes a short summary statement: "I have taken other characteristics, in particular the bureaucratization of labor in the organisation of the labor process of unproductive workers in order to show the significance of the distribution of authority. It is these elements which determine the class position of the new petty bourgeoisie. The new petty bourgeoisie interiorizes the social division of labor imposed by the bourgeoisie throughout the whole of society. Each level of the new petty bourgeoisie exercises specific authority and ideological domination over the working class, which takes on particular characteristics within the factory division of labor since the workers do not exert any kind of authority or ideological dominance over other workers, for example, over unskilled workers, that has even remotely the same characteristics as that exercised by the different levels of the new petty bourgeoisie over the working class. These are the political and ideological elements in the social division of labor that I have taken to show the class specificity of the new petty bourgeoisie. It is important to stress that these are the elements that have nothing to do with the so-called 'class for itself.'"

26. Serge Mallet, *The New Working Class* (Bristol: Spokesman, 1975). Andre Gorz, *Strategy for Labor: A Radical Proposal* (Boston: Beacon, 1967).

27. Erik Olin Wright, *Classes* (London: Verso, 1985), 7.

28. Ibid., 87.

29. Ibid., 56.

30. G. A. Cohen, *Karl Marx's Theory of History: A Defence* (Princeton, N.J.: Princeton University Press, 1978).

31. Wright, *Classes*, 65.

32. Ibid., 66.

33. Erik Olin Wright, "A General Framework for the Analysis of Class Structure," in Erik Olin Wright, ed., *The Debate on Classes* (London: Verso, 1989), 10.

34. Wright, *Classes*, 67.

35. Ibid., 69–70.

36. Ibid., 76.

37. Ibid., 78.

38. Ibid., 79.

39. Ibid., 89.

40. Ibid., 91.

41. Ibid., 132.

42. Ibid., 285–86.

43. Ibid., 290.

44. Guglielmo Carchedi, *Class Analysis and Social Research* (New York: Basil Blackwell, 1987), 7.

45. Ibid., 60.

46. Wright, *Classes*, 182. Requoted by Guglielmo Carchedi, "Classes and Class Analysis," in Wright, *Debate on Classes*, 106.

47. Ibid., 107.

48. Ibid., 108–9.

49. Wright, *Classes*, 154.

50. Ibid., 147.

51. Ibid., 115.

52. Ibid., 125.

53. Carchedi, *Class Analysis*, 79–80.

54. Ibid., 80.

55. Ibid., 80–81.

56. Ibid., 82–83.

57. Ibid., 85–86.

58. Ibid., 177.

59. Ibid., 191.

60. Ibid., 108.

61. Merton, *Social Theory*, 66.

62. Ibid., 42.

63. William Julius Wilson, *The Declining Significance of Race: Blacks and Changing American Institutions* (Chicago: University of Chicago Press, 1980).

64. Barbara Ehrenreich, *Fear of Falling* (New York: HarperCollins, 1989).

65. Alvin Gouldner, *The Future of Intellectuals and the Rise of the New Class* (New York: Seabury, 1979), 30–31.

66. Ibid., 44–45.

67. Ibid., 45–46.

68. Eric Lichten, *Class, Power and Austerity: The New York City Fiscal Crisis* (South Hadley, Mass.: Bergin and Garvey, 1986).

69. Stanley Aronowitz, "On Intellectuals," in Bruce Robbins, ed., *Intellectuals: Aesthetics, Politics, Academics* (Minneapolis: University of Minnesota Press), 54.

70. Editorial Collective, "Introduction," *Zerowork* 1 (December 1975): 2. "The political strategy of the working class in the last cycle of struggles upset the Keynesian plan for develop-

ment. It is this cycle that the struggle for income through work changes to a struggle for income independent of work. The working class strategy for full employment that had provoked the Keynesian solution of the Thirties became in the last cycle of struggle a general strategy of the refusal of work. The strategy that pits income against work is the main characteristic of struggle in all articulations of the social factory. The transformation marks a new level of working class power and must be the starting point of any revolutionary organization. The strategy of refusal of work overturns previous conceptions of where the power of the working class lies and junks all the organizational formulae appropriate to the previous phases of the class relation."

71. William DiFazio, *Longshoremen: Community and Resistance on the Brooklyn Waterfront* (South Hadley, Mass.: Bergin and Garvey, 1985), vii.

72. Alain Touraine, *Return of the Actor: Social Theory in Postindustrial Society* (Minneapolis: University of Minnesota Press, 1988).

10. Quantum Measures: Capital Investment and Job Reduction

1. David Lodge, *Nice Work* (New York: Macmillan, 1986), 84–85.

2. Terri Mizrahi, *Getting Rid of Patients: Contradictions in the Socialization of Physicians* (New Brunswick, N.J.: Rutgers University Press, 1984).

3. Lawrence Mead, *The New Politics of Poverty: The Nonworking Poor in America* (New York: Basic Books, 1992).

4. On service pay, see Mead, *New Politics of Poverty*, 76. On average factory pay, see the U.S. Bureau of Labor Statistics *Labor Yearbook*, 1990.

5. Oscar Lewis, *Five Families* (New York: Basic Books, 1959); *Pedro Martinez* (New York: Basic Books, 1961); *La Vida* (New York: Random House, 1964).

6. Nathan Glazer and Daniel Patrick Moynihan, *Beyond the Melting Pot* (New York: Free Press, 1966). But the book is based on an earlier report by Moynihan for the U.S. Department of Labor, Office of Planning and Research: *The Negro Family, the Case for National Action* (Washington, D.C.: U.S. Government Printing Office, 1965). This report, together with Lewis's anthropological studies in Mexico and especially Puerto Rico, became the basis of the current conservative view that disorganization, rather than either discrimination or economic vicissitudes, led to the devaluation of the work ethic among the poor.

7. Jean Elshtain, *Public Man, Private Woman* (Princeton, N.J.: Princeton University Press, 1981); Elizabeth Fox-Genovese, *Feminism without Illusions* (Chapel Hill: University of North Carolina Press, 1991).

8. Robert Bellah et al., *Habits of the Heart* (Berkeley: University of California Press, 1987).

9. William Julius Wilson, *The Truly Disadvantaged: The Inner City, the Underclass and Public Policy* (Chicago: University of Chicago Press, 1987).

10. Juliet Schor, *The Overworked American* (New York: Basic Books, 1991).

11. Arlie Hochschild and Anne Machung, *The Second Shift: Working Parents and the Revolution at Home* (New York: Viking, 1989).

12. Mead, *New Politics of Poverty*, 77.

13. Ibid., 19–21.

14. Ibid., 76.

15. Benjamin Hunnicut, *Work without End: Abandoning Shorter Hours for the Right to Work* (Philadelphia: Temple University Press, 1988).

16. "Fewer Jobs Filled as Factories Rely on Overtime Pay," *New York Times*, May 16, 1993.

17. *Wall Street Journal*, December 3, 1992.

11. The Jobless Future?

1. David Lodge, *Nice Work* (New York: Macmillan, 1986), 85–86.

2. Hannah Arendt, *The Human Condition* (Chicago: University of Chicago Press, 1958).

3. Ibid., 137.

4. Ibid., 146.

5. Ibid., 152.

6. There is a long polemic in the *Grundrisse* against Fourier's utopian program to transform work into play. Apparently these passages escaped Arendt's notice.

7. Arendt, *Human Condition*, 207–19, 325.

8. Jürgen Habermas, *Theory of Communicative Action,* vol. 1 (Boston: Beacon, 1984).

9. Jürgen Habermas, *Moral Consciousness and Communicative Action* (Cambridge, Mass.: MIT Press, 1991).

10. William DiFazio, *Longshoremen* (South Hadley, Mass.: Bergin and Garvey, 1985).

11. *Solidarity,* a publication of the United Auto Workers (AFL-CIO), May 1993.

12. Karl Marx, *Grundrisse* (New York: Vintage, 1973), 706.

13. "Companies Cutting Funds for Scientific Research," *New York Times,* June 30, 1993.

14. Karl Polanyi, *The Great Transformation* (Boston: Beacon, 1957), 33.

15. William Honan, "New Pressures on the University," *New York Times* Education Life, January 8, 1993, 18.

16. R. William Kapp, *The Social Costs of Private Enterprise* (New York: Schocken, 1951).

17. Paul Lafargue, *The Right to Be Lazy* (Chicago: Charles Kerr, 1907).

18. This is a major argument of Jürgen Habermas's theory of communicative action. See especially Habermas's "Toward a Reconstruction of Historical Materialism" in *Communication and the Evolution of Society* (Boston: Beacon, 1979), 131–38.

19. Ernest Mandel, *Late Capitalism* (London: New Left Books, 1979).

Index

san tool, 97; automatic factory and, 300; and elimination of labor, 20, 85, 194, 195, 299; indeterminate role of, 103; investment in, 6; labor process and, 89, 103, 191–96, 327; skills required by, 92; and transfer of knowledge, 20

Computer numerical control (CNC), 99, 298

Computer programmers, 16, 21, 87

Computers: art, authenticism, and, 136; and education, 131, 132–34, 135; as "interested" machines, 132; reading with, 135–37

Conant, James B., 57

Conglomerates, 16

Consumer society, 53, 189–91; advertising industry and, 206; social movements within, 186–87, 189–91; and spurious pleasures, 61

Consumption, 27–28; as cosmetics for catastrophe, 39; credit and, 28; as cultural ideal, 56; and productivity, 27; state support for, 307; suburban living and, 323

Cooley, Mike (Lucas Aerospace), 91, 122

Corey, Lewis: and "crisis" of middle class, 40

Cornfeld, Bernard, 16

Creationist science, 73

Credit: economic crisis delayed by use of, 50

Cremin, Lawrence, 245

Crick, Francis, 153

Critical Theory, 60–61, 68, 79

Cultural capital, 51, 211

Cultural contradictions, 44

Cultural criticism: as specialized knowledge, 183

Culture: degradation of, 45; as habituation, 31; and language, 31; national, 22; system, as socialization, 52; technology and, 45

Culture industry, 23, 45

Culture of critical discourse, 43, 127, 174, 175, 187

Culture of Narcissism, The (Lasch), 310

"Culture of poverty," 308–9, 311, 321

Currency: internationalization of, 22

Cybernetic revolution, 37; displacement of labor by, 82; and pleasure, 78

Cyberspace, 76

Czechoslovakia: cultural intellectuals in, 177

Dahrendorf, Ralf, 44

Darwin, Charles, 35, 36

Davis, Leon, 217

Debt, 26; mortgages and, 349; students and, 316

Democracy: complexity and, 72; corporate, 188; and expert authority, 69; information gap and, 18; liberal, 357; libertarian, 279; participatory, 18, 162; plebiscitary, 18; plural, 281; public sphere and, 162; radical, 279–82, 288; technology and, 63; technoculture and, 71, 162; undermined by consumer society, 61

Democratic party, 39, 212

Department of Environmental Protection, New York City: use of CADD by, 110–15, 123, 127, 131; use of McAuto by, 123

Department of Transportation, New Jersey: use of CADD by, 122, 127, 129–31

Depression (1930s), 40, 45, 46, 202

Derber, Charles, 223

Desk Set (film), 48

Dewey, John, 40, 63, 71, 180, 181

Dialectic: and emergence of novelty, 31

Dick, Phillip K., 31, 75

DiFazio, Sebastian, 269

DiFazio, William, 120, 234, 235, 236

Direct numerical control (DNC), 100

Disaccumulation, 26

Do Androids Dream of Electric Sheep? (Dick), 31

Doctors: unionism of, 224

Domestic content legislation, 349

Do the Right Thing (film), 268–70, 293

Dr. Strangelove (film), 183

Dreyfus, Hubert, 77, 132

Dreyfus, Stuart, 77, 132

Dubos, Rene, 62

Duchin, Faye, 2, 122

Du Pont, 9, 57, 338

Eastern Europe: intellectuals and, 176–77; students of science from, 148

Ecology movement, 62, 186–87, 189, 190; bioregionalism, 345; critique of technoscience, 60; Habermas, Arendt, and nature, 334; and social justice, 354; unions and, 345

Index

Index

Stanley Aronowitz is professor of sociology at the Graduate Center of the City University of New York. He is the author of *False Promises, Science as Power: Discourse and Ideology in Modern Society, The Crisis in Historical Materialism: Class, Politics and Culture in Marxist Theory, Roll over Beethoven,* and (with Henry Giroux) *Postmodern Education: Politics, Culture, and Social Criticism.*

William DiFazio is professor of sociology at St. John's University in Jamaica, New York. He is the author of *Longshoremen: Community and Resistance on the Brooklyn Waterfront.*